T0260793

Quantile Regression

The Wiley Series in Probability and Statistics is well established and authoritative. It covers many topics of current research interest in both pure and applied statistics and probability theory. Written by leading statisticians and institutions, the titles span both state-of-the-art developments in the field and classical methods.

Reflecting the wide range of current research in statistics, the series encompasses applied, methodological and theoretical statistics, ranging from applications and new techniques made possible by advances in computerized practice to rigorous treatment of theoretical approaches.

This series provides essential and invaluable reading for all statisticians, whether in academia, industry, government, or research.

Series Editors:
David J. Balding, *University College London, UK*
Noel A. Cressie, *University of Wollongong, Australia*
Garrett Fitzmaurice, *Havard School of Public Health, USA*
Harvey Goldstein, *University of Bristol, UK*
Geof Givens, *Colorado State University, USA*
Geert Molenberghs, *Katholieke Universiteit Leuven, Belgium*
David W. Scott, *Rice University, USA*
Ruey S. Tsay, *University of Chicago, USA*
Adrian F. M. Smith, *University of London, UK*

Related Titles

Nonparametric Finance by Jussi Klemela February 2018,

Machine Learning: Topics and Techniques by Steven W. Knox February 2018

Measuring Agreement: Models, Methods, and Applications by Pankaj K. Choudhary, Haikady N. Nagaraja November 2017

Engineering Biostatistics: An Introduction using MATLAB and WinBUGSby Brani Vidakovic October 2017

Fundamentals of Queueing Theory, 5th Editionby John F. Shortle, James M. Thompson, Donald Gross, Carl M. Harris October 2017,

Reinsurance: Actuarial and Statistical Aspects by Hansjoerg Albrecher, Jan Beirlant, Jozef L. Teugels September 2017,

Clinical Trials: A Methodologic Perspective, 3rd Edition by Steven Piantadosi August 2017

Advanced Analysis of Variance by Chihiro Hirotsu August 2017

Matrix Algebra Useful for Statistics, 2nd Editionby Shayle R. Searle, Andre I. Khuri-April 2017

Statistical Intervals: A Guide for Practitioners and Researchers, 2nd Edition by William Q. Meeker, Gerald J. Hahn, Luis A. Escobar March 2017

Time Series Analysis: Nonstationary and Noninvertible Distribution Theory, 2nd Edition by Katsuto Tanaka March 2017

Probability and Conditional Expectation: Fundamentals for the Empirical Sciencesby Rolf Steyer, Werner Nagel March 2017

Theory of Probability: A critical introductory treatment by Bruno de Finetti February 2017

Simulation and the Monte Carlo Method, 3rd Edition by Reuven Y. Rubinstein, Dirk P. Kroese October 2016,

Linear Models, 2nd Edition by Shayle R. Searle, Marvin H. J. Gruber October 2016
Robust Correlation: Theory and Applications by Georgy L. Shevlyakov, Hannu Oja August 2016

Statistical Shape Analysis: With Applications in R, 2nd Edition by Ian L. Dryden, Kanti V. Mardia July 2016,

Matrix Analysis for Statistics, 3rd Edition by James R. Schott June 2016

Statistics and Causality: Methods for Applied Empirical Research by Wolfgang Wiedermann (Editor), Alexander von Eye (Editor) May 2016

Time Series Analysis by Wilfredo Palma February 2016

Quantile Regression

Estimation and Simulation

Volume 2

Marilena Furno

Department of Agriculture,
University of Naples Federico II, Italy

Domenico Vistocco

Department of Economics and Law,
University of Cassino, Italy

This edition first published 2018
© 2018 John Wiley & Sons Ltd

The right of Prof Marilena Furno and Domenico Vistocco to be identified as the authors has been asserted in accordance with law.

Registered Office(s)
John Wiley & Sons, Inc., 111 River Street, Hoboken, NJ 07030, USA
John Wiley & Sons Ltd, The Atrium, Southern Gate, Chichester, West Sussex, PO19 8SQ, UK

Editorial Office
9600 Garsington Road, Oxford, OX4 2DQ, UK

For details of our global editorial offices, customer services, and more information about Wiley products visit us at www.wiley.com.

Wiley also publishes its books in a variety of electronic formats and by print-on-demand. Some content that appears in standard print versions of this book may not be available in other formats.

Library of Congress Cataloging-in-Publication Data

Davino, Cristina.
 Quantile regression : theory and applications / Cristina Davino, Marilena Furno, Domenico Vistocco.
  pages cm – (Wiley series in probability and statistics)
 Includes bibliographical references and index.
 ISBN 978-1-119-97528-1 (hardback)
1. Quantile regression. 2. Regression analysis. I. Furno, Marilena, 1957–. II. Vistocco, Domenico.
III. Title.
 QA278.2.D38 2013
 519.5'36–dc23
                                        2013023591

Cover design by Wiley
Cover image: Image used by permission of the Ministero dei Beni e delle Attività Culturali e del Turismo – Museo Archeologico di Napoli

Set in 10/12pt TimesLTStd by SPi Global, Chennai, India
Printed in Singapore by C.O.S. Printers Pte Ltd

10 9 8 7 6 5 4 3 2 1

Contents

Preface			*xi*
Acknowledgements			*xiii*
Introduction			*xv*
About the companion website			*xix*

1 Robust regression — 1

Introduction — 1

1.1 The Anscombe data and OLS — 1

1.2 The Ancombe data and quantile regression — 8

  1.2.1 Real data examples: the French data — 12

  1.2.2 The Netherlands example — 14

1.3 The influence function and the diagnostic tools — 17

  1.3.1 Diagnostic in the French and the Dutch data — 22

  1.3.2 Example with error contamination — 22

1.4 A summary of key points — 26

References — 26

Appendix: computer codes in Stata — 27

2 Quantile regression and related methods — 29

Introduction — 29

2.1 Expectiles — 30

  2.1.1 Expectiles and contaminated errors — 39

  2.1.2 French data: influential outlier in the dependent variable — 39

  2.1.3 The Netherlands example: outlier in the explanatory variable — 45

2.2 M-estimators — 49

  2.2.1 M-estimators with error contamination — 54

  2.2.2 The French data — 58

  2.2.3 The Netherlands example — 59

2.3 M-quantiles — 60

  2.3.1 M-quantiles estimates in the error-contaminated model — 64

  2.3.2 M-quantiles in the French and Dutch examples — 64

  2.3.3 Further applications: small-area estimation — 70

2.4 A summary of key points — 72

References — 73

Appendix: computer codes — 74

3   Resampling, subsampling, and quantile regression   81
       Introduction   81
    3.1   Elemental sets   81
    3.2   Bootstrap and elemental sets   89
    3.3   Bootstrap for extremal quantiles   94
          3.3.1   The French data set   97
          3.3.2   The Dutch data set   98
    3.4   Asymptotics for central-order quantiles   100
    3.5   Treatment effect and decomposition   101
          3.5.1   Quantile treatment effect and decomposition   107
    3.6   A summary of key points   117
          References   118
          Appendix: computer codes   120

4   A not so short introduction to linear programming   127
       Introduction   127
    4.1   The linear programming problem   127
          4.1.1   The standard form of a linear programming problem   129
          4.1.2   Assumptions of a linear programming problem   131
          4.1.3   The geometry of linear programming   132
    4.2   The simplex algorithm   141
          4.2.1   Basic solutions   141
          4.2.2   Optimality test   147
          4.2.3   Change of the basis: entering variable and leaving variable   148
          4.2.4   The canonical form of a linear programming problem   150
          4.2.5   The simplex algorithm   153
          4.2.6   The tableau version of the simplex algorithm   159
    4.3   The two–phase method   168
    4.4   Convergence and degeneration of the simplex algorithm   176
    4.5   The revised simplex algorithm   181
    4.6   A summary of key points   190
          References   190

5   Linear programming for quantile regression   191
       Introduction   191
    5.1   LP formulation of the L_1 simple regression problem   191
          5.1.1   A first formulation of the L_1 regression problem   193
          5.1.2   A more convenient formulation of the L_1 regression
                  problem   204
          5.1.3   The Barrodale–Roberts algorithm for L_1 regression   210
    5.2   LP formulation of the quantile regression problem   217
    5.3   Geometric interpretation of the median and quantile regression
          problem: the dual plot   218
    5.4   A summary of key points   228
          References   229

6   Correlation   233
      Introduction   233
6.1   Autoregressive models   233
6.2   Non-stationarity   242
      6.2.1   Examples of non-stationary series   243
6.3   Inference in the unit root model   248
      6.3.1   Related tests for unit root   252
6.4   Spurious regression   254
6.5   Cointegration   259
      6.5.1   Example of cointegrated variables   260
      6.5.2   Cointegration tests   261
6.6   Tests of changing coefficients   262
      6.6.1   Examples of changing coefficients   265
6.7   Conditionally heteroskedastic models   269
      6.7.1   Example of a conditional heteroskedastic model   272
6.8   A summary of key points   274
      References   274
      Appendix: Stata computer codes   275

Index   **283**

Preface

Quantile regression is an expanding area in theoretical and applied statistics, as evidenced from the increased number of papers on the topic in mainstream statistical journals as well as in many specialized journals devoted to different scientific disciplines. Such interest has grown quickly as can be seen from the infographic in Figure 1, where the number of citations for the search key "quantile regression" is reported (the Web of Science Thomson Citation Index has been used). In particular, the number of papers on the topic "quantile regression" are grouped in classes of five years, starting from the appearance of the seminal paper on the topic in 1978[1]. The first class groups the papers up to 1993, due to the reduced number of referenced paper in the early years. The size of the bubbles is proportional to the number of papers in each period. The interest in scientific journals goes hand in hand with the treatment of quantile regression in chapters or sections of more generalist books on statistical models and regression. Nevertheless, few are the textbooks entirely devoted to quantile regression. This book follows up a first volume and complements it addressing some advanced topics and deepening issues only partially covered in the earlier volume. This volume follows the same example based approach adopted for the first book. All the topics are treated both theoretically and using real data examples. Graphical representations are widely adopted to visually introduce several issues, and in particular to illustrate the geometric interpretation of the simplex method, which historically represents the turning point for the diffusion of quantile regression.

Marilena Furno and Domenico Vistocco

---

[1] Koenker R and Bassett G (1978) Regression quantiles. *Econometrica* 46, 33–50.

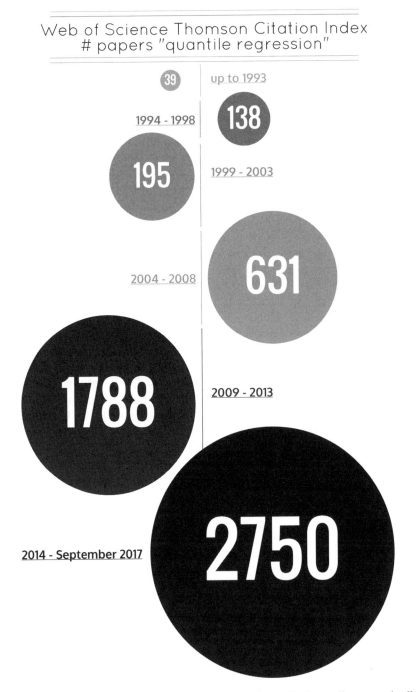

Figure 1 Evolution of the number of papers on the topic "quantile regression" in the Web of Science Thomson Citation Index.

Acknowledgements

The final structure of this book has benefited from the contributions, discussions, and comments of several colleagues.

In particular we would like to thank (in alphabetic order): Francesco Caracciolo, Maria Rosaria D'Esposito, Thomas Kirschstein, Roger Koenker, Michele La Rocca, Adriano Masone, Davide Passaretti, and Antonio Sforza, who helped us in revising previous drafts of the manuscript.

The authors are very grateful to the anonymous referees who evaluated the initial project: their comments helped us to improve the general structure of the book. We would also like to thank Wiley's staff for their support during the editorial process.

Finally we would like to thank Mariano Russo for his graphing skills with the cover figure. The latter depicts a stylized detail of a pompeian mosaic exhibited in the Museo Archeologico di Naples, reproduced by kind permission of Ministero dei Beni e delle Attività Culturali e del Turismo. Last but not least, our special thanks are for our families, who have contributed to make this book possible.

Introduction

Quantile regression, albeit officially introduced by Koenker and Basset in their seminal paper "Regression quantiles" appeared in Econometrica in 1978, has a long history, dating back to 1755, earlier than the widespread least squares regression. Boscovich's initial study on the ellipticity of the earth is indeed the first evidence of the least absolute criterion and anticipates by fifty years the official introduction of least squares by Legendre in 1805. Following Boscovich, Edgeworth proposed in 1888 a geometric approach to the median regression, naming it plural median, which can be considered the precursor of the simplex method. The plural median is the extension of the "methode de situation" proposed by Laplace in 1793 following Boscovich's idea. The studies of Boscovich, Laplace, and Edgeworth can be considered the prelude to the passionate work of Koenker[1] on quantile regression:

> I have spent a large fraction of my professional energy arguing [...] we can take a more comprehensive view of the statistical relationship between variables by expanding the scope of the linear model to include quantile regression.

The history of quantile regression is intertwined with that of linear programming and in particular with one of its main solving methods, the simplex. Albeit the problem of solving a system of linear inequalities dates back to Fourier in 1826, the first formulation of a problem equivalent to the general linear programming problem was given by Leonid Kantorovich, in 1939, with his studies on the optimal allocation of resources. The main turning point is in 1947, with the proposal of the simplex method by Dantzig. Even if nowadays there are alternatives in literature, the simplex is the first method for efficiently tackling linear programming problem. The studies on least absolute deviation are innumerable, and it would not be possible to mention all of them. The contribution of Wagner in 1959, published in the *Journal of the American Statistical Society*, is, however, of fundamental importance since it links statistics and linear programming. The formulation of the least absolute deviation problem in terms of linear programming is indeed the entry point to sound and efficient methods for solving the least absolute regression problem. An asymmetric loss function, leading to the definition of quantile regression, is the ensuing step. The main leading figures in this history are reported in the infographic of Figure 1. It is a little curious that the same first name, Roger, appears at the beginning and at the end of the proposed quantile regression timeline.

---

[1] Koenker R (2000) Galton, Edgeworth, Frisch, and prospects for quantile regression in econometrics. *Journal of Econometrics* 95, 347–374.

Figure 1   A quantile regression timeline.

This book complements its first volume on quantile regression with the following topics: robustness, M-estimators and M-quantile, expectiles and the treatment of correlated data. Two chapters are devoted to the linear programming formulation of the quantile regression problem and to the use of elemental sets. Even if recent literature offers alternative methods to compute quantile regression, we focus only on the simplex method for its historical importance in the development and spread of quantile regression.

Structure of the book

Chapter 1 – *Robust regression*, focuses on the robustness of quantile regressions, that is, on the reduced impact of anomalous values on the median regression estimator. Analogously to the univariate case, where the median is less affected by outliers than the mean, the conditional median regression is less attracted by outliers than the conditional mean regression, the OLS estimates. The chapter starts by pointing out the shortcomings of the OLS estimator in a small data set. Then the behavior of the median regression is analyzed and compared with the OLS results. In addition, some diagnostic measures to spot outliers are discussed.

Chapter 2 – *Quantile regression and related methods*, compares quantile regression with alternative estimators such as expectiles, M-estimators, and M-quantiles. Expectiles allow to estimate the equation in the tails of the error distribution, introducing an asymmetric weighting system within the OLS objective function in order to move the estimated equation away from the conditional mean. Its advantage is the ease of computation, while its drawback is a lack of robustness. M-estimators consider a weighting system that allows to control the impact of outliers at the conditional mean, thus introducing robustness in the OLS framework, but the estimated line can be computed at the conditional mean and not in the tails. The M-quantiles combine both of the above mentioned weighting systems, defining an estimator robust to anomalous values as the M-estimators and apt to estimate the regression at many points of the error distribution, not only at the mean, as the expectiles.

Chapter 3 – *Resampling, subsampling, and quantile regression*, considers bootstrap and elemental sets. The elemental sets are small subsets of data entering the computational routines of quantile regressions, as will be highlighted in Chapter 5. Bootstrap is implemented to compute extreme quantile regressions, like the 5-th or the 98-th quantile. The behavior of extreme quantiles differs from the central quantile regression estimator, and a bootstrap-based approach allows to correct bias and to implement inference. In addition, quantile treatment-effect estimators heavily rely on the bootstrap approach.

Chapter 4 – *A not so short introduction to linear programming*, offers a general introduction to linear programming focusing on the simplex method. The chapter streamlines the main concepts rather than focusing on mathematical details, meanwhile seeking to keep the presentation rigorous. The use of several graphical representations offers the geometrical insights of linear programming in general and of the simplex method in particular. The presence of a chapter entirely devoted to linear programming is intended to make the text self-contained. The chapter contains only the tools to invite the readers to look behind the scenes of quantile regression in order to understand what is behind the software command to compute the quantile regression. Readers with a background in linear programming may wish to skip to the subsequent chapter.

Chapter 5 – *Linear programming for quantile regression*, focuses on the computational aspects of quantile regressions. The chapter first presents the case of median regression and then moves to the more general quantile regression setting. The median regression problem is faced using both the standard simplex algorithm and its variant due to Barrodale and Roberts. This variant is implemented in most statistical software for solving the median regression problem and it was adapted by Koenker and d'Orey for solving quantile regression. Finally, the dual plot, which represents the data and the solutions in the parameter space, is exploited to offer insights for the geometric interpretation of median and quantile regression. It also relates to the elemental sets discussed in the second chapter, and introduces the quantile regression process.

Chapter 6 – *Correlation*, considers quantile regressions for time series. Issues like non-stationarity, spurious regressions, cointegration, conditional heteroskedasticity are analyzed via quantile regression.

Although the book is the result of the joint work of both authors, each one has individually contributed to the chapters. In particular Marilena Furno wrote Chapters 1, 2, 3, and 6. Domenico Vistocco wrote Chapters 4 and 5.

About the companion website

The Stata codes that replicate the results of the textbook examples are here reported. Different Stata versions provide differing results, particularly in case of artificially generated data. The differences would not contradict the discussion in the textbook. However, besides each file generating artificial data there is a companion file including the list of the artificial data followed by all the other codes. They would make it easier to replicate the results and the graphs in the text.

The first three chapters of the book use the same data sets. Therefore, the files are grouped by type of data and not by chapters: Anscombe, Health, Artificial data. Within each file, for each group of codes is reported the section of the book where the results they provide are discussed.

In Chapter 2 some estimators like expectiles, M-quantiles and M-estimators for the Tukey, Huber, Hampel routines, are more easily computed in R than in Stata. The R codes for these estimators are reported in the appendix. When needed, like for the expectiles graphs, the R results are inserted in the Stata data sets so the reader can plot them in Stata.

For the bootstrap examples of Chapter 3 the Stata codes are reported. The results change each time these files are implemented.

The last chapter for time series considers different data sets and they are grouped within the Chapter 6 folder.

1

Robust regression

Introduction

This chapter considers the robustness of quantile regression with respect to outliers. A small sample model presented by Anscombe (1973) together with two real data examples are analyzed. The equations are estimated by OLS and by the median regression estimator, in order to compare their behavior in the presence of outliers. The impact of an outlying observation on a selected estimator can be measured by the influence function, and its sample approximation allows to evaluate the robustness of an estimator. The difference between the influence function of the OLS and of the quantile regression estimators is discussed, together with some other diagnostic measures defined to detect outliers.

1.1   The Anscombe data and OLS

In the linear regression model $y_i = \beta_0 + \beta_1 x_i + e_i$, the realizations of the variables x_i and y_i, in a sample of size n with independent and identically distributed (i.i.d.) errors, allow to compute the $p = 2$ unknown coefficients β_0 and β_1. The ordinary least squares (OLS) estimator is the vector $\beta^T = (\beta_0, \beta_1)$ that minimizes the sum of squared errors, $\sum_{i=1,n} e_i^2 = \sum_{i=1,n} (y_i - \beta_0 - \beta_1 x_i)^2$. The minimization process yields the OLS estimators $\beta_1 = \frac{cov(x,y)}{var(x)}$ and $\beta_0 = \bar{y} - \beta_1 \bar{x}$, where $\bar{y}$ and $\bar{x}$ are the sample means. These estimators are the best linear unbiased (BLU) estimators, and OLS coincides with maximum likelihood in case of normally distributed errors. However OLS is not the sole criterion to compute the unknown vector of regression coefficients, and normality is not the unique error distribution. Other criteria are available, and they turn out to be very useful in the presence of outliers and when

Quantile Regression: Estimation and Simulation, Volume 2. Marilena Furno and Domenico Vistocco.
© 2018 John Wiley & Sons Ltd. Published 2018 by John Wiley & Sons Ltd.
Companion website: www.wiley.com/go/furno/quantileregression

Table 1.1   Anscombe data sets and its modifications.

	original data set					modified data set				
X_1	Y_1	Y_2	Y_3	Y_4	X_2	Y_1^*	Y_3^*	Y_4^*	X_2^*	Y_1^{**}
10	8.04	9.14	7.46	6.58	8	8.04	7.46	6.58	8	8.04
8	6.95	8.14	6.77	5.76	8	6.95	6.77	5.76	8	6.95
13	7.58	8.74	**12.74**	7.71	8	7.58	8.5	7.71	8	7.58
9	8.81	8.77	7.11	8	8	15	7.11	8	8	15
11	8.33	9.26	7.81	8.47	8	8.33	7.81	8.47	8	8.33
14	9.96	8.10	8.84	7.04	8	9.96	8.84	7.04	8	9.96
6	7.24	6.13	6.08	5.25	8	7.24	6.08	5.25	8	7.24
4	4.26	3.10	5.39	**12.5**	**19**	4.26	5.39	8.5	8	4.26
12	10.84	9.13	8.15	5.56	8	10.84	8.15	5.56	8	14.84
7	4.82	7.26	6.42	7.91	8	4.82	6.42	7.91	8	4.82
5	5.68	4.74	5.73	6.89	8	5.68	5.73	6.89	8	5.68

Note: The first six columns present the Anscombe data sets, where the numbers in bold are anomalous values. The last five columns present the modified data sets, where few observations have been altered. These observations are underlined in the table. In Y_3^*, Y_4^* and X_2^* the outliers are brought closer to the other observations. In Y_1^* and Y_1^{**} outliers are introduced in the original clean data set.

the errors are realization of non-normal distributions. The small data set in Table 1.1 allows to explore some of the drawbacks of OLS that motivate the definition of different objective functions, that is different criteria defining the estimators of β, like in the quantile and the robust regression estimators.

Anscombe (1973) builds an artificial data set comprising $n = 11$ observations of four dependent variables, Y_1, Y_2, Y_3, and Y_4, and two independent variables, X_1 and X_2. This data set is reported in the first six columns of Table 1.1, while the remaining columns modify some observations of the original variables. The variables in the first six columns define four simple linear regression models where the OLS estimate of the intercept is always equal to $\widehat{\beta}_0 = 3$ and the OLS estimated slope is always equal to $\widehat{\beta}_1 = 0.5$. These estimates are significantly different from zero, and the goodness of fit index is equal to $R^2 = 0.66$ in each of the four models. Figure 1.1 presents the plots of these models: the top-left graph shows a regression model where OLS well summarizes the data set $[Y_1\ X_1]$. In the other three models, however, the OLS estimates poorly describe the majority of the data in the sample.

In the top-right graph, the data $[Y_2\ X_1]$ follow a non-linear pattern, which is incorrectly estimated by a linear regression. Here the assumption of linearity is wrong, and the results are totally unreliable since the model is misspecified.

In the two bottom graphs, the OLS line is attracted by one anomalous value, that is by one observation that is far from the majority of the data. In the bottom-left graph, the $[Y_3\ X_1]$ data set is characterized by one observation greater than all the others with respect to the dependent variable Y_3. This is a case of one anomalous observation in the dependent variable, reported in bold in the table, where the third observation

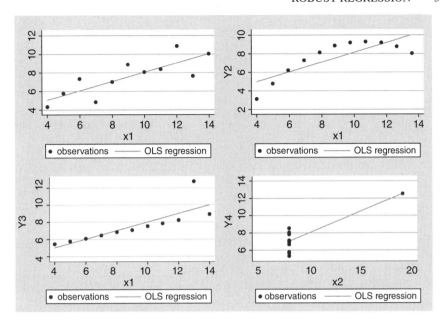

Figure 1.1   OLS estimates in the four Anscombe data sets $[Y_1\ X_1]$, $[Y_2\ X_1]$, $[Y_3\ X_1]$ and $[Y_4\ X_2]$. The four linear regressions yield the same OLS estimated coefficients, $\widehat{\beta}_0 = 3$ and $\widehat{\beta}_1 = 0.5$, and the same goodness of fit value, $R^2 = 0.66$, in all the four regressions. The sample size is $n = 11$.

$(Y_3\ X_1)_3 = (\textbf{12.74}\ 13)$ presents the largest value of Y_3, the farthest from its median, $\text{Me}(Y_3) = 7.11$, and from its mean $\overline{Y}_3 = 7.5$. In this case the outlier attracts the OLS regression, causing a larger OLS estimated slope and a smaller OLS intercept. This example shows how one outlying observation can cause bias in the OLS estimates. If this observation is replaced by an observation closer to the rest of the data, for instance by the point $(Y_3^*\ X_1)_3 = (\underline{8.5}\ 13)$ as reported in the eighth column of the table, the variance of the dependent variable drops from $\widehat{\sigma}^2(Y_3) = 4.12$ of the original series to $\widehat{\sigma}^2(Y_3^*) = 1.31$ of the modified series, all the observations are on the same line, the goodness of fit index attains its maximum value, $R^2 = 1$, and the unbiased OLS estimated coefficients are $\widehat{\beta}_0^* = 4$ and $\widehat{\beta}_1^* = 0.34$. The results for the modified data set $[Y_3^*\ X_1]$ are depicted in the top-right graph of Figure 1.2.

In the bottom-right graph of Figure 1.1, instead, OLS computes a non-existing proportionality between Y_4 and X_2. The proportionality between these two variables is driven by the presence of one outlying observation. In this case one observation, given by $(Y_4\ X_2)_8 = (\textbf{12.5}\ \textbf{19})$ and reported in bold in the table, is an outlier in both dimensions, the dependent and the explanatory variable. If the eighth observation of Y_4 is brought in line with all the other values of the dependent variable, that is if $(Y_4\ X_2)_8$ is replaced by $(Y_4^*\ X_2)_8 = (\underline{8.5}\ \textbf{19})$ as reported in the ninth column of Table 1.1, this observation is anomalous with respect to the independent variable alone. This case, for the data set $[Y_4^*\ X_2]$, is depicted in the bottom-left graph of Figure 1.2. Even

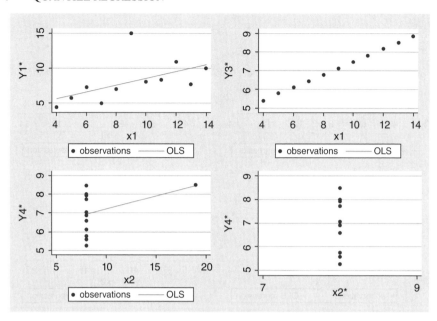

Figure 1.2   OLS estimates of the modified data sets $[Y_1^* \; X_1]$ and $[Y_3^* \; X_1]$ in the top graphs, $[Y_4^* \; X_2]$ and $[Y_4^* \; X_2^*]$ at the bottom section of the figure. The sample size is $n = 11$.

now there is no true link between the two variables; nevertheless OLS computes a non-zero slope driven by the outlying value in $(X_2)_8$. When the eighth observation $(Y_4 \, X_2)_8 = (\mathbf{12.5 \; 19})$ is replaced by $(Y_4^* \, X_2^*)_8 = (\underline{8.5} \; \underline{8})$, so that also the independent variable is brought in line with the rest of the sample, it becomes quite clear that Y_4^* does not depend on X_2^* and that the previously estimated model is meaningless, as can be seen in the bottom-right graph of Figure 1.2 for the data set $[Y_4^* \, X_2^*]$.

These examples show that there are different kinds of outliers: in the dependent variable, in the independent variable, or in both. The OLS estimator is attracted by these observations, and this causes a bias in the OLS estimated coefficients.

The bottom graphs of Figure 1.1 illustrate the impact of the so-called leverage points, which are outliers generally located on one side of the scatterplot of the data. Their sideway position enhances the attraction, and thus the bias, of the OLS estimated line. The bias can be related to the definition of the OLS estimator, which is based on the sample mean of the variables. The mean is not a robust statistic, as it is highly influenced by anomalous values, and its lack of robustness is transmitted to the OLS estimator of the regression coefficients.

There are, however, cases of non-influential outliers, i.e., of anomalous values that do not attract the OLS estimator and do not cause bias. This case is presented in the top-left panel of Figure 1.2. In this graph the data set $[Y_1 \; X_1]$ is modified to include one outlier in Y_1. In particular, by changing the fourth observation $(Y_1 \; X_1)_4 = (8.81 \; 9)$ into $(Y_1^* \; X_1)_4 = (\underline{15} \; 9)$ – as reported in the seventh column of Table 1.1 – the

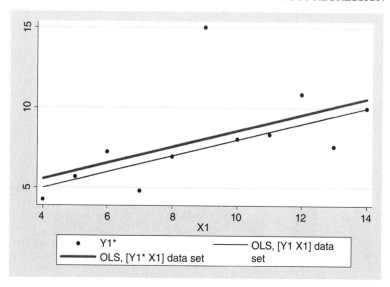

Figure 1.3 OLS estimated regressions in the presence of a non-influential outlier, as in the data set $[Y_1^* \ X_1]$ characterized by one outlier in the dependent variable. The anomalous value in Y_1^* has a central position in the graph and causes a change in the OLS intercept but not in the OLS estimated slope. The lower line is estimated in the $[Y_1 \ X_1]$ data set, without outliers. The upper line is estimated in the $[Y_1^* \ X_1]$ modified data set, with one non-influential outlier in the dependent variable affecting the intercept but not the slope in the OLS estimates. The sample size is $n = 11$.

estimated slope remains the same, $\widehat{\beta}_1^* = 0.5$, while the intercept increases to $\widehat{\beta}_0^* = 3.56$. The comparison of these OLS estimates is depicted in Figure 1.3: the lower line in this graph is estimated in the $[Y_1 \ X_1]$ data set without outlier and yields the values $\widehat{\beta}_0 = 3.0$ and $\widehat{\beta}_1 = 0.5$. The upper line is estimated in the modified data set $[Y_1^* \ X_1]$ with one outlier in the fourth observation of Y_1^*. The stability of the OLS slope in this example is linked to the location of the outlying point, which assumes a central position with respect to the explanatory variable, close to the mean of X_1. Thus a non-influential outlier is generally located at the center of the scatterplot and has an impact only on the intercept, without modifying the OLS slope.

The bias of the OLS estimator in the presence of influential outliers has prompted the definition of a wide class of estimators that, by curbing the impact of outliers, provide more reliable – robust – results. The payout of robust estimators is a reduced efficiency with respect to OLS in data sets without outlying observations. This is particularly true in case of normal error distributions, since under normality, OLS coincides with maximum likelihood and provides BLU estimators. However, in the presence of anomalous data, the idea of normally distributed errors must be discarded. Indeed the presence of outliers in a data set can be modeled by assuming non-normal errors, like Student-t, χ^2, double exponential, contaminated distributions, or any other distribution characterized by greater probability in the tails with respect to the normal

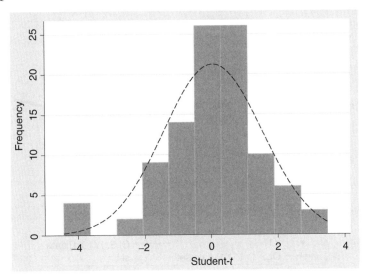

Figure 1.4 Comparison of the histogram of a Student-t with two degrees of freedom and a standard normal density, represented by the dashed line, in a sample of size $n = 50$. With respect to the standard normal, the Student-t histogram presents a small peak in the left tail, thus yielding values far from the rest of the data with frequency higher than in the standard normal case.

case. A greater probability in the tails implies a greater probability of realizations far from the center of the distribution, that is, a greater probability of outliers in the data. Figure 1.4 compares the realizations of a Student-t distribution with 2 degrees of freedom and a standard normal, represented by the dashed line, in a sample of $n = 50$ observations. The realizations of the Student-t distribution present a small peak in the left tail. This peak shows that data far from the center occur with a frequency greater than in the case of a normal density. Analogously, Figure 1.5 presents histogram of the realizations of a contaminated normal distribution f. This distribution is defined as the linear combination of two normal distributions centered on the same mean, in this example centered on zero, but having different variances. The outliers are realizations of the distribution with higher variance. In Figure 1.5 a standard normal density, f_a, generates 95% of the observations while the remaining 5% are realizations of f_b, a contaminating normal distribution having zero mean and a larger standard error, $\sigma_b = 4$. In this example the degree of contamination, i.e., the percentage of observations coming from the contaminating distribution f_b, is 5%. In a sample of size $n = 50$, this amounts to 2 or 3 anomalous data. The figure reports the histogram of $f = (1 - .05)f_a + .05 f_b$, which is more dispersed than the normal distribution, depicted by the solid line, skewed and with a small peak in the left tail. It is of course possible to have all sorts of contaminated distributions, with densities that differ not only in variance but also in shape, like in the case of the linear combination of a normal and a χ^2 or a Student-t distribution. Figure 1.6 reports the linear combination

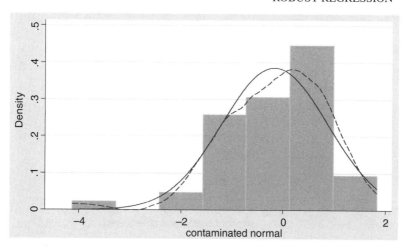

Figure 1.5 Histogram of the realizations of a contaminated normal distribution, where the observations close to the mean are generated by a standard normal while the observations in the tails, the outliers in the left tail, to be precise, are generated by a contaminating normal having zero mean and $\sigma = 4$. The degree of contamination, i.e., the percentage of observations generated by the distribution characterized by a larger variance, is 5% in a sample of size $n = 50$. The dashed line is the Epanechnikov (1969) kernel density, while the solid line is the normal density.

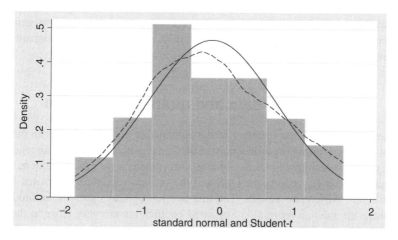

Figure 1.6 Histogram of the realizations of a standard normal distribution contaminated by a Student-t with 4 degrees of freedom. The degree of contamination is 5% in a sample of size $n = 50$. The dashed line is the Epanechnikov (1969) kernel density, which is skewed and thick tailed when compared to the solid line representing the normal density.

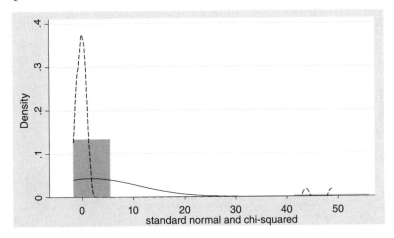

Figure 1.7   Histogram of the realizations of a standard normal distribution contaminated by a χ^2 distribution. The degree of contamination is 5%, in a sample of size $n = 50$, and contamination occurs only in the right tail. The dashed line is the Epanechnikov (1969) kernel density, characterized by two small peaks in the far right, while the solid line is the normal density.

of a standard normal and a Student-t with 4 degrees of freedom. The latter, depicted by the dashed line, is characterized by tails larger than normal, particularly on the right-hand side. Figure 1.7 presents a standard normal contaminated by a χ^2 distribution, which is a case of asymmetric contamination since it generates outliers exclusively in the right tail. The figure shows very large outliers to the right, generated by the contaminating χ^2 distribution.[1] In addition, various degrees of contamination can be considered.

1.2   The Ancombe data and quantile regression

This section considers the behavior of the quantile regression estimator in the presence of outliers, looking in particular at the median regression. The median regression can be directly compared to OLS, which in turn coincides with the conditional mean regression. Table 1.2 reports the OLS and the median regression estimates for the four different Anscombe models. The first two top columns present the model $[Y_1 \ X_1]$ where the median regression estimated coefficients are very close to the OLS values. In this case the median regression coefficients are within the 95% OLS confidence intervals: $0.75 < \hat{\beta}_0(.5) = 3.24 < 5.25$ and $0.26 < \hat{\beta}_1(.5) = 0.48 < 0.74$. The weakness of quantile regression is in the standard errors, which at the median are about twice the OLS standard errors, $se(\hat{\beta}_0(.5)) = 2.043 > se(\hat{\beta}_0)_{OLS} = 1.125$

---

[1] An example of asymmetric contamination, and more in general of asymmetric behavior, is provided by variables like market prices, which rarely decrease, or stocks, which rapidly increase during an economic crisis.

Table 1.2 Comparison of OLS and median regression results in the four Anscombe data sets.

	Y_1 median	Y_1 OLS	Y_2 median	Y_2 OLS	Y_3 median	Y_3 OLS
X_1	0.480	0.500	0.500	0.500	0.345	0.500
se	(.199)	(.118)	(.197)	(.118)	(.001)	(.118)
intercept	3.240	3.000	3.130	3.001	4.010	3.002
se	(2.043)	(1.125)	(1.874)	(1.125)	(.007)	(1.124)

	Y_4 median	Y_4 OLS
X_2	0.510	0.508
se	(.070)	(.107)
intercept	2.810	2.857
se	(1.323)	(1.019)

Note: Standard errors in parenthesis, sample size $n = 11$.

and $se(\hat{\beta}_1(.5)) = 0.199 > se(\hat{\beta}_1)_{OLS} = 0.118$. This occurs since, as mentioned, under normality, OLS is the BLUE and thus more efficient than the quantile regression estimator.

Model $[Y_2\ X_1]$ is incorrect in both OLS and median regression since the true equation is non-linear and any attempt to describe these observations by a linear model causes misspecification. The second pair of columns in the top section of Table 1.2 present these results, with OLS and median regression providing very similar estimates. The top-right graph in Figure 1.1 presents this data set together with the OLS estimated regression, which coincides with the median regression.

The $[Y_3\ X_1]$ data set is characterized by one influential outlier in the dependent variable. Here the median regression improves upon OLS since the median fitted line is not attracted by the anomalous value in the third observation of Y_3. Furthermore, the precision of the median regression estimator improves upon OLS, and the standard errors of the quantile regression coefficients are much smaller than the OLS analogues: $se(\hat{\beta}_0(.5)) = 0.007 < se(\hat{\beta}_0)_{OLS} = 1.124$ and $se(\hat{\beta}_1(.5)) = 0.001 < se(\hat{\beta}_1)_{OLS} = 0.118$. The right-hand side graph in Figure 1.8 presents the behavior of the median regression in this case, showing that the estimated slope is unaffected by the anomalous value and that the estimated median regression well represents the majority of the data.

The case of outliers in both dependent and independent variables, analyzed in the data set $[Y_4\ X_2]$, yields median regression results comparable to OLS, and quantile regression in this case is just as unreliable as OLS. This can be seen in the two bottom columns of Table 1.2 and in the left plot of Figure 1.9, where the presence of one outlier in $(Y_4\ X_2)_8$ yields very close results in the OLS and the median regression: none of them is robust with respect to outliers in the explanatory variables.

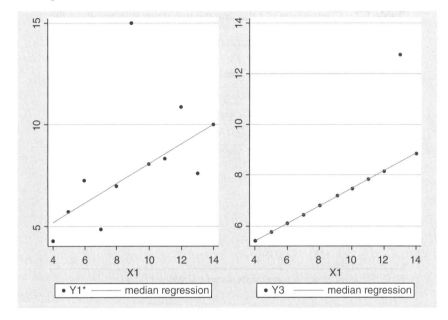

Figure 1.8   Median regression estimates in the presence of outliers. In the left graph is the $[Y_1^* \; X_1]$ modified data set with one outlier in the dependent variable placed in a central position. The right panel depicts the $[Y_3 \; X_1]$ data set with one outlier located on the side. In both cases there is no impact of the outlier on the coefficients as estimated by the median regression. The OLS estimates, instead, present an increase in the intercept in the $[Y_1^* \; X_1]$ data set, as shown in Figure 1.3, and an increased slope in the $[Y_3 \; X_1]$ data set, as can be seen in the bottom left panel of Figure 1.1. Sample size $n = 11$.

Summarizing, the median regression is robust with respect to outliers in the dependent variable, as in $[Y_3 \; X_1]$, but is not robust to outliers in the explanatory variables, as in $[Y_4 \; X_2]$. More insights can be gained by the analysis of the modified data sets.

In $[Y_1^* \; X_1]$ the location of the outlier in $(Y_1^* \; X_1)_4$ is close to the mean of the independent variable X_1, i.e., the outlier is in a central position and is not a leverage point. The first two top columns of Table 1.3 present the estimated coefficients. This is a case of one non-influential outlier, where its central position modifies only the OLS intercept. Indeed in Figure 1.3 the OLS fitted line shifts upward but yields an unchanged slope. This can be compared with the left-hand side graph of Figure 1.8, which shows the same median regression in the two data sets, $[Y_1 \; X_1]$ and $[Y_1^* \; X_1]$. The second column in the top section of Table 1.3 shows that the estimated line does not shift at all, yielding the same intercept and the same slope of Table 1.2, $\hat{\beta}_0(.5) = \hat{\beta}_0^*(.5) = 3.24$ and $\hat{\beta}_1(.5) = \hat{\beta}_1^*(.5) = 0.48$. In addition, the median regression standard errors are slightly smaller than in the OLS case, $se(\hat{\beta}_0^*)_{OLS} = 2.441 > se(\hat{\beta}_0^*(.5)) = 2.043$ and $se(\hat{\beta}_1^*)_{OLS} = 0.256 > se(\hat{\beta}_1^*(.5)) = 0.199$, showing that one non-influential outlier worsens the precision of the OLS estimator.

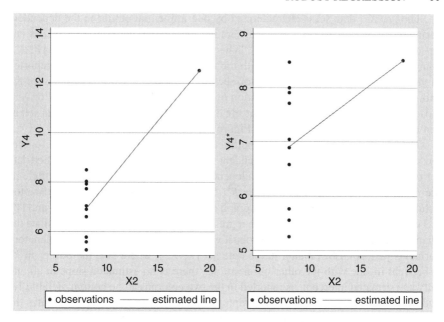

| • observations | ——— estimated line | | • observations | ——— estimated line |

Figure 1.9 OLS and median regression estimates in the $[Y_4\ X_2]$ and $[Y_4^*\ X_2]$ Anscombe data sets overlap. In the left plot there is one outlier in both the dependent and the independent variable, while the right plot presents one outlier only in the explanatory variable X_2. Sample size $n = 11$.

Table 1.3 Comparison of OLS and median regression results in the modified Anscombe data sets.

	Y_1^* OLS	Y_1, Y_1^* median	Y_3^* OLS	Y_3, Y_3^* median	Y_4^* OLS	Y_4^* median
X_1	0.500	0.480	0.346	0.345		
se	(0.256)	(0.199)	(0.0003)	(0.001)		
X_2					0.144	0.146
se					(0.107)	(0.070)
intercept	3.563	3.240	4.005	4.010	5.765	5.719
se	(2.441)	(2.043)	(0.003)	(0.007)	(1.019)	(1.323)

	Y_4^* OLS	Y_4^* median
X_2^*	0 (omitted)	0 (omitted)
se	0	0
intercept	7.061	7.04
se	(.351)	(.628)

Note: Standard errors in parentheses, sample size $n = 11$.

The $[Y_3\ X_1]$ data set is characterized by one influential outlier in the dependent variable that affects the OLS estimates but not the median regression results, as can be seen in the right plot of Figure 1.8. When the outlier is removed, as in the $[Y_3^*\ X_1]$ data set, the OLS results coincide with the median regression estimates, as reported in the second pair of columns in the top section of Table 1.3. Once accounted for the anomalous value in the third observation of Y_3, the standard errors are equal or very close to zero since all the observations are on the same line in the $[Y_3^*\ X_1]$ data set, as can be seen in the top-right graph of Figure 1.2.

The $[Y_4\ X_2]$ data set has been modified in two steps, first bringing the outlier closer to the rest of the data in the dependent variable, so that the $[Y_4^*\ X_2]$ modified data set presents one outlier only in the independent variable. The estimated coefficients are $\hat{\beta}^*_{0,OLS} = \hat{\beta}^*_0(.5) = 5.7$ and $\hat{\beta}^*_{1,OLS} = \hat{\beta}^*_1(.5) = 0.14$, as reported in the last two top columns of Table 1.3 and in the right graph of Figure 1.9. In both the $[Y_4\ X_2]$ and $[Y_4^*\ X_2]$ data sets the OLS and the median regression estimates are very similar, since the quantile regression estimator is not robust with respect to outliers in the explanatory variables just as OLS. Finally, in the data set $[Y_4^*\ X_2^*]$, where also the outlier in X_2 is brought in line with the other observations, there is no estimated slope at all and only an estimated intercept, as reported in the two columns at the bottom of Table 1.3 and as depicted in the bottom-right graph of Figure 1.2. The intercept computes the selected quantile of the dependent variable, the conditional median and, in the OLS case, the conditional mean.

Summarizing:

- under normality, as in the data set $[Y_1\ X_1]$, the OLS estimator is BLU and provides the smallest variance;

- in case of non-normality with outliers in the dependent variable, as in $[Y_3\ X_1]$ and $[Y_1^*\ X_1]$, the quantile regression estimator is unbiased and has smaller variance;

- in case of incorrect model specification, as in $[Y_2\ X_1]$, and in case of outliers in the explanatory variables, as in $[Y_4\ X_2]$ and $[Y_4^*\ X_2]$, both OLS and quantile regression provide unreliable results.

1.2.1   Real data examples: the French data

This section considers a real data set example provided by the 2004 Survey of Health, Ageing and Retirement in Europe (SHARE).[2] SHARE samples the population aged 50 or above in 11 European countries. The survey involved 19286 households and 32022 individuals, covering a wide range of topics, such as physical health, socioeconomic status, and social interactions. The data set is quite large, but focusing on one country at a time and on a specific issue, the sample size becomes more manageable and apt to compare the behavior of OLS and median regression in the presence of outliers. The focus is on body mass index as predictor of walking speed. Pooling together all the 11 European countries of the SHARE data set, the link between walking speed and body mass index is negative and statistically relevant. When the

---

[2] The website to download the SHARE data set is http://www.share-project.org/

model is individually estimated in each country, only some of them yield statistically relevant negative slopes.

Consider for instance French data on walking speed, *wspeed*, and body mass index, *BMI*, which provide a subsample of $n = 393$ observations, presented in Figure 1.10. The graph shows the presence of a large walking speed value coinciding with a value of the *BMI* variable that characterizes decidedly overweight individuals. This observation describes an individual with $BMI = 36.62851$ and $wspeed = 3.333333$, where $BMI > 30$ signals obesity and the median walking speed in this sample is $\text{Me}(wspeed) = 0.62$. This observation is highly suspicious and in need of further scrutiny. It provides an example of one anomalous value in the dependent variable, similar to the Anscombe $[Y_3 \; X_1]$ example.

Next consider a linear regression model relating walking speed to the *BMI* values, where it is reasonable to expect an inverse proportionality between these two variables. Figure 1.11 presents the OLS and the median regression, showing that the two estimated slopes present a discrepancy: $\hat{\beta}_{1,OLS} = 0.0033$ and $\hat{\beta}_1(.5) = -0.00056$. The OLS line is attracted by the large anomalous value of *wspeed* located to the top right of the scatterplot, and the OLS slope becomes positive. This observation is highly influential. It can be brought in line with the rest of the sample by replacing it with the median value of walking speed, given by $\text{Me}(wspeed) = 0.62$. Actually, since the sample size is large and the observation is possibly a coding error, it could also be simply discarded. The OLS regression implemented in the cleaned sample, i.e., computed by replacing the outlier 3.333 with the median value of the variable 0.62,

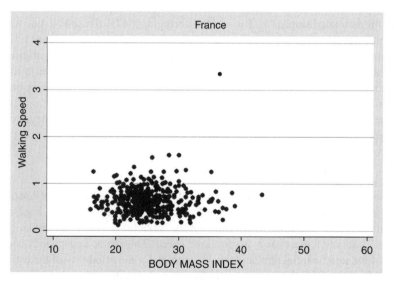

Figure 1.10   French data on walking speed and body mass index in a sample of size $n = 393$. This data set presents one outlier in the dependent variable describing an individual with the fastest walking speed of the sample and with a quite large *BMI* value, well inside the obesity range. This makes the observation quite suspicious, leading to consider it as a possible coding error.

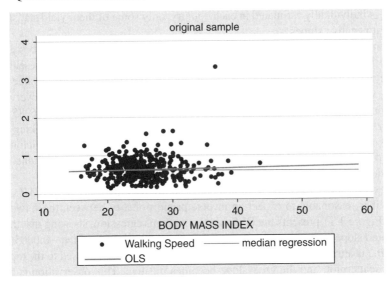

Figure 1.11 OLS and median regression in the French data set, sample size $n = 393$. The two regressions do not have the same slope: OLS yields a positive estimate since the upper line is attracted by the influential outlier in the dependent variable. The median regression, due to its robustness to outliers in the dependent variable, yields a negative slope.

yields the estimated slope $\widehat{\beta}^*_{1,OLS} = -0.0009$, bringing the OLS results in line with the median regression computed in the original sample, as can be seen in Figure 1.12. The estimated slope at the median is $\widehat{\beta}_1(.5) = \widehat{\beta}^*_1(.5) = -0.00056$ in both the original and the cleaned sample. The estimated median lines coincide, and the outlier in *wspeed* does not modify the median regression estimated slope. This is the case since quantile regressions are robust to outliers in the dependent variable.

The final results show that in the French data set there is a small negative link between walking speed and body mass index.

1.2.2   The Netherlands example

Next consider the same regression model, walking speed as a function of *BMI*, computed for the Dutch data, in a subset of size 307. The scatterplot in Figure 1.13 presents at least one outlier in the explanatory variable, with one individual having $BMI = 73.39$ and a reasonably low walking speed. This is not a suspicious observation, but it could be influential and thus dangerous for the reliability of the estimated coefficients. The OLS slope estimated in the original sample, of size $n = 307$, does not really differ from the median regression estimated slope, with $\widehat{\beta}_{1,OLS} = -0.0058$ and $\widehat{\beta}_1(.5) = -0.0065$. The OLS and the median regression are depicted in Figure 1.14.

If the outlying value in *BMI*, equal to 73.39, is replaced with the median value of the variable, Me(*BMI*) = 25.60, the estimated coefficients do not change much, yielding $\widehat{\beta}^*_{1,OLS} = -0.0054$ and $\widehat{\beta}^*_1(.5) = -0.0063$. Figure 1.15 reports the estimated

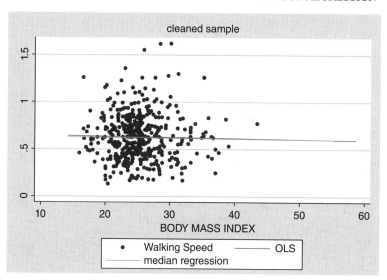

Figure 1.12 OLS and median regression in the cleaned French data set, $n = 393$. The data are cleaned by replacing the outlier with the median value of the *wspeed* variable. The OLS estimated slope becomes small and negative, just as in the median regression estimated in the original, not cleaned sample.

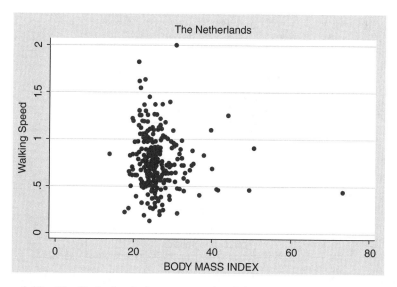

Figure 1.13 The Netherlands data set, sample of size $n = 307$. The data present one very large value of *BMI* and provide an example of at least one very large outlier in the explanatory variable.

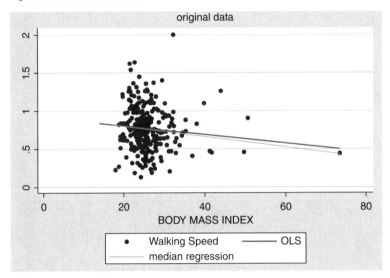

Figure 1.14   Dutch data, OLS and median regression, sample size $n = 307$. The OLS and the median regression estimated coefficients do not significantly differ from one another.

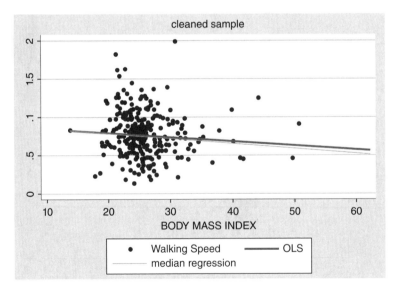

Figure 1.15   Dutch cleaned data set, OLS and median regression. The data are cleaned by replacing the outlier with the median of the explanatory variable *BMI*. The OLS regression estimates in the cleaned data set does not differ much from the median regression estimates computed in the original sample.

lines for the cleaned sample. The results are not very different between the original and the cleaned sample, and the outlier can be considered as non-influential.

In the Netherlands example the link between walking speed and *BMI* is small, negative, and statistically different from zero in the original sample.

1.3 The influence function and the diagnostic tools

In the previous sections, the robustness of quantile regression and of OLS has been analyzed by looking at their behavior in data sets where the position of one outlier and its impact on the estimates could be easily spotted. More generally, the appropriate tool to analyze the robustness of an estimator is provided by the influence function, IF, introduced by Hampel (1974). The IF measures the impact of one observation on the selected estimator, in a sample generated by a given distribution F_a. Consider next Δx_i, which assigns probability mass one to the additional point x_i. The IF measures the impact of x_i on the statistic T. It compares T evaluated at the contaminated distribution $(1 - t)F_a + t\Delta x_i$ with the same statistic T evaluated at the uncontaminated F_a. The IF is given by

$$IF(x_i, T, F_a) = lim_{t \to 0} \frac{T((1 - t)F_a + t\Delta x_i) - T(F_a)}{t}$$

A large IF signals that the addition of x_i greatly modifies the estimator and T is not robust. Vice versa, a small IF shows a negligible impact of x_i on T thus signaling that T is a robust estimator.

In the case of a linear regression model $y_i = \mathbf{x}_i^T \beta + e_i$, where $\mathbf{x}_i^T$ is the p-row vector comprising the ith observation for all the p explanatory variables, the IF is given by the ratio between the first-order condition defining the estimator and the expectation of the second-order condition. In the simple regression model with $p = 2$ and $\mathbf{x}_i^T = \begin{bmatrix} 1 & x_i \end{bmatrix}$, OLS is characterized by the following IF

$$IF_{OLS} = \frac{e_i \mathbf{x}_i}{E(\mathbf{x}_i \mathbf{x}_i^T)}$$

The IF_{OLS} can be split in two different factors, the scalar term of the Influence of Residuals, $IR_{OLS} = e_i$, and the Influence of Position in the factor space, $IP_{OLS} = \frac{\mathbf{x}_i}{E(\mathbf{x}_i \mathbf{x}_i^T)}$. The former singles out the impact on the estimator of each regression residual, while the latter shows the impact of the position of the ith observation of the explanatory variables:

$$IF_{OLS} = e_i \frac{\mathbf{x}_i}{E(\mathbf{x}_i \mathbf{x}_i^T)} = IR_{OLS} IP_{OLS}$$

These two components show that a large e_i increases by the same amount IR_{OLS} and that an outlier in the explanatory variables increases the IP_{OLS} term. Thus the OLS estimator is not robust to outliers in the dependent variable, as mirrored by IR_{OLS}, nor to outliers in the explanatory variables as measured by IP_{OLS}.

For the quantile regression, the IF is defined as

$$IF_\theta = \frac{\psi(e_i)\mathbf{x}_i}{\int f(x_i^T \widehat{\beta}(\theta))\mathbf{x}_i \mathbf{x}_i^T dG(x)} = \psi(e_i)\mathbf{x}_i Q^{-1} = IR_\theta\, IP_\theta$$

where $Q = \int f(x_i^T \widehat{\beta}(\theta))\mathbf{x}_i \mathbf{x}_i^T dG(x)$. In this equation $IR_\theta = \psi(e_i) = (\theta - 1(e_i < 0))$, while $IP_\theta = \mathbf{x}_i Q^{-1}$. The most important difference between IF_{OLS} and IF_θ is in the IR term, in the treatment of the residuals. In OLS any large value of $\widehat{e}_i = y_i - \widehat{\beta}_0 - \widehat{\beta}_1 x_i$ has an equal impact on IF_{OLS}, since the IR_{OLS} term is free to assume any value. In quantile regression, the impact of a residual on the estimator depends upon its sign, positive or negative according to its position above or below the estimated line, and not to its value measuring the distance between the observation and the estimated line. A very large e_i does not influence much the quantile regression estimator since only its position with respect to the estimated equation matters. The different behavior of IR is what makes the difference between the OLS and the quantile regression results in the Anscombe example for the $[Y_3\ X_1]$, the $[Y_1^*\ X_1]$, and in the French data sets. For instance in the $[Y_3\ X_1]$ sample, the IF for the outlying observation $(Y_3\ X_1)_3$ is function of $IR_{OLS,3} = \widehat{e}_3 = 3.241$, where $\widehat{e}_i$ is the OLS residual. The latter is larger than $IR_{\theta,3} = \psi(\tilde{e}_3) = \psi(4.245) = \theta$, where $\tilde{e}_i$ is the residual from the median regression and $\theta = .5$. Analogously in the $[Y_1^*\ X_1]$ modified Anscombe data set, characterized by one outlier in the fourth observation of the dependent variable, the $IR_{OLS,4} = \widehat{e}_4 = 6.936 >> IR_{\theta,4} = \psi(\tilde{e}_4) = \psi(7.44) = \theta = .5$. Finally, in the French data, the OLS residual for the outlying observation yields $IR_{OLS,i} = \widehat{e}_i = 2.66$, while for the median regression it is $IR_{\theta,i} = \psi(\tilde{e}_i) = \psi(2.72) = \theta = .5$. At the median IR_θ is always equal to $\pm.5$, while at a selected quantile θ it is $IR_\theta = \theta - 1(e_i < 0)$.[3]

It is worth stressing that the value of the quantile regression residuals for the influential outliers, the third residual in $[Y_3\ X_1]$, the fourth in $[Y_1^*\ X_1]$, and the residual linked to the large walking speed value in the French data are actually larger than their OLS analogues. This occurs because the OLS residuals are not a good diagnostic tool to detect outliers. When the OLS estimated line is attracted by an influential outlier, the corresponding residual is smaller than it should be. Looking at the ith OLS residual and replacing $\widehat{\beta}_0$ and $\widehat{\beta}_1$ by the OLS estimators – both of which depend on the non-robust sample mean – the OLS residual can be written as (Huber, 1981)

$$\widehat{e}_i = y_i - \widehat{y}_i = y_i - \widehat{\beta}_0 - \widehat{\beta}_1 x_i = y_i - \bar{y} + \widehat{\beta}_1 \bar{x} - \widehat{\beta}_1 x_i = y_i - \bar{y} - \widehat{\beta}_1(x_i - \bar{x})$$

$$= y_i - \bar{y} - \frac{cov(x,y)}{var(x)}(x_i - \bar{x})$$

$$= y_i - \bar{y} - \frac{(x_i - \bar{x})^2(y_i - \bar{y}) + \sum_{j \neq i}(x_j - \bar{x})(x_i - \bar{x})(y_j - \bar{y})}{\sum (x_i - \bar{x})^2}$$

$$= \left[1 - \frac{(x_i - \bar{x})^2}{\sum (x_i - \bar{x})^2}\right](y_i - \bar{y}) - \frac{\sum_{j \neq i}(x_j - \bar{x})(x_i - \bar{x})(y_j - \bar{y})}{\sum (x_i - \bar{x})^2}$$

The last equality shows that the ith OLS residual is small when the term $\left[1 - \frac{(x_i - \bar{x})^2}{\sum (x_i - \bar{x})^2}\right] = 0$. This occurs when $\frac{(x_i - \bar{x})^2}{\sum (x_i - \bar{x})^2} \simeq 1$, that is when x_i is very far from

---

[3] Since $0 < \theta < 1$, its value can slightly curb the quantile regression IP_θ.

its sample mean $\bar{x}$, and its deviation from the mean greatly contributes to the total deviation of X. When this is the case, the numerator, $(x_i - \bar{x})^2$ is approximately equal to the denominator, which represents the dispersion of the entire X. Thus the farther is x_i from its mean, the smaller is the OLS residual, and the OLS fitted line is attracted by the influential outlier. This is the reason why it is not advisable to consider the OLS residuals to detect outliers, but it is preferable to use alternative diagnostic measures.

Various diagnostic tools have been defined in the literature. In order to find outliers in the explanatory variables, the key statistic is $h_{ii} = x_i^T(X^TX)^{-1}x_i$, which is closely related to the influence of position IP. To detect outliers in both dependent and independent variables, the appropriate statistic is given by the standardized residuals, defined as $std(e_i) = \dfrac{e_i}{\sigma\sqrt{1-h_{ii}}}$. Alternatively, the studentized residuals can be considered, given by $stu(e_i) = \dfrac{e_i}{\sigma_{(i)}\sqrt{1-h_{ii}}}$. The difference between $std(e_i)$ and $stu(e_i)$ is in the denominator, where $\sigma_{(i)}$ is computed after the exclusion of the ith observation.[4] These two statistics are distributed respectively as standard normal and Student-t distribution so that a formal test on each observation could be implemented. The test verifies if each data point is or is not a rightful observation, respectively under the null and the alternative hypothesis. However, it suffices to scrutinize only the very large standardized or studentized residual.[5]

Back to the examples, the IP term in the influence function of the quantile regression estimators explains why this estimator is not robust to outliers in the explanatory variables: indeed there is no element in $IP_\theta = x_iQ^{-1}$ to bound outliers in the explanatory variables. This is the reason why OLS and median regression provide similar results in the Anscombe model $[Y_4\ X_2]$ and in its transformed version $[Y_4^*\ X_2]$. Looking at the h_{ii} values in both data sets, all the observations in X_2 yield $h_{11} = 0.1$ but the eighth, which yields $h_{88} = 1$. The large value in the explanatory variable yields similar results in both the OLS and the median regression estimators since IP is not bounded.

One simple way to approximate the IF is provided by the change in the estimated coefficients caused by the exclusion of one observation. This involves the comparison of $\widehat{\beta}$ and $\widehat{\beta}_{(i)}$, where the subscript in parenthesis signals that the regression coefficient has been estimated in a sample of size $n - 1$, where the ith observation has been excluded. For instance, in the Anscombe data set $[Y_3\ X_1]$, the exclusion of the largest observation, the third one given by $(Y_3\ X_1)_3 = (\mathbf{12.74}\ 13)_3$, yields the OLS estimated slope $\widehat{\beta}_{1(3)} = 0.34$ and the difference is $\widehat{\beta}_1 - \widehat{\beta}_{1(3)} = 0.5 - 0.34 = 0.16$. When instead the smallest observation is excluded from the sample, like the data point in position eight, which is equal to $(Y_3\ X_1)_8 = (5.39\ 4)_8$, the OLS slope is $\widehat{\beta}_{1(8)} = 0.52$ and $\widehat{\beta}_1 - \widehat{\beta}_{1(8)} = 0.5 - 0.52 = -0.02$. This observation has a much smaller impact than the third one on the OLS estimates, i.e., it is less influential.

---

[4] A subscript within parenthesis implies that the particular observation, or set of observations, in parenthesis has been excluded when computing the statistic under analysis.

[5] For further references on outlier detection, see Chatterjee and Hadi (1988) or Hoaglin, Mosteller, and Tukey (1983).

In the median regression, the drop of the third observation does not cause any change, $\hat{\beta}_1(.5) - \hat{\beta}_{1(3)}(.5) = 0$, and the exclusion of the smallest observation yields $\hat{\beta}_{1(8)}(.5) = .346$ with $\hat{\beta}_1(.5) - \hat{\beta}_{1(8)}(.5) = .345 - .346 = -.001$, which is quite small. The change in coefficient, $\hat{\beta} - \hat{\beta}_{(i)}$, allows to approximate the IF and to compute the Sample Influence Function, SIF, defined as SIF$=(n-1)(\hat{\beta} - \hat{\beta}_{(i)})$. For the third observation of the $[Y_3\, X_1]$ data set, SIF$(\beta_{1,OLS})_3 = (n-1)\,(\hat{\beta}_1 - \hat{\beta}_{1(3)}) = 1.6$ while for the eighth one, SIF$(\beta_{1,OLS})_8 = (n-1)\,(\hat{\beta}_1 - \hat{\beta}_{1(8)}) = -0.2$.

In the Anscombe $[Y_4\, X_2]$ data set, the exclusion of the eighth outlying observation from the sample yields the estimated slope $\hat{\beta}_{1(8)} = \hat{\beta}_{1(8)}(.5) = 0$ in both OLS and the median regression. The impact of this observation is quite large, since its presence causes the slope coefficient to move from a positive value to zero, $\hat{\beta}_1 - \hat{\beta}_{1(8)} = 0.5 - 0 = 0.5$. For the intercept, the SIF grows even further, since $\hat{\beta}_0 = \hat{\beta}_0(.5) = 2.8$ in the full sample turns into $\hat{\beta}_{0(8)} = \hat{\beta}_{0(8)}(.5) = 6.9$ in the smaller sample, yielding the difference $\hat{\beta}_0 - \hat{\beta}_{0(8)} = -4.1$. The final value of the sample influence function is given by SIF$(\beta_{1,OLS})_8 = $ SIF$(\beta_1(.5))_8 = (n-1)\,(\hat{\beta}_1 - \hat{\beta}_{1(8)}) = 5$ and SIF$(\beta_{0,OLS})_8 = $ SIF$(\beta_0(.5))_8 = (n-1)(\hat{\beta}_0 - \hat{\beta}_{0(8)}) = -41$ in both OLS and median regression estimators. This signals the eighth observation as highly influential.

In the $[Y_4^*\, X_2]$ data set, since the outlier is anomalous only in the independent variable, the change in the estimated coefficients due to the exclusion of the eighth observation is smaller than in the $[Y_4\, X_2]$ case. The estimated slope from $\hat{\beta}_1 = \hat{\beta}_1(.5) = 0.14$ turns into $\hat{\beta}_{1(8)} = \hat{\beta}_{1(8)}(.5) = 0$ in both OLS and median regression. The impact of the anomalous value is $\hat{\beta}_1 - \hat{\beta}_{1(8)} = 0.14$ for the slope, and $\hat{\beta}_0 - \hat{\beta}_{0(8)} = 5.7 - 6.9 = -1.2$ for the intercept in both OLS and median regression estimators. In this case the sample influence function is SIF$(\beta_{1,OLS})_8 = $ SIF$(\beta_1(.5))_8 = (n-1)(\hat{\beta}_1 - \hat{\beta}_{1(8)}) = 1.4$ and SIF$(\beta_{0,OLS})_8 = $ SIF$(\beta_0(.5))_8 = (n-1)(\hat{\beta}_0 - \hat{\beta}_{0(8)}) = -12$.

The Anscombe data sets provide examples for the case of a single outlier, but real data may have more than one anomalous value. In case of multiple outliers, assuming there are m of them in a sample of size n, the sample influence function becomes SIF$(\beta)_I = \frac{n-m}{m}(\hat{\beta} - \hat{\beta}_{(I)})$, where the index (I) refers to the set of m excluded observations, and the coefficient $\hat{\beta}_{(I)}$is computed by excluding all of them. For instance, in the $[Y_1^*\, X_1]$ modified Anscombe data set, one additional outlier can be introduced. Besides changing the fourth observation, also the ninth observation of the dependent variable can be modified from $(Y_1^*\, X_1)_9 = (10.84\ 12)_9$ into $(Y_1^{**}\, X_1)_9 = \underline{(14.84\ 12)_9}$ as reported in the last column of Table 1.1. In the full sample, the estimated regression coefficients are $\hat{\beta}_{OLS} = \begin{bmatrix} \hat{\beta}_0 = 2.94 \\ \hat{\beta}_1 = 0.61 \end{bmatrix}$ and $\hat{\beta}(.5) = \begin{bmatrix} \hat{\beta}_0(.5) = 3.24 \\ \hat{\beta}_1(.5) = 0.48 \end{bmatrix}$. Figure 1.16 reports the OLS and the median regression estimated lines in the sample of size $n = 11$ comprising two outliers. Next both the fourth and the ninth observation are excluded from the sample. The OLS and the median regression estimated coefficients in the subset without outliers are very similar to one another, $\hat{\beta}_{OLS(4,9)} = \begin{bmatrix} \hat{\beta}_{0(4,9)} = 3.21 \\ \hat{\beta}_{1(4,9)} = 0.43 \end{bmatrix}$ and $\hat{\beta}(.5)_{(4,9)} = \begin{bmatrix} \hat{\beta}_{0(4,9)}(.5) = 3.27 \\ \hat{\beta}_{1(4,9)}(.5) = 0.46 \end{bmatrix}$, with OLS getting closer to the median regression

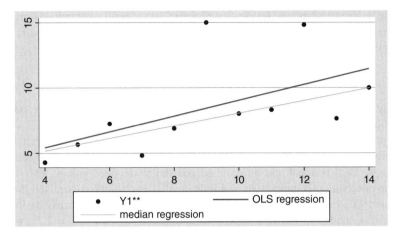

Figure 1.16 OLS and median regression when there are two anomalous values in the dependent variable, the 4th and the 9th observation of the $[Y_1^{**}\ X_1]$ data set, in a sample of size $n = 11$. The scatterplot clearly shows the two outliers, located at the top of the graph. Together they cause an increase in the OLS estimated slope, as depicted by the upper line in the graph. This is not the case for the median regression, which yields the same estimates in the $[Y_1\ X_1]$, $[Y_1^*\ X_1]$ and $[Y_1^{**}\ X_1]$ data sets.

estimates. The OLS sample influence function is

$$\text{SIF}(\hat{\beta}_{OLS})_{4,9} = \frac{11 - 2}{2}(\hat{\beta}_{OLS} - \hat{\beta}_{OLS(4,9)})$$

$$= 4.5 \begin{bmatrix} \hat{\beta}_0 - \hat{\beta}_{0(4,9)} = 2.94 - 3.21 = -0.27 \\ \hat{\beta}_1 - \hat{\beta}_{1(4,9)} = 0.61 - 0.43 = 0.18 \end{bmatrix} = \begin{bmatrix} -1.21 \\ 0.81 \end{bmatrix}$$

while for the median regression, it is

$$\text{SIF}(\hat{\beta}(.5))_{4,9} = \frac{11 - 2}{2}(\hat{\beta}(.5) - \hat{\beta}(.5)_{(4,9)})$$

$$= 4.5 \begin{bmatrix} \hat{\beta}_0(.5) - \hat{\beta}_{0(4,9)}(.5) = 3.24 - 3.27 = -0.03 \\ \hat{\beta}_1(.5) - \hat{\beta}_{1(4,9)}(.5) = 0.48 - 0.46 = 0.02 \end{bmatrix} = \begin{bmatrix} -0.13 \\ 0.09 \end{bmatrix}$$

The two outliers are highly influential in OLS. A more than double value of $\text{SIF}(\hat{\beta}_{OLS})_{4,9}$ compared to $\text{SIF}(\hat{\beta}(.5))_{4,9}$ shows once again the robustness of the quantile regression with respect to outliers in the dependent variable.

As a final check, in the $[Y_1^{**}\ X_1]$ data set, the vector of standardized residuals from the median regression can be computed:

$$std(\tilde{e}_i) = \left[\frac{\tilde{e}_i}{\sigma \sqrt{1 - h_{ii}}} \right]^T$$

$$= \begin{bmatrix} 1.13e^{-08} & -.18 & -4.72 & 9.83 & -.26 & 0 & 5.46 & 0 & 9.42 & -3.25 & 0 \end{bmatrix}$$

The largest standardized residuals are located at the fourth and ninth observations, signaling that these two data points, which are the ones intentionally modified, are influential outliers.

1.3.1 Diagnostic in the French and the Dutch data

Starting with the French data, in the original sample, the OLS residual corresponding to the outlying observation with the highest walking speed and large *BMI* is $\hat{e}_i = 2.66$, while the standardized and studentized residuals for this observation are much larger, respectively $std(\hat{e}_i) = 9.36$ and $stu(\hat{e}_i) = 10.61$. There is quite a discrepancy between the simple OLS residual and its standardized and studentized values. This signals that the observation is anomalous and influential. The outlier has attracted the OLS estimated line causing a smaller OLS residual and a positive estimated slope. In the median regression, for this same observation, the residual is $\tilde{e}_i = 2.71$, and the standardized and the studentized residuals are similar, respectively $std(\tilde{e}_i) = 2.72$ and $stu(\tilde{e}_i) = 2.74$. There is not much difference among these three values since the median regression is not attracted by outliers in the dependent variable. For the slope, the SIF_i computing the impact of this observation on the OLS estimator, is $SIF(\hat{\beta}_{OLS})_i = (n-1)(\hat{\beta}_{OLS} - \hat{\beta}_{OLS(i)}) = 392(.0033 + .0009) = 1.64$, to be compared with its analogue for the median regression, $(n-1)(\hat{\beta}(.5) - \hat{\beta}_{(i)}(.5)) = 0$.

Next consider the Dutch data set, which is characterized by one outlier in the explanatory variable. The best tool to identify this kind of outlier is $h_{ii} = x_i^T (X^T X)^{-1} x_i$, which reaches the highest value of $h_{ii} = 0.280$ for $BMI = 73.39$, followed by $h_{ii} = 0.12$ for $BMI = 57.34$ and $h_{ii} = 0.11$ for $BMI = 55.53$. These are the three data points located to the right of Figure 1.13. The largest value in h_{ii} clearly signals the largest outlier in the explanatory variable. However, the corresponding SIF_i in this example is quite small. Indeed, for the OLS slope it is $SIF(\hat{\beta}_{OLS})_i = (n-1)(\hat{\beta}_{OLS} - \hat{\beta}_{OLS(i)}) = 306(-.0058 + .0054) = -0.122$, and for the median regression slope is $SIF(\hat{\beta}(.5))_i = 306(-.0065 + .0069) = 0.122$. Thus the *BMI* outlier is not particularly influential; indeed the corresponding OLS residual, together with the standardized and studentized residual, are respectively $\hat{e}_i = -0.048$, $std(\hat{e}_i) = -0.20$, and $stu(\hat{e}_i) = -0.19$. These values are all quite small and become even smaller in the median regression case: $\tilde{e}_i = -6.07e^{-9}$, $std(\tilde{e}_i) = -6.13e^{-9}$ and $stu(\tilde{e}_i) = -6.12e^{-9}$.

1.3.2 Example with error contamination

This section implements a controlled experiment by analyzing an artificial data set with outliers generated by a contaminated error distribution. In a sample of size $n = 100$, the explanatory variable x_i is given by the realizations of a uniform distribution defined in the [0, 1] interval, and the error term e_i is provided by the realizations of a 10% contaminated normal where a standard normal is contaminated by a zero mean normal with $\sigma = 6$. Figure 1.17 presents the histogram of the realizations of e_i.

Once known x_i and e_i, the dependent variable y_i is defined as $y_i = 3 + 0.9x_i + e_i$. Next the OLS and the median regression estimators are implemented in the artificial

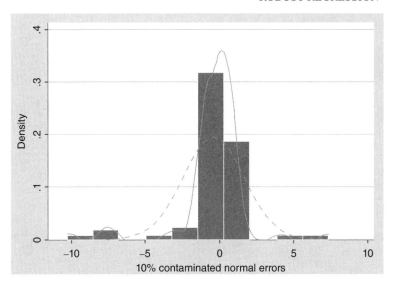

Figure 1.17 Histogram of the realizations of a 10% contaminated normal error distribution. The dashed line is the normal density while the solid line is the Epanechnikov (1969) kernel density of the contaminated normal, sample size $n = 100$. The kernel density shows the presence of values far from the mean, generated by the contaminating distribution.

data set $(y_i \ x_i)$, and their behavior in the presence of outliers can be fully evaluated since the parameters defining y_i are known in advance. The mean and the median conditional regressions results are reported in the first two columns of Table 1.4, while Figure 1.18 presents the scatterplot of the data. The graph clearly shows the presence of outliers. Although the OLS estimates are close to the true coefficients, $\beta_0 = 3$ and $\beta_1 = 0.9$, the standard error of the slope is so large that $\widehat{\beta}_1$ is not statistically different from zero. The standardized OLS residuals, defined as $std(\widehat{e}_i) = \dfrac{\widehat{e}_i}{\sigma\sqrt{1-h_{ii}}}$, and the OLS studentized residuals, $stu(\widehat{e}_i) = \dfrac{\widehat{e}_i}{\sigma_{(i)}\sqrt{1-h_{ii}}}$, allow to point out the following outliers in the dependent variable: $y_{16} = 7.89$, $y_{36} = 10.79$, $y_5 = -4.28$, $y_{18} = -7.02$,

Table 1.4 OLS and median regression in the case of contaminated error distribution.

	OLS	median	OLS$_{(I)}$	median$_{(I)}$
slope	.940	1.753	1.263	1.696
se	(.71)	(.36)	(.32)	(.41)
constant	2.60	2.372	2.64	2.427
se	(.42)	(.21)	(.19)	(.24)

Note: Standard errors in parenthesis, sample size $n = 100$.

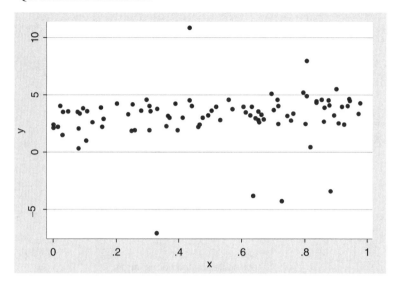

Figure 1.18   Data generated as $y_i = 3 + 0.9x_i + e_i$ where e_i is the contaminated error distribution depicted in Figure 1.17, sample of size $n = 100$. The presence of outliers in the dependent variable, caused by the contaminated error distribution, is clearly visible.

$y_{48} = -3.83$, $y_{52} = -3.46$, as detected by values of $std(\hat{e}_i)$ outside the ±1.96 interval. These values are reported in Table 1.5 for both the OLS and the median regression, respectively, $\hat{e}_i$ and $\tilde{e}_i$ in the table. This table shows that the standardized and studentized residuals of the median regression point out two additional outliers undetected in OLS, $y_{45} = 0.25$ and $y_{97} = 0.38$. Once the anomalous values are detected, it is

Table 1.5   Largest values of the studentized and standardized residuals in both OLS and median regression, $n = 100$.

y_i	OLS $stu(\hat{e}_i)$	OLS $std(\hat{e}_i)$	median $stu(\tilde{e}_i)$	median $std(\tilde{e}_i)$
7.89	2.26	2.22	4.54	4.15
10.79	4.08	3.79	12.17	7.69
−4.28	−3.96	−3.69	−13.53	−8.00
−7.02	−5.52	−4.85		−10.06
0.25			−2.36	−2.31
−3.83	−3.63	−3.43	−11.00	−7.37
−3.46	−3.58	−3.38	−11.39	−7.49
0.38			−3.68	−3.47

Note: The empty cells in the table are for the observations that are not detected as outliers.

possible to compute their impact on an estimator by means of the sample influence function, $\text{SIF}(\beta)_I = \frac{n-m}{m}(\widehat{\beta} - \widehat{\beta}_{(I)})$. The last two columns of Table 1.4 report the OLS and median regression estimated coefficients when the outlying values are eliminated, respectively the six observations detected by the standardized and studentized OLS residuals and the eight observations detected by the standardized and studentized residuals of the median regression. For the OLS estimator, the SIF $(\beta)_I$ is

$$\text{SIF}(\widehat{\beta}_{OLS})_I = \frac{(n-m)}{m}(\widehat{\beta}_{OLS} - \widehat{\beta}_{OLS(I)})$$

$$= \frac{100-6}{6} \begin{bmatrix} \widehat{\beta}_0 - \widehat{\beta}_{0(I)} = 2.60 - 2.64 = -0.04 \\ \widehat{\beta}_1 - \widehat{\beta}_{1(I)} = 0.94 - 1.26 = -0.32 \end{bmatrix} = \begin{bmatrix} -0.63 \\ -5.01 \end{bmatrix}$$

while for the median regression, it is

$$\text{SIF}(\widehat{\beta}(.5))_I = (\widehat{\beta}(.5) - \widehat{\beta}(.5)_{(I)})$$

$$= \frac{100-8}{8} \begin{bmatrix} \widehat{\beta}_0(.5) - \widehat{\beta}_0(.5)_{(I)} = 2.37 - 2.42 = -0.05 \\ \widehat{\beta}_1(.5) - \widehat{\beta}_1(.5)_{(I)} = 1.75 - 1.69 = 0.06 \end{bmatrix} = \begin{bmatrix} -0.57 \\ 0.69 \end{bmatrix}$$

The comparison of $\text{SIF}(\widehat{\beta}_{OLS})_I$ and $\text{SIF}(\widehat{\beta}(.5))_I$ shows that the impact of the set of observations in I is larger in the OLS model than in the median regression case, and the difference is quite marked for the slope coefficient, although the number of excluded observations in the OLS regression is smaller than in the median regression case. Finally, by computing the quantile regression at $\theta = .10, .25, .50, .75, .90$, all the eight outliers detected by the standardized and the studentized quantile residuals are clearly visible, as shown in Figure 1.19. The estimated coefficients at the θ

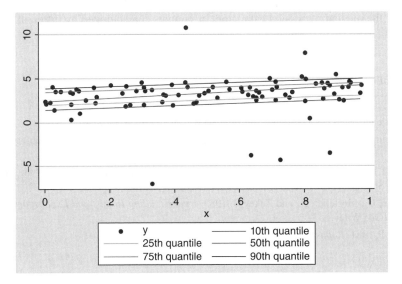

Figure 1.19   Estimated quantile regressions for the model with 10% contamination in the errors, sample of size $n = 100$.

Table 1.6   Quantile regression estimates, contaminated errors.

θ	10th	25th	50th	75th	90th
slope	1.21	1.12	1.75	1.08	1.31
se	(5.8)	(.54)	(.36)	(.39)	(3.7)
constant	1.40	1.99	2.37	3.44	3.81
se	(3.4)	(.32)	(.21)	(.23)	(2.2)

Note: Standard errors in parenthesis, sample size $n = 100$.

quantiles are reported in Table 1.6, where both the lower and the higher quantile regression estimates are not significantly different from zero.

1.4   A summary of key points

Section 1.1 focuses on the shortcomings of the OLS estimator in the linear regression model by analyzing an artificial data set, discussed by Anscombe (1973), where OLS yields the same estimated coefficients in four different models, but the result is correct only in one of them. Section 1.2 considers the behavior of the median regression in the Anscombe data and in two additional models analyzing real data, in order to show the robustness of the quantile regression approach. Section 1.3 considers the influence function, a statistic that measures the impact of outliers on an estimator, together with its empirical approximation, the sample influence function. Examples with real and artificial data sets allow to compare the influence function of the OLS and of the median regression estimators in samples characterized by different kinds of outliers.

References

Anscombe, F. 1973. "Graphs in statistical analysis." The American Statistician 27, 17–21.

Chatterjee, S., and Hadi, A. 1988. *Sensitivity Analysis in Linear Regression*. Wiley.

Epanechnikov, V. 1969. "Nonparametric estimation of a multidimensional probability density." Theory of Probability and its Applications 14 (1), 153–158.

Hampel, F. 1974. "The influence curve and its role in robust estimation." Journal of the American Statistical Association 69, 383–393.

Hoaglin, D., Mosteller, F., and Tukey, J. 1983. *Understanding Robust and Exploratory Data Analysis*. Wiley.

Huber, P. 1981. *Robust Statistics*. Wiley.

Appendix: computer codes in Stata

A) generate contaminated distributions	
`gen norm1=invnorm(uniform())`	generate N(0,1)
`gen norm4 = 4*invnorm(uniform())`	generate N(0,16)
`gen conta = cond(norm1 >-1.96,norm1,norm4)`	left tail contamination
`gen contam = cond(norm1 > 1.96,norm4,conta)`	right tail contamination

B) compute diagnostic measures to detect outliers	
`reg y x (or qreg y x)`	OLS (or quantile regression)
`predict name, resid`	compute regression residuals
`predict name1, hat`	compute h_{ii}
`predict name2, rstand`	compute standardized residuals
`predict name3, rstud`	compute studentized residuals

2

Quantile regression and related methods

Introduction

The detailed analysis of a regression model at various points of the conditional distribution, allowed by quantile regressions, can be imported into the least squares framework. This implies investigating the dependence among variables not only at the conditional mean but also in the tails, just as occurs in the quantile regression framework. An asymmetric weighting system is included within the least squares objective function to define the asymmetric least squares or expectile estimator. The problem with expectiles is their lack of robustness, since the squared errors in the objective function enhance instead of curbing the impact of anomalous values, analogously to what was discussed in the previous chapter for the standard OLS estimator. To gain robustness in the least squares approach, a weighting system bounding the outliers defines a class of robust estimators, the M-estimators. The combination of these two approaches, asymmetric weights to compute expectiles together with weights to control outliers and gain robustness of the M-estimators, leads to the definition of the M-quantiles. The M-quantile estimator merges the weighting system curbing outliers of the M-estimators and the asymmetric weights defining the location of the expectiles. The purpose is to robustly compute the regression coefficients at different points of the conditional distribution of the dependent variable. The M-quantiles estimators are frequently used in small-area estimation. Examples analyzing real and artificial data sets point out the characteristics of the above estimators.

Quantile Regression: Estimation and Simulation, Volume 2. Marilena Furno and Domenico Vistocco.
© 2018 John Wiley & Sons Ltd. Published 2018 by John Wiley & Sons Ltd.
Companion website: www.wiley.com/go/furno/quantileregression

2.1   Expectiles

Newey and Powell (1987) consider the extension of the quantile regression estimator to the least squares framework, defining the asymmetric least squares estimator, also known as expectile. In the linear regression $y_i = \beta_0 + \beta_1 x_i + e_i$, the entire class of measures of location estimated by the quantile regressions can be computed in the least squares framework as well, by minimizing the following objective function

$$\sum_{i=1,n} |\theta - 1(e_i < 0)| e_i^2$$

where the parameter θ sets the expectile, that is, the location of the regression line away from the mean. The advantage of the above definition with respect to quantile regression is in its differentiability. The above objective function is based on squared values – the L_2 norm – and is easily differentiable. The normal equations can be defined as a weighted least squares estimator where the weights select the position of the estimated line with respect to the conditional distribution of the dependent variable, away from the center, in the left or the right tail:

$$\sum_{i=1,n} w_{exp,i} e_i^2$$

$$w_{exp,i} = \theta \qquad \text{for } e_i \geq 0$$

$$1 - \theta \qquad \text{elsewhere}$$

By choosing $\theta = .50$, the estimated expectile coincides with the standard OLS regression passing through the conditional mean. Table 2.1 compares the estimated coefficients of the expectile and the quantile regression at $\theta = .10, .25, .50, .75, .90$, in the Anscombe data set $[Y_1 \, X_1]$. The slopes estimated by the expectiles are slightly smaller than the quantiles results everywhere but at $\theta = .50$. Figures 2.1 and 2.2

Table 2.1   Anscombe data set $[Y_1 \, X_1]$, expectile and quantile regressions estimated coefficients.

expectiles	10th	25th	50th	75th	90th
slope	0.44	0.48	0.50	0.52	0.56
se	(0.19)	(0.18)	(0.13)	(0.17)	(0.21)
constant	2.53	2.68	3.00	3.29	3.49
se	(1.20)	(1.33)	(1.13)	(1.50)	(1.73)

quantiles	10th	25th	50th	75th	90th
slope	0.46	0.57	0.48	0.67	0.60
se	(0.06)	(0.21)	(0.20)	(0.19)	(0.14)
constant	1.60	1.98	3.24	2.72	3.63
se	(0.64)	(1.93)	(2.04)	(1.47)	(1.63)

Note: Standard errors in parenthesis, sample size $n = 11$.

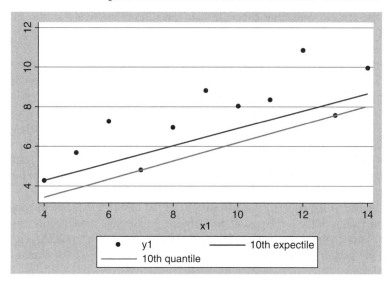

Figure 2.1 Estimates of the 10th quantile and of the 10th expectile regressions in the Anscombe data set $[Y_1\ X_1]$, sample size $n = 11$. The quantile fitted line goes through $p = 2$ observations.

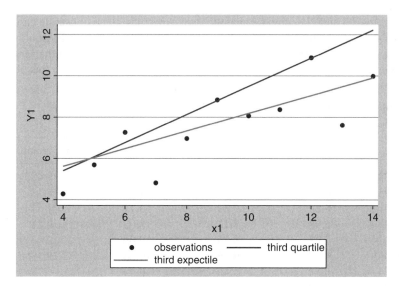

Figure 2.2 Comparison of the estimated third quartile and the estimated third expectile regressions in the Anscombe data set $[Y_1\ X_1]$, sample size $n = 11$. The quantile fitted line goes through $p = 2$ observations.

report the graphs comparing expectiles and quantiles respectively at the values of $\theta = .10, .75$. In both graphs the quantile regression fitted lines pass through $p = 2$ observations, providing for these observations a perfect fit.

Table 2.2 compares the fitted values of the quantile and expectile regressions at $\theta = .25, .75$, showing that while the quantile regression estimated lines go through exactly $p = 2$ observations – which are the values highlighted in the table – this is not the case for the estimated expectiles.

The disadvantage of expectiles is in their interpretation. While quantile regressions are directly interpretable as the exactly computed lines passing through p observations and splitting the sample according to the selected conditional quantile, θ below and $1 - \theta$ above the fitted line, this is not the case in the expectiles for different reasons:

i) the use of the squared residual instead of the absolute value in the objective function, that is, of the L_2 instead of the L_1 norm, implies that the expectile estimated line does not go through exactly p observations as occurs for the quantile regression so that the perfect fit interpretation of the quantile regressions is not shared by the expectiles;

ii) the fitted expectile does not break up residuals into θ elements below and $1 - \theta$ above the fitted expectile: θ does not represent a quantile, but it is basically a skewness coefficient needed to set the position of the estimated regression. While the quantile regressions can be interpreted as order statistics in the regression setting, this is not the case for the expectiles.

Figures 2.3 and 2.4 compare quantiles and expectiles in the univariate case. The former depicts the quantiles of the realizations of a standard normal distribution in a sample of size $n = 10000$, while Figure 2.4 provides the expectiles for the same distribution. The two figures are not identical, showing that quantiles and expectiles are

Table 2.2   Fitted values of the expectile and the quantile regressions computed at $\theta = .25, .75$.

Y_1	25th expectile	25th quantile	75th expectile	75th quantile
8.04	7.46	7.68	8.54	9.48
6.95	6.50	6.54	7.49	8.13
7.58	8.89	9.39	10.11	11.51
8.81	6.98	7.11	8.02	**8.81**
8.33	7.93	8.25	9.06	10.16
9.96	9.37	**9.96**	10.64	12.19
7.24	5.54	5.4	6.44	6.78
4.26	4.59	**4.26**	5.39	5.42
10.84	8.41	8.82	9.59	**10.84**
4.82	6.02	5.97	6.96	7.45
5.68	5.06	4.83	5.91	6.10

Note: In bold are the fitted values coinciding with the observed values of the dependent variable in the Anscombe $[Y_1 \, X_1]$ data set.

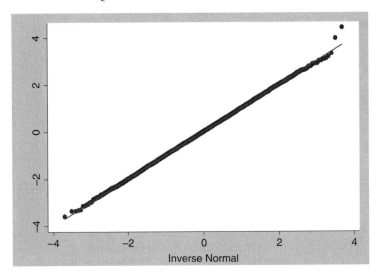

Figure 2.3   The graph plots the quantiles of the realizations of a standard normal, sample size $n = 10000$.

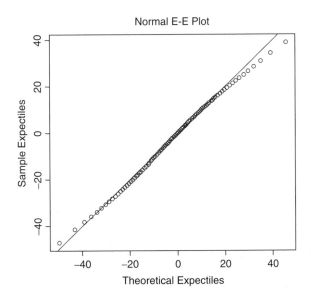

Figure 2.4   The graph plots the expectiles of the realizations of a standard normal, sample size $n = 10000$. The comparison with Figure 2.3 shows the difference between quantiles and expectiles, particularly marked in the tails.

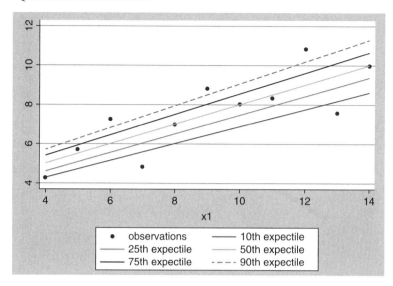

Figure 2.5   Estimated expectiles in the Anscombe model $[Y_1\ X_1]$, sample size $n = 11$.

indeed different, although closely related. Yang and Zou (2014) show the existence of a one-to-one mapping between them. Jones (1994) provides the link between expectiles and quantiles, showing that expectiles are quantiles of a distribution related to the original one.

Next consider the behavior of the expectiles in the presence of outliers. Figure 2.5 depicts the estimated expectiles for the Anscombe data $[Y_1\ X_1]$ in the sample of size $n = 11$, while Figures 2.6 and 2.8 present the estimated expectiles in the data sets $[Y_1^*\ X_1]$ and $[Y_1^{**}\ X_1]$, which modify the original $[Y_1\ X_1]$ sample to include respectively one and two outliers in the dependent variable. Figures 2.7 and 2.9 depict the estimated quantiles for these same data sets. Tables 2.3 and 2.4 report the expectile and the quantile regressions estimates. Focusing on the $[Y_1^*\ X_1]$ sample, the comparison of the graphs in Figures 2.5 and 2.6 and of the results in the top section of Table 2.1 with those in Table 2.3 shows that the estimated expectiles do change in the presence of one outlier, just as occurs in the OLS case of Figure 1.3 in Chapter 1. One non-influential anomalous value in Y_1^*, located at the center of the graph, moves upward the estimates of the intercept at all expectiles in the $[Y_1^*\ X_1]$ data set, while the quantile regression estimates of the intercept in Figure 2.7 do not change much across quantiles. In Table 2.1, the expectile estimated intercept ranges from 2.53 to 3.49 for the $[Y_1\ X_1]$ data set, while in Table 2.3, for the $[Y_1^*\ X_1]$ data set, the estimated constant goes from 2.63 to 6.48. Vice versa, the quantile regression estimates of the intercept in the $[Y_1^*\ X_1]$ data set coincide almost everywhere with the results for original data $[Y_1\ X_1]$.

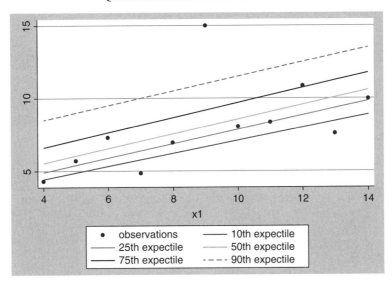

Figure 2.6 Estimated expectiles in the modified Anscombe data set $[Y_1^* \; X_1]$ with one outlier in the dependent variable, sample size $n = 11$. The outlier attracts all the expectiles causing an increasing change of the intercept across expectiles.

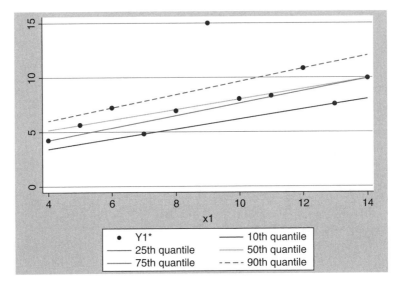

Figure 2.7 Quantile regressions in the Anscombe $[Y_1^* \; X_1]$ data set. Contrarily to the expectile estimates, here the increasing change in the intercept does not occur. It can be noticed that the first quartile converges to the median at large values of X_1.

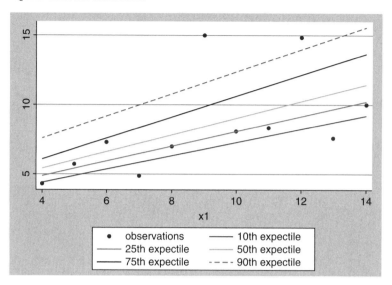

Figure 2.8   Estimated expectiles in the modified data set $[Y_1^{**} \ X_1]$, sample size $n = 11$. The presence of two outliers modifies all the expectiles causing an increase in the estimated slopes across expectiles and yielding a fan-shaped graph.

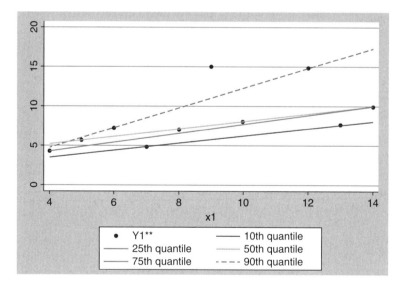

Figure 2.9   Quantile regression estimates in the Anscombe $[Y_1^{**} \ X_1]$ data set. Only the top line is attracted by the outliers. It can be noticed that the median regression crosses the third quartile at the low values of X_1 and converges to the first quartile at large values of X_1. The former is a case of quantile crossing.

Table 2.3 Estimated expectile and quantile regressions in the data set with one outlier $[Y_1^* \, X_1]$.

expectiles	10th	25th	50th	75th	90th
slope	0.44	0.49	0.50	0.51	0.50
se	(0.26)	(0.24)	(0.17)	(0.24)	(0.47)
constant	2.63	2.94	3.56	4.57	6.48
se	(1.63)	(1.86)	(1.63)	(3.11)	(5.65)
quantiles	10th	25th	50th	75th	90th
slope	0.46	0.57	0.48	0.60	0.60
se	(0.06)	(0.20)	(0.20)	(0.45)	(0.29)
constant	1.60	1.98	3.24	3.64	3.64
se	(0.64)	(1.80)	(2.04)	(4.55)	(3.94)

Note: Standard errors in parenthesis, sample size $n = 11$.

Table 2.4 Estimated expectile and quantile regressions in the data set with two outliers $[Y_1^{**} \, X_1]$.

expectiles	10th	25th	50th	75th	90th
slope	0.48	0.53	0.61	0.76	0.80
se	(0.34)	(0.16)	(0.25)	(0.40)	(0.49)
constant	2.47	2.70	2.94	3.04	4.36
se	(2.05)	(1.44)	(1.99)	(3.55)	(5.38)
quantiles	10th	25th	50th	75th	90th
slope	0.46	0.57	0.48	1.26	1.26
se	(0.06)	(0.20)	(0.20)	(0.45)	(0.17)
constant	1.60	1.98	3.24	−0.36	−0.36
se	(0.64)	(1.80)	(2.04)	(4.49)	(2.33)

Note: Standard errors in parenthesis, sample size $n = 11$.

In the $[Y_1^{**} \, X_1]$ example with two influential outliers located in the top right of Figure 2.8, the comparison of the results in Table 2.1 with those in Table 2.4 and of the graphs in Figures 2.5 and 2.8 shows the increase of the slope across expectiles due to the presence of two influential outliers, extending to the tails what occurs in the simple OLS of Figure 1.16 in the previous chapter. Indeed the estimated slope ranges from 0.44 to 0.56 across expectiles in Table 2.1 for $[Y_1 \, X_1]$, while it goes from 0.48 to 0.80 in Table 2.4 for the $[Y_1^{**} \, X_1]$ data set. To the contrary, the quantile regression estimated slope changes only at the top quantiles, as can be seen in Figure 2.9 and in the bottom section of Table 2.4. While in Figure 2.8 the expectiles are fan shaped, in Figure 2.9

the lower quantiles up to the median are close to one another and only the upper line is attracted by the outliers and diverges. As mentioned, the main characteristic of expectiles is the adoption of the L_2 norm, and in the L_2 norm is relevant not only the position of the observation with respect to the estimated line, as in the L_1 norm, but also its numerical value. This causes the lack of robustness of the expectile with respect to the quantile regression estimator, and this is particularly evident at $\theta = .50$, where the expectile coincides with the OLS estimates. Indeed the results at the 50th expectile of Tables 2.1, 2.3, and 2.4 coincide with the OLS results of Tables 1.2 and 1.3 and with the results for $[Y_1^{**} \ X_1]$, the two outliers of the Anscombe example, reported at the end of Section 1.3. When the least squares analysis is extended beyond the mean of the conditional distribution, the lack of robustness affects all the selected expectiles. Figures 2.6 and 2.8 show how all the expectiles are attracted by the position of one or two outliers in the modified data sets $[Y_1^* \ X_1]$ and $[Y_1^{**} \ X_1]$ when compared with the estimated expectiles in Figure 2.5 for the $[Y_1 \ X_1]$ sample.

The lack of robustness of the expectiles reduces their capability to signal asymmetry and heteroskedasticity in the error distribution. While quantile regressions are attracted by outliers only at the extreme quantiles, expectiles are influenced by few anomalous values even at the lower values of θ. A set of fan-shaped estimated quantiles signals heteroskedasticity, but fan-shaped expectiles can be caused by few anomalous values that attract all the expectiles and cause a changing slope across all the expectiles, as in the $[Y_1^{**} \ X_1]$ case. Figure 2.8 shows that in the expectile case all the estimated lines present an increasing slope, while in the quantile regression case of Figure 2.9 only the top estimated line is attracted by the two outliers in Y_1^{**} and the other slopes do not change much. Analogously, while unequal interquantile differences in the intercept signal skewness, this is not necessarily true for expectiles in the presence of few outliers. For instance, unequal increments in the intercept at equally spaced expectiles occur in the $[Y_1^* \ X_1]$ data set, as shown in Figure 2.6, while this is not the case with the quantile regression estimates of Figure 2.7.

A final comment on expectiles considers the computational methods. There are different ways to compute expectiles: using splines (Schnabel and Eilers, 2009) or fitting the mean expectile, that is, OLS and then computing the other expectiles as deviation from the mean expectile (Sobotka et al., 2013). This idea goes back to the work by He (1997) on quantile regressions. It has the twofold advantage of avoiding crossing of the expectile lines and of reducing the number of fits. The disadvantage is a possible suboptimal fit in heteroscedastic models. Finally Figures 2.7 and 2.9, depicting the estimated quantiles for the $[Y_1^* \ X_1]$ and $[Y_1^{**} \ X_1]$ data sets, show one drawback of quantile regressions: since quantile regressions are not computed as deviation from the median, the fitted lines may intersect. In Figure 2.9, for instance, the first quartile reaches the median fitted line at high values of X_1, and the top quantile intersects the median regression at low values of X_1. Thus, at low values of X_1 an observation on the 75th quantile is below the median, and this is a quite puzzling result. The small sample size and/or the presence of large outliers can cause this problem.

Efron (1992) extends the asymmetric least squares approach to maximum likelihood, defining the asymmetric maximum likelihood estimator. The latter is implemented in case of non-normal distributions, particularly those distributions belonging to the exponential family, such as the Poisson. In particular, in the Poisson case, this estimator provides good approximations of the percentiles, similarly to the quantile regression estimator. In case of normality, the asymmetric maximum likelihood coincides with the expectiles, just as symmetric maximum likelihood coincides with OLS. Computationally, the standard maximum likelihood estimates provide the starting values, which are updated by weighted least squares. The estimated coefficients are computed by

$$\beta_w = (X^T W(\theta) X)^{-1} X^T W(\theta) Y$$

where θ is the asymmetry parameter setting the position of the regression away from the mean, X is the matrix of explanatory variables, Y is the dependent variable, and $W(\theta)$ is the matrix of weights having diagonal elements $w = 1$ if the residual from the previous iteration is negative or equal to zero and $w > 1$ if the residual is positive. The iterations update β_w and $W(\theta)$ until convergence. The final solution coincides with the standard maximum likelihood implemented in an augmented data set. In the augmented data set, the observations below the solution have unit weight and appear only once, while the data points above the solution are repeated w times. Obviously, when $w = 1$ throughout, the results are the standard maximum likelihood estimates computed in the original data set.

2.1.1   Expectiles and contaminated errors

The small sample size of the Anscombe data allows to point out the characteristics of the expectile estimator. Here its behavior is analyzed in a larger sample, like the one considered in Section 1.3.2. The dependent variable is generated as $y_i = 3 + 0.9x_i + e_i$, the errors are generated by a 10% contaminated normal distribution, while the explanatory variable x_i is given by the realizations of a uniform distribution in the [0, 1] interval in a sample of size $n = 100$. Figures 2.10 and 2.11 consider the quantile and the expectile plot of the dependent variable. In both graphs the tails diverge from the straight line – particularly in the quantile plot – thus signaling the presence of outliers. The model $y_i = \beta_0 + \beta_1 x_i + e_i$ is estimated at the 10th, 25th, 50th, 75th, and 90th expectiles. The results are reported in Table 2.5 and in the graph of Figure 2.12. While the quantile regression estimates of this model, reported in Table 1.6, were all statistically relevant but at the 10th and 90th quantiles, here only the two top expectiles yield statistically significant estimated slopes. Figure 2.12, when compared with Figure 1.19, shows that some of the outliers are not well detected since the expectiles are attracted by them and for $\theta < .75$, their slopes are flatter than at the corresponding quantiles.

2.1.2   French data: influential outlier in the dependent variable

As discussed in Section 1.2.1, this data set presents a very implausible observation characterized by the highest walking speed and a large *BMI*. This is an outlier in the dependent variable and is most likely a coding error. To check the normality of the variables, the quantile plot compares the theoretical quantiles of the normal

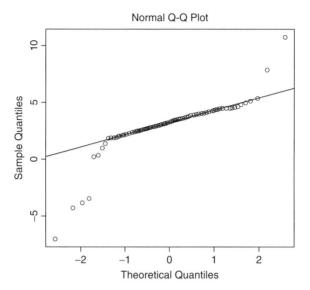

Figure 2.10 Quantile plot of the dependent variable in the data set with 10% contaminated error distribution. In the tails, the sample quantiles are quite far from the theoretical ones, clearly signaling the presence of at least six outliers.

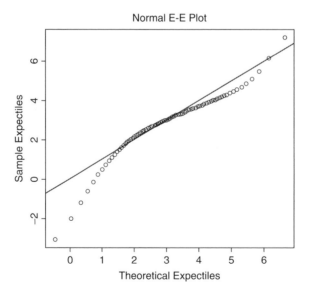

Figure 2.11 Expectile plot of the dependent variable in the data set with 10% contaminated error distribution. In the tails, the sample expectiles are far from the theoretical ones. The divergence is smaller than in the quantile plot case.

Table 2.5   Estimated expectiles, 10% contamination in the errors.

θ	10th	25th	50th	75th	90th
slope	−0.414	0.578	0.94	1.07	1.32
se	(1.97)	(0.98)	(0.59)	(0.45)	(0.68)
intercept	1.32	2.06	2.60	3.12	3.55
se	(1.18)	(0.53)	(0.33)	(0.28)	(0.44)

Note: Standard errors in parenthesis, sample size $n = 100$.

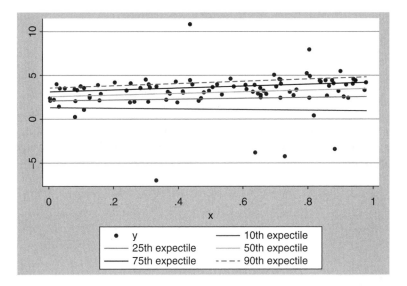

Figure 2.12   Estimated expectiles in the data set characterised by 10% contamination in the error distribution, in a sample of size $n = 100$. The graph shows the presence of some outlying observations generated by the contaminating normal with larger variance, $\sigma^2 = 36$. The lower expectiles are attracted by the outliers and are flatter than the corresponding quantiles in Figure 1.19. As a result some outliers are not well detected, such as the observation ($y = .38\,x = .8$), which is quite close to the 10th estimated expectile.

distribution with the empirical quantiles of the variable under analysis. An analogous plot is computed for the expectiles, comparing the expectiles of a normal variate with the empirical expectiles of the data. The graphs in Figures 2.13 to 2.16 report the expectile and the quantile plots for *wspeed* and *BMI* in the French data set. It can be seen that the presence of an outlier in *wspeed* is by far more evident in the quantile than in the expectile plot. This is the case since while one outlier affects all the expectiles, only the extreme quantile, the top one in this sample – or the bottom one, depending on the position of the outlier – is attracted by an outlier, and this makes its presence more evident.

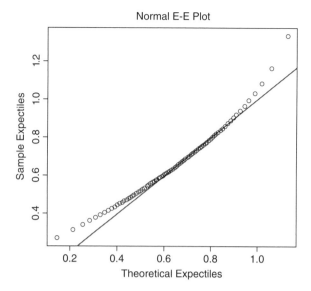

Figure 2.13   Expectile plot of *wspeed*, French data set. The expectiles of this variable are compared to the expectiles of the normal distribution. Both tails diverge from the straight line.

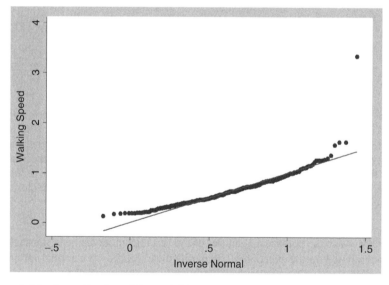

Figure 2.14   Quantile plot of the variable *wspeed* in the French data set. Besides the divergence from normality in the tails, the top quantile is quite far from the others and from the theoretical normal quantile. This signals the presence of outliers attracting the top quantile. The comparison with the expectile plot shows how the presence of an outlier is more evident in the quantile than in the expectile graph.

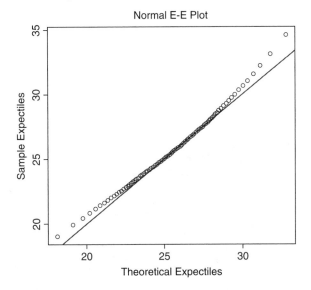

Figure 2.15 Expectile plot of *BMI* in the French data set.

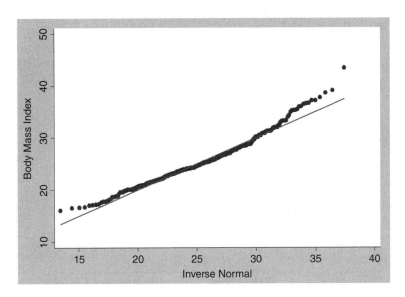

Figure 2.16 Quantile plot of *BMI* in the French data set.

The previous chapter has discussed how the outlying data points attract the OLS but not the median regression. Indeed OLS yields a positive slope, while the median regression presents a negative slope. This same attraction effect can be found in all the estimated expectiles. Figure 2.17, depicting the estimated expectiles, shows a fan-shaped set of estimated lines, where the slopes are all positive, although quite small, and the 90th expectile presents the largest of all slopes. Figure 2.18, which

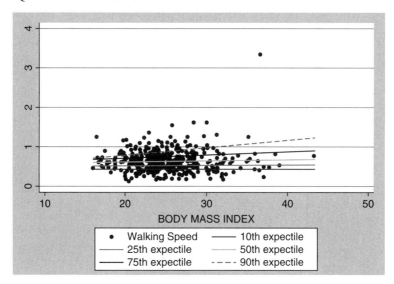

Figure 2.17   Estimated expectiles at $\theta = .10, .25, .50, .75, .90$, French data set of size $n = 393$. The outlier in the dependent variable at the top left of the graph attracts all the expectiles, thus computing an implausible positive slope between walking speed and body mass index.

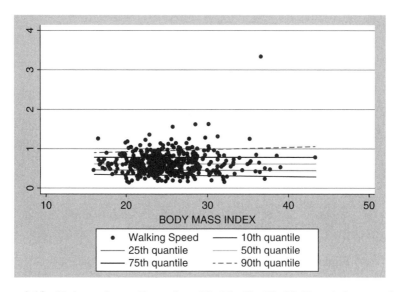

Figure 2.18   Estimated quantiles at $\theta = .10, .25, .50, .75, .90$, French data set of size $n = 393$. The outlier in the dependent variable at the top left does not attract all the quantile regressions. The latter provides a negative slope at all but the 90th quantile.

Table 2.6 Estimated slope at the selected expectiles and quantiles, French data set.

θ	10th	25th	50th	75th	90th
expectile	0.0005	0.0011	0.0033	0.0084	0.0196
quantile	−0.0029	−0.0007	−0.00057	−0.00023	0.0059

Note: Sample size $n = 393$.

collects the estimated quantiles, presents all negatively sloped lines but at the 90th quantile, thus showing the greater robustness of the quantile estimator with respect to outliers in the dependent variable. Table 2.6 compares the slope estimated at the selected expectiles and quantiles.

2.1.3   The Netherlands example: outlier in the explanatory variable

The expectile and the quantile plots for *wspeed* and *BMI* of the Dutch sample, reported in Figures 2.19 to 2.22, show once again how the presence of outliers is more evident in the quantile than in the expectile plots. In the Dutch data set there is one non-influential outlier in the explanatory variable, *BMI*, and the OLS results do not differ much from the ones of the median regression. This similar behavior carries

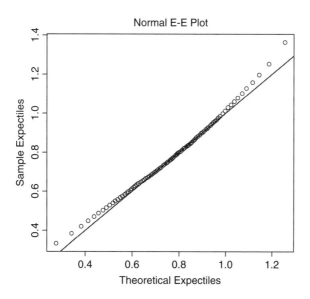

Figure 2.19   Expectile plot of *wspeed* in the Dutch data set.

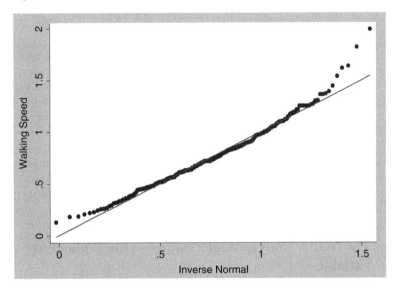

Figure 2.20   Quantile plot of *wspeed* in the Dutch data set.

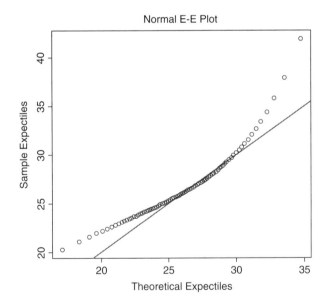

Figure 2.21   Expectile plot of *BMI* in the Dutch data set.

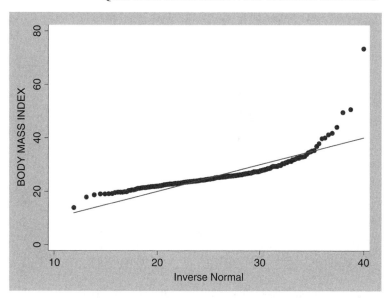

Figure 2.22   Quantile plot of *BMI* in the Dutch data set. The anomalous value in *BMI* is more evident in this graph than in the previous graph depicting the expectile plot.

Table 2.7   Estimated slope at the selected expectiles and quantiles, Dutch data set.

θ	10th	25th	50th	75th	90th
expectile	−0.0024	−0.0042	−0.0058	−0.0070	−0.0078
quantile	−0.0001	−0.0052	−0.0065	−0.0089	−0.0093

Note: Sample size $n = 307$.

on to the expectiles computed at $\theta \neq .50$, as can be seen in Table 2.7, which reports the slopes estimated by the two estimators. Figures 2.23 and 2.24 depict respectively the estimated expectiles and the estimated quantiles. It should be noted that, due to the selected method to compute the expectiles, which calculates an expectile as deviation from the conditional mean, i.e., the OLS regression, crossing of the fitted lines is avoided in Figure 2.23 while this is not the case in the quantile regressions of Figure 2.24. Quantile crossing is a problem occurring in case of misspecification, collinearity in the explanatory variables, small sample size, or, as in this example, in the presence of large outliers attracting the extreme quantiles.

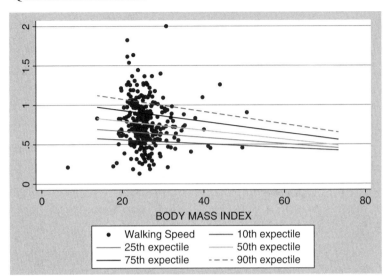

Figure 2.23 Estimated expectiles at $\theta = .10, .25, .50, .75, .90$, Dutch data set of size $n = 307$. Due to the selected computational method, the expectile fitted lines do not cross.

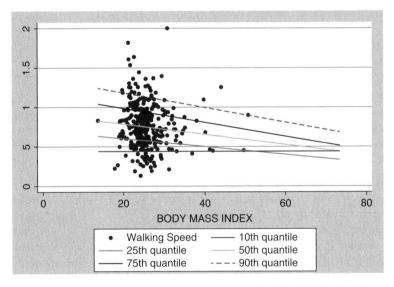

Figure 2.24 Estimated quantile regressions at $\theta = .10, .25, .50, .75, .90$, Dutch data set of size $n = 307$. In this graph the 10th quantile intersects the 25th estimated line due to the large outlier in *BMI* affecting the lower quantile. This is a case of quantile crossing.

2.2   M-estimators

Section 1.3 shows how the OLS estimator is characterized by an unbounded influence function. The IF_{OLS} does not control residuals nor outliers in the explanatory variables. In the linear regression model $y_i = \beta_0 + \beta_1 x_i + e_i$, the influence function of the OLS estimator is $\text{IF}_{OLS} = \frac{e_i \mathbf{x}_i}{E \mathbf{x}_i \mathbf{x}_i^T}$, where $\mathbf{x}_i^T = \begin{bmatrix} 1 & x_i \end{bmatrix}$ is the ith observation for the $p = 2$ explanatory variables. Huber (1964, 1981) proposes to induce robustness in the OLS estimator by bounding large residuals, thus defining the class of M-estimators for the linear regression model. The purpose is to provide a good fit to the majority of the data by controlling the impact of outliers. The introduction of a bound implies to modify the least squares objective function in order to control outliers. The new objective function is defined as

$$\sum \rho \left(\frac{y_i - \beta_0 - \beta_1 x_i}{\sigma} \right)$$

where the ρ function is given by

$$\rho(e_i) = \begin{cases} \dfrac{1}{2} u_i^2 & \text{for } |u_i| \le c \\ c|u_i| - \dfrac{1}{2} c^2 & \text{for } |u_i| > c \end{cases}$$

The u_i are the standardized errors, $u_i = \frac{y_i - \beta_0 - \beta_1 x_i}{\sigma} = \frac{e_i}{\sigma}$, and σ is generally estimated by $\text{MAD} = \frac{median|e_i|}{.6745}$. The above objective function keeps the L_2 norm: it minimizes the sum of squared errors, as OLS, meanwhile controlling their value through the ρ function. The arbitrary constant c is the term that bounds outliers. A very large value of c yields the standard OLS estimator, while a very small value of c drastically curbs outliers. A frequent choice for this tuning constant is $c = 1.345$. As in OLS, the M-estimator computes the regression passing through the center of the conditional distribution of the dependent variable.

The partial derivatives for intercept and slope defining the M-estimators of the regression coefficients are

$$\sum \psi \left(\frac{y_i - \beta_0 - \beta_1 x_i}{\sigma} \right) = 0$$

$$\sum \psi \left(\frac{y_i - \beta_0 - \beta_1 x_i}{\sigma} \right) x_i = 0$$

where the $\psi(u_i)$ function is defined as (Huber proposal 2)

$$\psi(u_i) = \begin{cases} -c & \text{for } u_i < -c \\ u_i & \text{for } -c \le u_i \le c \\ c & \text{for } u_i > c \end{cases}$$

The $\psi(u_i)$ function replaces the too large/small residuals with the c value so that outliers are not allowed to have a value larger/smaller than $\pm c$, and their impact on the M-estimator is under control. This provides an influence function bounded in the residuals, the IR term of IF. Indeed, for the slope coefficient, the IF of the M-estimator is given by $\text{IF}_M = \frac{\psi(u_i)x_i}{E\psi'(u_i)}$, where $\psi'(u_i)$ is the first derivative of $\psi(u_i)$, while for the intercept it is $\text{IF}_M = \frac{\psi(u_i)}{E\psi'(u_i)}$. The IF_M of the slope shows that outliers in the explanatory variables are not curbed, since $\psi(u_i)$ controls only residuals, the IR term, but not the IP component of IF_M, which remains unbounded. Thus, the slope in the Huber M-estimator is still influenced by anomalous values in the explanatory variables.

The $\rho(u_i)$ functions of the M-estimators can be expressed in terms of weights, and this estimator can be defined as a weighted least squares estimator,

$$\sum \rho \left(\frac{y_i - \beta_0 - \beta_1 x_i}{\sigma} \right) = \sum w_i \left(\frac{y_i - \beta_0 - \beta_1 x_i}{\sigma} \right)^2 = \sum w_i u_i^2$$

with partial derivatives for intercept and slope given by

$$\sum w_i \left(\frac{y_i - \beta_0 - \beta_1 x_i}{\sigma} \right) = 0$$

$$\sum w_i \left(\frac{y_i - \beta_0 - \beta_1 x_i}{\sigma} \right) x_i = 0.$$

The weighting system in the Huber M-estimator is defined as

$$w_{i,H} = \min \left(1, \frac{c}{|u_i|} \right)$$

when $|u_i| < c$, then $w_{i,H} = 1$, while for large standardized residuals, $w_{i,H} = \frac{c}{|u_i|} < 1$, and the ith observation is downweighted. The estimation procedure starts implementing the simple OLS. The OLS residuals are standardized, and the weights are computed. The second round implements a weighted least squares regression, and the new standardized residuals allow to recompute the weights. These steps are repeated until convergence is reached.

A different M-estimator, presented by Krasker and Welsh (1982), introduces a bound not only on the residuals but also on the explanatory variables, in order to control both the IR and the IP component of IF. The Krasker and Welsh M-estimator is defined by the following weighting system

$$w_{i,KW} = \min \left(1, \frac{c}{(\mathbf{x}_i^T A^{-1} \mathbf{x}_i)^{1/2} |u_i|} \right).$$

where the term $(\mathbf{x}_i^T A^{-1} \mathbf{x}_i)$ is a robust measure of distance in the $\mathbf{x}_i$'s, standardized by $A = E(w_{i,KW}^2 (u_i)^2 \mathbf{x}_i \mathbf{x}_i^T)$. The residuals, the weights, and the robust measure of distance in the $\mathbf{x}$'s are iteratively computed. In this estimator, the bound c controls both outlying residuals and/or outliers in the explanatory variables.

There are many possible alternative definitions of the $\psi(u_i)$ function, each identifying a different M-estimator. The Beaton and Tukey (1974) biweight M-estimator is defined by the function

$$\psi(u_i) = u_i(c^2 - u_i^2)^2 1_{[-c,c]}(u_i)$$

or by the weights

$$w_{i,T} = \min\left[0, \left(1 - \left(\frac{c}{|u_i|}\right)^2\right)^2\right]$$

and a frequently used value of the bound is $c = 4.685$. The Hampel M-estimator (Andrews et al., 1972) is defined by

$$\psi(u_i) = \begin{cases} u_i & \text{for} \quad 0 \le |u_i| < a \\ a\ sgn(u_i) & \text{for} \quad a \le |u_i| < b \\ a\ \dfrac{c - |u_i|}{c - b}\ sgn(u_i) & \text{for} \quad b \le |u_i| \le c \\ 0 & \text{for} \quad |u_i| > c \end{cases}$$

with $0 < a < b < c < \infty$, $c - b \ge 2a$, and $sgn(.)$ is the sign function.

By appropriately defining the weighting system, all the $\psi(u_i)$ functions can be rewritten in terms of weighted least squares, and the weights provide a very good diagnostic tool. The smaller weights signal outlying data points that have been heavily downweighted or even excluded by the sample in order to control their impact on the estimated coefficients.

Both the Tukey and the Hampel M-estimators, also known as redescending estimators, comprise the possibility of observations having assigned a zero weight, which implies the exclusion of these observations from the sample. In these estimators, the weights can assume any value in the [0, 1] interval.

In addition, the simple exclusion of the anomalous values from the sample yields the trimmed estimator, which coincides with an M-estimator assigning only two values: unit weight to the majority of the data and zero weight to the large outliers.

As mentioned, the M-estimators are computed iteratively. Convergence is guaranteed for convex ρ functions and for redescending estimators. Often the Huber ρ function is considered for the initial iterations while a redescending estimator is implemented for the final convergence, and this is actually the way the Huber-Tukey M-estimator results of Table 2.8 have been computed, where the redescending estimator implemented in the final iterations is the biweight function. This allows to assign a zero weight to those data points that is preferable to exclude from the sample.

Finally, the class of M-estimators comprises the mode regression estimator. The regression passing through the conditional mode is defined by maximizing $n^{-1} \sum \delta^{-1} K(e_i/\delta)$, with e_i being the regression errors, δ the bandwidth, and K the kernel function (Lee, 1993, Kemp and Santos Silva, 2012). The kernel function assumes the same role of the robust functions $\rho(\ )$ and $\psi(\ )$ in the M-estimators,

Table 2.8   Results in the Anscombe data, $n = 11$.

	$[Y_1\, X_1]$	$[Y_1^*\, X_1]$	$[Y_1^{**}\, X_1]$
Huber-Tukey			
slope	0.50	0.49	0.44
se	(0.14)	(0.14)	(0.12)
constant	2.97	2.87	3.17
se	(1.34)	(1.34)	(1.17)
Huber estimator			
slope	0.50	0.50	0.50
se	(0.15)	(0.19)	(0.19)
constant	2.98	2.99	2.99
se	(1.44)	(1.80)	(1.80)
Hampel estimator			
slope	0.50	0.49	0.46
se	(0.12)	(0.13)	(0.13)
constant	3.00	2.89	3.07
se	(1.12)	(1.29)	(1.33)
median regression			
slope	0.48	0.48	0.48
se	(0.19)	(0.19)	(0.19)
constant	3.24	3.24	3.24
se	(2.04)	(2.04)	(2.04)
OLS estimator			
slope	0.50	0.50	0.61
se	(0.12)	(0.25)	(0.29)
constant	3.00	3.56	2.94
se	(1.12)	(2.44)	(2.84)

Note: Standard errors in parenthesis, sample size $n = 11$.

i.e., bounding outliers. In case of truncated data, the conditional mode provides a consistent estimate of the non-truncated conditional mean.

Next consider the Anscombe model to analyze the behavior of the M-estimators. Table 2.8 reports the results of some of the M-estimators so far discussed of the median regression and of OLS for the $[Y_1\, X_1]$ Anscombe data without outliers, together with the estimates in the modified data sets with one outlier, $[Y_1^*\, X_1]$, and with two outliers in the dependent variable $[Y_1^{**}\, X_1]$. In this table the median regression does not change when one and then two outliers are introduced in the original data set, since the modified observations preserve their original position above the median. Also the slope computed by the Huber estimator does not change from one sample to the other. Vice versa, in the OLS, in the Hampel, and in the Huber-Tukey estimator the slope does change. In $[Y_1^{**}\ X_1]$ the OLS slope increases since the OLS line is attracted by the two outliers. In the Hampel and in the Huber-Tukey case, instead, the slope decreases as the number of outliers in the sample increases. In these two M-estimators the bound curbs more severely the outliers, and thus the slope decreases as their number increases. Figures 2.25 and 2.26 compare the OLS, the M-estimators, and the median regression fitted lines in the modified Anscombe data $[Y_1^*\ X_1]$ and $[Y_1^{**}\ X_1]$. In these figures, the Huber-Tukey estimated slope decreases from one data set to the other to balance the outliers, while the OLS estimates increase since OLS is attracted by the outliers in the dependent variable.

Table 2.9 reports the weights of the M-estimators introduced to curb outliers. They turn out to be a very useful diagnostic tool. The smallest weights, those close or equal to zero, signal the outlying observations. For instance in the Anscombe data $[Y_1^{**}\ X_1]$, modified to include the two outliers at the 4th and the 9th observation, the Huber M-estimator yields the following less-than-unit weights: $w_{3,H} = .756; w_{4,H} = .206; w_{9,H} = .266; w_{10,H} = .883$. The two anomalous data are heavily downweighted, but also the nearby observations have been somewhat downweighted. In this same data set the Huber-Tukey M-estimator yields the following weights: $w_{4,HT} = 0; w_{9,HT} = .015$, and the nearby observations are not much downweighted, $w_{3,HT} = .923; w_{10,HT} = .910$. Thus the non-exclusion of large outliers in the Huber ρ function causes the downweighting of the nearest data points, which are, instead, legitimate observations. The simple Tukey weights w_T do not

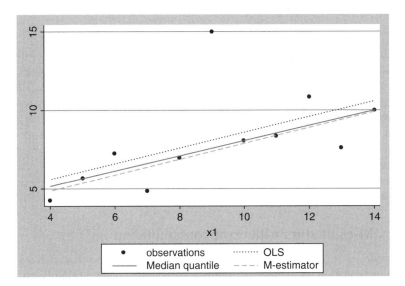

Figure 2.25 Comparison of results in the Anscombe model modified to include one outlier in the dependent variable, data set $[Y_1^*\ X_1]$, sample size $n = 11$. The two bottom lines are the median regression and the Huber-Tukey M-estimator fitted lines. The OLS estimator provides the greatest intercept since the fitted line is attracted by the outlier.

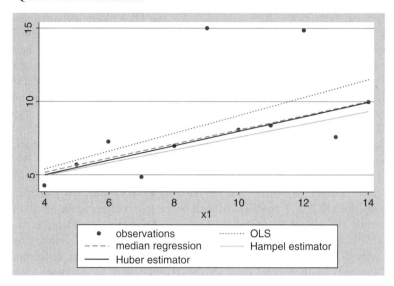

Figure 2.26   Comparison of results in the Anscombe model modified to include two outliers in the explanatory variable, data set $[Y_1^{**}\ X_1]$, sample size $n = 11$. Median regression and Huber and Hampel estimators yield similar results. The Huber estimator does not change its estimated slope with respect to the $[Y_1^*\ X_1]$ sample. The OLS slope is the only one to increase from one sample to the other; the fitted line is attracted by the two outliers and is steeper than all the others.

differ much from the combined Huber-Tukey weights w_{HT}. They both downweight by a small amount the good observations while annihilating the outlying ones.

There is indeed a trade-off in the definition of the weighting systems. The zero weights exclude outliers, avoiding their impact on the estimates, but they also give up completely their informative content.[1] Alternatively, assigning them a very small weight to keep them in the sample causes the nearby legitimate observations to be downweighted. The Hampel weights, w_{Ha}, in the $[Y_1^{**}\ X_1]$ sample are capable to pinpoint and downweight only the outliers. However, in the $[Y_1^*\ X_1]$ case the 4th outlying observation is not downweighted at all while the 9th legitimate observation receives a weight smaller than one.

2.2.1   M-estimators with error contamination

The model $y_i = \beta_0 + \beta_1 x_i + e_i$ of Section 1.3.2, with errors characterized by a 10% contamination and with a uniformly distributed X, is here robustly estimated considering the Huber estimator and the combination of the Huber and Tukey $\rho(\ )$ functions. The results provided by these estimators are reported in the first

---

[1] In a regression with many explanatory variables, $p > 2$, the ith outlier may be damaging only for some of the x_{ij}'s. By dropping the entire observation, the information of the good components of x_i are lost.

Table 2.9   Robust weights in the $[Y_1 \ X_1]$, $[Y_1^* \ X_1]$ and $[Y_1^{**} \ X_1]$ data sets.

Huber-Tukey weights $w_{i,HT}$			Huber weights $w_{i,H}$		
$[Y_1 \ X_1]$	$[Y_1^* \ X_1]$	$[Y_1^{**} \ X_1]$	$[Y_1 \ X_1]$	$[Y_1^* \ X_1]$	$[Y_1^{**} \ X_1]$
.999	.997	.991	1	1	1
.999	.999	.997	1	1	1
.673	.750	.923	.752	.755	.756
.853	0	0	1	.206	.206
.995	.999	.996	1	1	1
.999	.999	.984	1	1	1
.865	.847	.915	1	1	1
.949	.969	.979	1	1	1
.723	.699	.015	.836	.879	.266
.751	.809	.910	.874	.883	.883
.997	.991	.996	1	1	1

Tukey weights w_T			Hampel weights w_{Ha}		
$[Y_1 \ X_1]$	$[Y_1^* \ X_1]$	$[Y_1^{**} \ X_1]$	$[Y_1 \ X_1]$	$[Y_1^* \ X_1]$	$[Y_1^{**} \ X_1]$
.999	.996	.982	1	1	1
.999	.999	.995	1	1	1
.731	.673	.843	1	1	1
.877	0	0	1	1	.0093
.997	.999	.992	1	1	1
.999	.998	.969	1	1	1
.887	.797	.829	1	1	1
.958	.958	.958	1	1	1
.766	.606	0	1	.912	.136
.794	.748	.820	1	1	1
.997	.988	.992	1	1	1

two columns of Table 2.10, under the heading Huber and Huber-Tukey. The third column of this table reports the OLS estimates, the second to last column shows the results provided by the trimmed estimator, and the last column reports the estimated median regression. The robustly estimated coefficients are statistically relevant in the M-estimators, as occurs in the median regression computed in Section 1.3.2. Vice versa, the OLS estimated slope is small and with a large standard error so that the OLS slope is not statistically different from zero. As mentioned, one relevant feature of the M-estimators is the set of final weights implemented to compute the robust regression coefficients. Table 2.11 reports the weights smaller than 0.7, signaling the outlying observations that have been downweighted in order to reduce their contribution to the estimated coefficients. For some observations the impact has

Table 2.10   Comparison of M-estimators and OLS results.

	Huber	Huber-Tukey	OLS	6%-trimmed	median
slope	1.183	1.299	.940	1.413	1.753
se	(.35)	(.33)	(.71)	(.36)	(.36)
constant	2.678	2.661	2.599	2.605	2.372
se	(.21)	(.20)	(.42)	(.21)	(.21)

Note: Standard errors in parenthesis, sample size $n = 100$.

Table 2.11   M-estimators, weights smaller than 0.7.

x_i	y_i	$w_{i,HT}$	$w_{i,H}$
.80	7.89	.072	.34
.43	10.79	0	.19
.73	−4.28	0	.18
.33	−7.02	0	.14
.08	.250	.540	.57
.64	−3.83	0	.20
.88	−3.46	0	.20
.81	.38	.283	.44

Note: The weights $w_{i,H}$ define the Huber M-estimator, while $w_{i,HT}$ refer to the weights generated by the Huber ρ function in the initial iterations followed by the Tukey ρ function afterward.

been reduced by less than 30%. As can be seen comparing Table 1.5 and Table 2.11, the observations downweighted by the M-estimators coincide with the observations detected as outliers by the standardized and studentized residuals of the median regression. In particular, the Huber-Tukey M-estimator completely excludes five observations, assigning them a zero weight, $w_{i,HT} = 0$. The remaining three outliers are very heavily downweighted, contributing by 54%, 28%, or by only 7% to the final estimates. The Huber weights, $w_{i,H}$, would not exclude any of the five data points but would assign them very low values, generally around .2. In the table, $w_{i,H}$ is generally larger than $w_{i,HT}$.

The weights of all the M-estimators provide a built-in diagnostic tool, thus bypassing the need to compute additional diagnostic measures such as the standardized or the studentized residuals. Figure 2.27 presents the weights of the Huber-Tukey M-estimator, where five of them are on the zero line, and another one is very close to it. Figure 2.28 compares the estimates provided by OLS, median regression and Huber-Tukey M-estimator in the contaminated errors model. Finally, once the outliers are detected, it is very easy to compute the trimmed regression, which amounts to implement OLS after the exclusion of the anomalous values. In this case the exclusions comprise the five observations having zero weight in

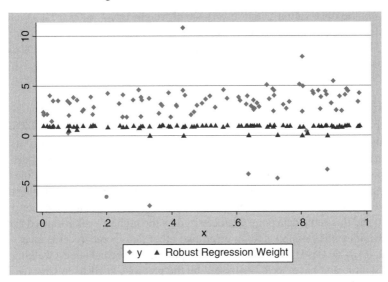

Figure 2.27   The triangles represent the weights generated by the Huber-Tukey M-estimator in the model with contaminated errors, in a sample of size $n = 100$. The six largest outliers receive a weight equal or very close to zero, while the majority of the observations receive $w_{i,HT}$ equal or close to one.

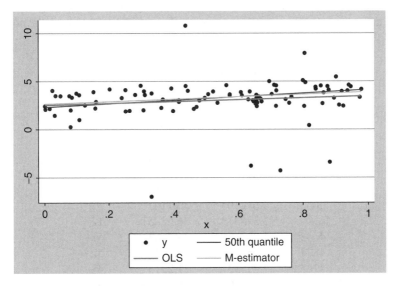

Figure 2.28   Comparison of median, Huber-Tukey M-estimator, and OLS in the model with contaminated errors, in a sample of size $n = 100$. The OLS line is flatter, and the slope has a larger standard error, while the other two estimated lines are close to one another.

$w_{i,HT}$ and the observation with the very small weight $w_{i,HT} = .072$. This yields the estimates reported in the second to last column of Table 2.10. The results are close to the Huber-Tukey estimates, with slightly larger standard errors. Since six out of 100 observations have been excluded, this is a 6%-trimmed estimator.

2.2.2   The French data

In the French example, the robustly estimated slope is negative, $\widehat{\beta}_{1,HT} = -0.00233$ in the Huber-Tukey case and $\widehat{\beta}_{1,H} = -0.0014$ in the Huber estimator, just as occurs in the median regression. The weights $w_{i,HT}$ smaller than .5 of the Huber-Tukey M-estimator are reported in the first column of Table 2.12, while in the last column are the Huber weights, $w_{i,H}$. In the central columns are the values of the dependent and independent variables. There is only one observation excluded from the sample by a zero $w_{i,HT}$ weight, and this is the data point characterized by the highest walking speed and high BMI. Besides this possible coding error, two other data points receive a very small weight, $w_{i,HT} = .06$. These two observations present the second largest walking speed of the sample and have been overlooked by the previous analysis in Chapter 1. The robust weighting system brings them to the attention of the analyst. These observations become clearly visible in the top section of the graph of Figure 1.12, once the possible coding error has been dropped. They were masked by the outlier. In Table 2.12, the downweighted data are those observations having a walking speed above 1.2, and the downweight becomes heavier in case of a walking speed above 1.5. The last column of the table reports the Huber weights, which once again are generally greater than $w_{i,HT}$, and assign a nonzero weight even to the largest outliers.[2] Indeed, the possible

Table 2.12   The smallest robust weights $w_{i,HT}$ and $w_{i,H}$ in the French data set.

$w_{i,HT}$	wspeed	BMI	$w_{i,H}$
.1238	1.552795	25.81663	.35
.4561	1.269036	29.70041	.50
.4935	1.25	25.63117	.52
.3684	1.351351	22.94812	.45
.0633	1.612903	30.11621	.33
0	3.333333	36.62851	.12
.4643	1.25	35.2653	.51
.4871	1.25	27.73438	.52
.4168	1.295337	31.2213	.48
.0661	1.612903	28.51563	.33

---

[2] The downweighting of the close-by observations is not reported in the table since these weights are greater than 0.5.

coding error, describing an individual obese and with the highest walking speed of the entire sample, receives a small but positive weight, $w_{i,H} = .12$, while it is excluded by the Huber-Tukey estimator, $w_{i,HT} = 0$.

2.2.3   The Netherlands example

The Dutch data set presents a robustly estimated slope coinciding with the OLS result, $\hat{\beta}_{1,HT} = -0.0058$ and $\hat{\beta}_{1,H} = -0.0060$. The weighting system provided by the Huber-Tukey estimator yields few weights smaller than .5, as reported in the first column of Table 2.13. The zero weight is assigned to the largest walking speed value, and the lower weights are assigned to the observations with $wspeed > 1.3$. The largest value of the explanatory variable, $BMI = 73.39$, receives a weight very close to one, $w_{i,HT} = .9987$, and is not reported in the table. This is the case since the M-estimator does not curb outliers in the explanatory variables, just as occurs in OLS, in the expectile, and in the quantile regression estimators. However, this observation, although being a large outlier in BMI, is not influential. Indeed, it has a low sample influence function, $\text{SIF}(\hat{\beta}_{1,OLS}) = -0.122$, $\text{SIF}(\hat{\beta}_1(.5)) = 0.122$, and $\text{SIF}(\hat{\beta}_{1,HT}) = 306(-0.0058 + 0.0055) = -0.091$. The last column in Table 2.13 reports the Huber estimator weights, which do not comprise any zero weight and which are generally larger than the Huber-Tukey weights. In this set of weights, w_H, the non-influential outlier in $BMI = 73.39$ receives a unit weight $w_{i,H} = 1$. The Dutch data set draws the attention to robustness with respect to outliers in the explanatory variables. Among the M-estimators, the bounded influence estimator defines weights that take into account both kinds of outliers, in the errors and in the explanatory variables. For the quantile regression estimator, Neykov et al. (2012) discuss the trimmed quantile regression, computed in a subset of the sample. By selecting the trimming constant k, that is the number of excluded observations from the sample, the trimmed quantile regression estimator is defined as $\min_\beta [\min_{I_k} \sum_{i \in I_{(k)}} \rho(e_i)]$, where $\rho(e_i)$ is the usual quantile regression function, $\rho(e_i) = e_i[\theta - 1(e_i < 0)]$, implemented in the subset $I_{(k)}$. This allows to exclude outliers in both dependent and explanatory variables.

Table 2.13   The smallest robust weights, $w_{i,HT}$ and $w_{i,H}$, in the Dutch data set.

$w_{i,HT}$	$wspeed$	BMI	$w_{i,H}$
.4458	1.440922	23.88946	.52
0	1.992032	30.80125	.28
.4715	1.388889	29.38476	.53
.3416	1.533742	21.79931	.46
.2463	1.612903	21.43461	.42
.0487	1.818182	21.29529	.34
.2124	1.633987	22.77319	.41

2.3   M-quantiles

M-quantiles merge together the M-estimators and the expectiles approach. Both methods are implemented within the least squares model, in the L_2 norm. While quantile regressions consider the L_1 norm, thus taking into account only the sign of the residuals, that is, the position of an observation above or below the estimated line, expectiles and M-estimators take into account both the sign and the value of each residual. This affects the robustness of expectiles but not of the M-estimators, which preserve robustness by introducing a weighting system to control the outlying observations. The M-estimator, which looks for the robust conditional mean regression, is generalized by Breckling and Chambers (1988) to compute robust regressions passing through different points of the conditional distribution besides the conditional mean. This yields the robust regression analogues of the expectiles, named asymmetric M-estimator or M-quantiles. The M-quantile estimator is defined by considering a $\psi_\theta(\ )$ function, which besides bounding outliers is also asymmetric, thus allowing to compute a robust regression away from the conditional mean. Although it can be considered as the analogue of a quantile regression within the least squares framework, it does not provide an alternative measure of quantiles since, just as in the expectiles approach, the estimated line – or plane – does not divide the observations in two groups, θ of them below and $1 - \theta$ above the estimated M-quantile line/plane. As for the expectiles, θ is only a skewness parameter, which sets the position of the estimated line away from the center. For $\theta=.5$ the M-quantile coincides with the M-estimator.

In detail, the objective function of the M-quantile estimator computed at θ is defined as

$$\sum_{i=1,n} \rho(u_i)[(1 - \theta)I(u_i \leq 0) + \theta I(u_i > 0)]$$

The $\rho(.)$ function is any of the robust regression functions defined in Section 2.2. The first derivative is $\sum_{i=1,n} \psi_\theta(u_i)\mathbf{x}_i = 0$, with $\mathbf{x}_i = [1 \ x_i]$ and

$$\psi_\theta(u_i) = \begin{cases} (1 - \theta)\psi(u_i) & \text{for} \quad u_i \leq 0 \\ \theta\psi(u_i) & \text{elsewhere} \end{cases}$$

The u_i are the standardized residual $u_i = \frac{e_i}{\sigma}$, σ is computed by MAD $= \frac{median|e_i|}{0.6745}$, and $\psi_\theta(u_i)$ premultiplies by the asymmetric weights θ and $(1 - \theta)$ the $\psi(u_i)$ function defining the Huber M-estimator or any other of the $\psi(\ )$ functions belonging to the class of the M-estimators. For instance in the Huber M-estimator case, it is (Pratesi et al., 2009)

$$\psi_\theta(u_i) = \begin{cases} -(1 - \theta)c & \text{for} \quad u_i < -c \\ (1 - \theta)u_i & \text{for} \quad -c \leq u_i < 0 \\ \theta u_i & \text{for} \quad 0 \leq u_i \leq c \\ \theta c & \text{for} \quad u_i > c \end{cases}$$

and for $\theta = .5$, the function $\psi_{.5}(u_i)$ yields the Huber M-estimator. As for all the M-estimators, a small c increases robustness, while a very large c yields the OLS estimator. For $\theta \neq .5$ and a very large c, the M-quantiles coincide with the expectiles. Considering the Huber robust weights $w_{i,H} = min(1, \frac{c}{|u_i|})$, their combination with θ and $(1 - \theta)$ location parameters defines the M-quantile weights $w_{i,M}$

$$
w_{i,M} = \begin{cases} -(1 - \theta)\dfrac{c}{|u_i|} & \text{for} & u_i < -c \\ (1 - \theta) & \text{for} & -c \le u_i < 0 \\ \theta & \text{for} & 0 \le u_i \le c \\ \theta\dfrac{c}{|u_i|} & \text{for} & u_i > c \end{cases}
$$

Table 2.14 reports the M-quantile estimated coefficients in the three Anscombe data sets, $[Y_1\ X_1]$, $[Y_1^*\ X_1]$ and $[Y_1^{**}\ X_1]$. The Huber-Tukey $\rho(\ )$ function has been here considered as the central M-estimator that robustly computes the conditional mean regression. Therefore, the regression model is first robustly computed at $\theta=.5$ by the simple Huber-Tukey robust estimator of Section 2.2. The regressions away from the robust conditional mean are estimated as deviations from the center using the above $w_{i,M}$ weights in the equation $\sqrt{w_{i,M}}Y_{1i} = \beta_0 + \beta_1 \sqrt{w_{i,M}}X_{1i} + e_i$. Analogously

Table 2.14   Huber-Tukey M-quantile estimates for the Anscombe data set $[Y_1\ X_1]$ and the modified data $[Y_1^*\ X_1]$ and $[Y_1^{**}\ X_1]$.

θ	10th	25th	50th	75th	90th
$[Y_1\ X_1]$					
slope	0.51	0.51	0.50	0.52	0.56
se	(0.08)	(0.09)	(0.14)	(0.11)	(0.10)
constant	2.26	2.56	2.98	3.21	3.17
se	(0.81)	(0.94)	(1.3)	(1.2)	(1.1)
$[Y_1^*\ X_1]$					
slope	0.48	0.49	0.49	0.49	0.49
se	(0.10)	(0.11)	(0.14)	(0.14)	(0.14)
constant	2.36	2.59	2.87	3.38	3.61
se	(0.91)	(1.1)	(1.3)	(1.4)	(1.3)
$[Y_1^{**}\ X_1]$					
slope	0.42	0.45	0.44	0.48	0.48
se	(0.10)	(0.14)	(0.12)	(0.17)	(0.17)
constant	2.60	2.77	3.17	3.46	3.67
se	(0.92)	(1.3)	(1.2)	(1.7)	(1.7)

Note: Standard errors in parenthesis, sample size $n = 11$.

in the data set $[Y_1^* \; X_1]$, the robust regression passing through the conditional mean is computed by the Huber-Tukey estimator, while the lines away from the conditional mean are computed as deviations from it in the equation $\sqrt{w_{i,M}} Y_{1i}^* = \beta_0 + \beta_1 \sqrt{w_{i,M}} X_{1i} + e_i$. The same procedure is implemented for the $[Y_1^{**} \; X_1]$ data set.[3]

The selection of a different M-estimator for the initial robust regression of the conditional mean, which provides the weights bounding outliers, would yield different M-quantiles estimates.

The results of Table 2.14 are depicted in Figures 2.29 to 2.31. They can be compared with the estimates and the graphs of the expectiles in Section 2.1. Figure 2.29 shows M-quantile estimated lines presenting a pattern very similar to the expectiles of Figure 2.5, since in $[Y_1 \; X_1]$ there are no outliers.

The robustness of the M-quantiles becomes clear when looking at the estimates of the intercept in the $[Y_1^* \; X_1]$ data set and at the estimated slope in the $[Y_1^{**} \; X_1]$ example. The constant term computed at the different θ in $[Y_1^* \; X_1]$ ranges from 2.3 to 3.6 across θ for the M-quantiles and from 2.6 to 6.5 for the expectiles. In Figure 2.30 the M-quantiles are not much affected by the outlier in the dependent variable, while

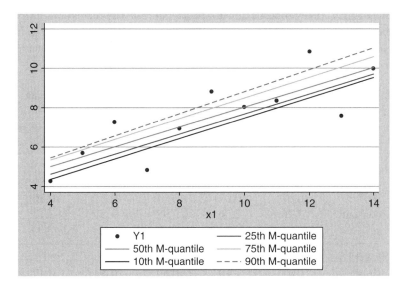

Figure 2.29   M-quantile estimates, $[Y_1 \; X_1]$ data set, sample size $n = 11$. In this data set, the M-quantiles are similar to the estimated expectiles of Figure 2.5, since there are no outliers.

---

[3] Here the M-quantiles are computed as deviation from the robust conditional mean without any iteration. Alternatively, they can be computed by iterative weighted least squares (WLS). Away from the conditional mean, the first WLS iteration combines the θ and $(1 - \theta)$ parameters with the robust weights at the conditional mean defined in Section 2.2, then the new set of residuals are used to recompute the robust weights and the WLS regression is re-estimated. These steps are repeated until convergence. The M-quantile results discussed in Section 2.3 are computed without iterations. Once estimated the robust regression at the mean, the asymmetric θ and $(1 - \theta)$ are implemented to diverge from the robust conditional mean.

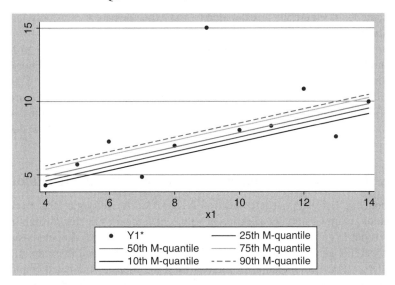

Figure 2.30   M-quantiles in the $[Y_1^* \ X_1]$ data set, sample size $n=11$. The M-quantile estimates are computed as deviations from the Huber-Tukey robust conditional mean estimator. This reduces the impact of the outlier and the M-quantiles have close-by intercepts with respect to the expectiles of Figure 2.6. Here the intercept ranges from 2.4 to 3.6, while in Section 2.1 it ranges from 2.6 to 6.5.

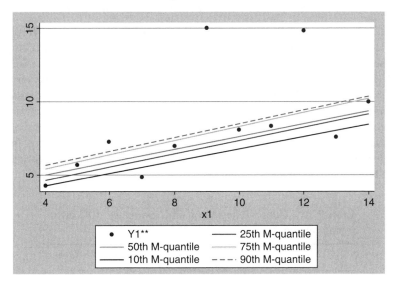

Figure 2.31   M-quantiles, $[Y_1^{**} \ X_1]$ data set, sample size $n=11$. The M-quantiles are not affected at all by the two outliers. This is due to the approach here implemented, which computes the M-quantiles by shifting toward the tails the robust conditional mean. While in Figure 2.8 the estimated lines are fan shaped, here this is not the case. In Section 2.1 the slope ranges from 0.48 to 0.80, while here it ranges from 0.42 to 0.48.

the intercepts of Figure 2.6 as estimated by the expectiles are increasingly far apart since the expectiles are attracted by the outlier.

For the data set $[Y_1^{**} \ X_1]$ the slope ranges from 0.42 to 0.48 in Table 2.14 for the M-quantiles and from 0.48 to 0.80 in Table 2.4 reporting the expectiles. Figure 2.31 shows how the M-quantiles are not attracted at all by the two outliers in $[Y_1^{**} \ X_1]$, since they are computed by shifting up or down the robust conditional mean. Figure 2.31 differs from Figure 2.8, where all the expectiles are instead attracted by the two outliers and yield a fan-shaped graph. Due to the robust weights, here the slope does not change much across θ.

2.3.1   M-quantiles estimates in the error-contaminated model

The model $y_i = \beta_0 + \beta_1 x_i + e_i$ of Section 2.1.1, characterized by a 10% error contamination, is now estimated at different values of θ meanwhile downweighting the data in order to curb outliers. Table 2.15 reports the M-quantiles computed using the robust weights generated by the Huber-Tukey M-estimator, and the regressions at the different θ are computed as deviations from the robustly estimated conditional mean in the model $\sqrt{w_{i,M}} y_i = \beta_0 + \beta_1 \sqrt{w_{i,M}} x_i + e_i$. Figure 2.32 depicts the estimated Huber-Tukey M-quantiles. With respect to the simple expectiles, which are statistically relevant only at the two higher expectiles as reported in Table 2.5, the estimated coefficients are statistically different from zero at all the selected θ.

2.3.2   M-quantiles in the French and Dutch examples

Figure 2.33 to 2.36 report the quantile plots of the variables *wspeed* and *BMI* robustly weighted using $w_{i,HT}$ for the French and the Dutch data sets. These figures, compared with the quantile plots of the original variables, reported in Figures 2.14, 2.16, 2.20, and 2.22, show how the robust weights have modified the quantiles of the variables. For instance, while in Figure 2.14 the *wspeed* top quantiles were above the straight line representing the theoretical normal quantiles, in the weighted *wspeed* variable, the higher quantiles are below the straight line of Figure 2.33. Table 2.16 reports the M-quantiles computed using the Huber-Tukey weights to control outliers. In both samples the slope is negative at all θ. Figures 2.37 and 2.38 present the M-quantiles for the French and the Dutch samples.

Table 2.15   Huber-Tukey M-quantile estimates, 10% error contamination.

θ	10th	25th	50th	75th	90th
slope	1.013	1.119	1.299	1.156	1.089
se	(.40)	(.42)	(.33)	(.34)	(.28)
constant	2.130	2.318	2.661	3.091	3.361
se	(.23)	(.24)	(.20)	(.21)	(.17)

Note: Standard errors in parenthesis, sample size $n = 100$.

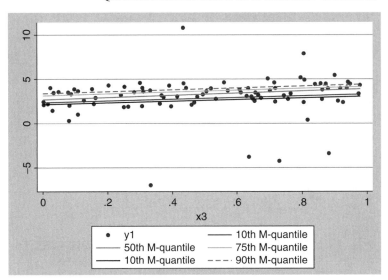

Figure 2.32 Huber-Tukey M-quantile estimates of the model with 10% contaminated normal error distribution in a sample of size $n = 100$. Compared with Figure 2.10, where the lower expectiles are attracted by the outliers so that some of them look like regular observations, here the outliers are all clearly visible.

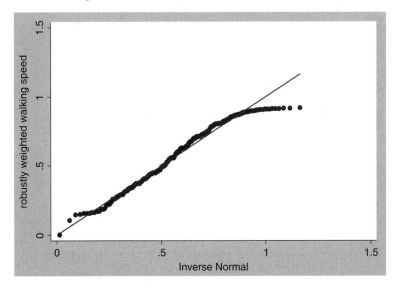

Figure 2.33 Quantile plot of the robustly weighted *wspeed*, France data set. Compared with Figure 2.14, which depicts the quantiles of the original data, the empirical quantiles of the weighted variable are below the theoretical normal quantile line at the top quantiles. The downweighting of outliers, particularly of the suspicious case of high walking speed of an individual with very large *BMI*, brings the upper quantiles below normality.

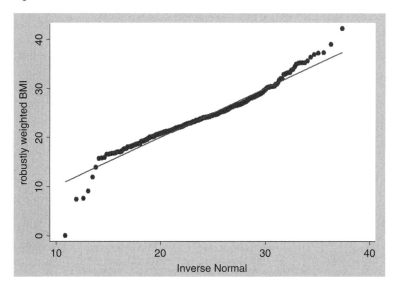

Figure 2.34   Quantile plot of the robustly weighted *BMI*, France data set. Compared with Figure 2.16, which depicts the quantiles of the original data, the empirical quantiles of the weighted variable are below the theoretical normal quantile line in the left tail. The downweighting has sizably affected the lower values of *BMI*.

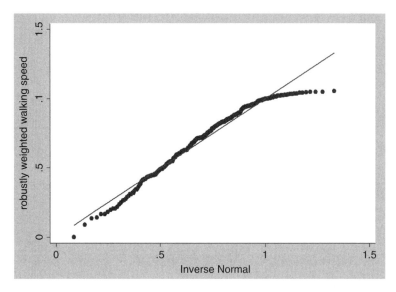

Figure 2.35   Quantile plot of the robustly weighted *wspeed*, the Netherlands data set. Compared with Figure 2.20, which depicts the quantiles of the original data, the empirical quantiles of the weighted variable are below the straight line both at the lower and at the higher quantiles.

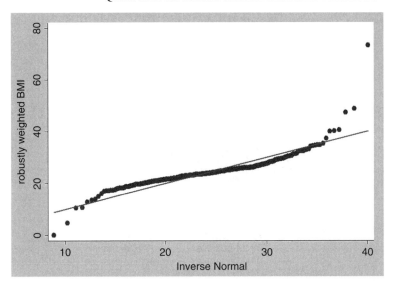

Figure 2.36 Quantile plot of the robustly weighted *BMI*, the Netherlands data set. Compared with Figure 2.22, which depicts the quantiles of the original data, the empirical quantiles of the weighted variable are below the theoretical normal quantile line in the left tail. The plot shows how the large but non-influential *BMI* value has not been downweighted.

Table 2.16   M-quantile estimates in the French and Dutch examples.

θ	10th	25th	50th	75th	90th
French data, $n = 393$					
intercept	0.542	0.580	0.672	0.735	0.789
slope	−0.003	−0.002	−0.002	−0.001	−0.001
Dutch data, $n = 307$					
intercept	0.696	0.776	0.898	1.02	1.09
slope	−0.003	−0.004	−0.006	−0.007	−0.007

To conclude, the health data set is considered in its entirety, pooling together *wspeed* and *BMI* for all the 11 European countries of the survey, in a sample of size $n = 2959$. Figure 2.39 depicts the scatterplot of the variables. The regression between walking speed and *BMI* in Europe yields a negative and statistically relevant slope coefficient whatever estimator is implemented, as shown in Table 2.17, with the sole exceptions of the 10th quantile estimated slope which is not statistically significant. Table 2.18 reports the observations receiving a zero weight in the

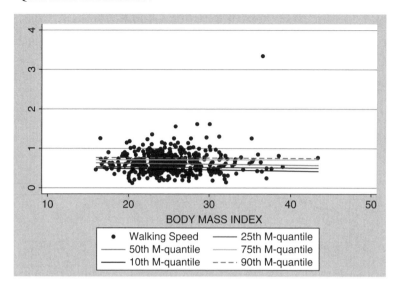

Figure 2.37 M-quantiles estimates, France data set, to be compared with Figure 2.17 depicting the expectiles. The M-quantiles are not attracted by the outlier as occurs in the fan shaped expectile graph.

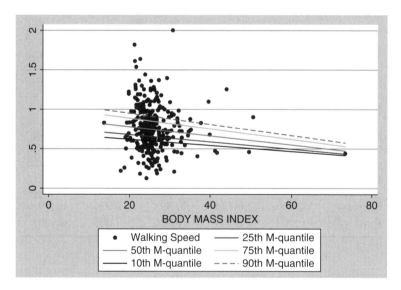

Figure 2.38 M-quantiles estimates, Dutch data set, to be compared with Figures 2.23 and 2.24, depicting respectively the expectiles and the quantiles fitted lines. The M-quantile estimates do not differ much from the expectiles of Figure 2.23, since the outlier in *BMI* is not influential. Compared to the quantile regressions of Figure 2.24, the M-quantile lines do not cross.

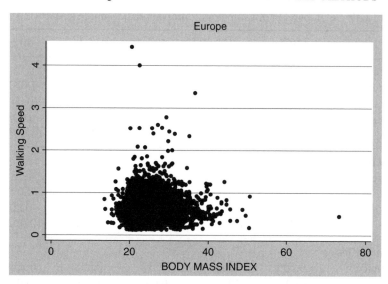

Figure 2.39 European data, sample of size $n = 2959$. The data set presents an extremely large outlier in *BMI*, *BMI* > 60, and at least three outliers in walking speed, with *wspeed* > 3.

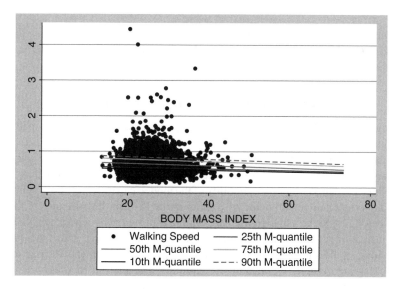

Figure 2.40 M-quantiles, European data, sample of size $n = 2959$.

Table 2.17   Comparison of estimators, European data, sample size $n = 2959$.

	10th	25th	50th	75th	90th
expectiles					
intercept	0.484	0.614	0.755	0.901	1.057
se	(0.03)	(0.03)	(0.03)	(0.05)	(0.09)
slope	−0.0013	−0.0022	−0.0029	−0.0033	−0.0033
se	(0.001)	(0.001)	(0.001)	(0.002)	(0.003)
quantiles					
intercept	0.317	0.534	0.729	0.935	1.25
se	(0.03)	(0.04)	(0.03)	(0.04)	(0.07)
slope	0.0003	−0.0026	−0.0035	−0.0038	−0.007
se	(0.001)	(0.001)	(0.001)	(0.001)	(0.002)
M-quantiles					
intercept	0.553	0.621	0.737	0.858	0.930
se	(0.02)	(0.02)	(0.03)	(0.03)	(0.02)
slope	−0.002	−0.0026	−0.0033	−0.0037	−0.0039
se	(0.0007)	(0.0009)	(0.001)	(0.001)	(0.0009)

Note: Sample size $n = 2959$.

M-quantile regressions. In this table the possible coding error in the French data set for *wspeed* $= 3.3$ receives a zero weight, while the anomalous value of *BMI* $= 73.39$ in the Dutch data set is not excluded from the sample.

2.3.3   Further applications: small-area estimation

A particularly fruitful implementation of M-quantiles can be found in small- area estimation problems, since they simplify the computation of inter-area differences (Chambers and Tzavidis, 2006; Pratesi et al., 2009) when the groups are small and the sample size of each group is tiny. The presence of a hierarchical structure in the data, which explains part of the variability of the model, can be taken into account, and the group effect can be computed by averaging the M-quantile estimates of the elements belonging to the same group. An example is provided by a work analyzing the performance of sixteen- year-old students (Tzavidis and Brown, 2010). M-quantile regressions on the students' scores at exams are function of students' characteristics such as language, ethnicity, and family background. The analysis focuses on assessing school performance. The impact of a specific school on students' achievements, that is, the group effect, is computed by merging the quantiles of students enrolled in the same school. The small-group effect (the school impact) can be computed by merging the estimates of all the elements belonging to the same group. This avoids the need of modeling the inter-area differences. For instance, in the European data on health analyzed in the previous section, the group effects would be the country- specific impact.

Table 2.18   Observations receiving a
zero weight in the M-quantile estimates,
European data, sample size $n = 2959$.

wspeed	BMI	country
4.4	20.5	Austria
4.4	22.1	Germany
2.0	23.8	Germany
2.5	20.2	Germany
4	22.4	Sweden
2.3	25.9	Sweden
1.9	30.8	Netherlands
2.2	29.7	Italy
3.3	36.6	France
2.7	29.3	Denmark
2.4	30.0	Greece
2.3	31.4	Greece
2.5	27.2	Greece
2.3	35.1	Belgium
1.9	29.7	Belgium
2.5	22.6	Belgium
2.5	28.2	Belgium
2.5	25.7	Belgium

Since in this example the sample size of each country is not really small, the data for each country are numerous enough to compute country-specific regressions. When this is not the case, that is, when the groups are small and the sample size of each group is inadequate, the group effect can be pulled out from the M-quantile estimates of the pooled data.

To measure the effect of group j, having sample size n_j, the statistic is

$$\widehat{m}_j = N_j^{-1} \left[\sum_{j \in s} y_{ij} + \sum_{j \notin s} x_{ij}^T \widehat{\beta} \right]$$

where the term $\sum_{j \in s} y_{ij}$ considers the n_j sample units in the j group, while the term $\sum_{j \notin s} x_{ij}^T \widehat{\beta}$ refers to the remaining $N_j - n_j$ units of group j.

For instance the French data set has $N_F = 3071$ observations, but the regression can be implemented only in a smaller sample, of size $n_F = 393$, due to missing data occurring mostly in the *wspeed* variable. The first term of the above equation yields $\sum_{j \in s} y_{ij} = 249.9$ for $s = n_F = 393$, while the second term computes at the 50th M-quantile the missing *wspeed* values as $\sum_{j \notin s} x_{ij}^T \widehat{\beta}(.5) = -233.8$. Thus at $\theta = .50$ the country effect for France is $\widehat{m}_F(.5) = \frac{[249.9 - 233.8]}{3071} = 0.005$.

In the Dutch subset there are $N_D = 2930$ observations, but due to missing data, the regression can be implemented only in a subset of size $n_D = 307$. In the subset $s = n_D = 307$, it is $\sum_{j \in s} y_{ij} = 234.6$ while for the missing data at the 50th M-quantile, it is $\sum_{j \notin s} x_{ij}^T \widehat{\beta}(.5) = -231.5$. Thus at $\theta = .50$ the Dutch impact is $\widehat{m}_D(.5) = 0.001$, and at $\theta = .50$ the French country effect is larger than the Dutch one.

This procedure can be repeated at each M-quantile, for $\theta = .10, .25, .75, .90$, to compute the country effect at the different points of the conditional distribution. At the 25th M-quantile the country effect for the Netherlands is $\widehat{m}_D(.25) = 0.018$ and for France is $\widehat{m}_F(.25) = 0.022$, while at the 75th M-quantile it is $\widehat{m}_D(.75) = -0.009$ in the Netherlands and $\widehat{m}_F(.75) = -0.004$ in France. Thus the country effect is positive at the lower M-quantiles and decreases to become negative at the top M-quantiles for both countries. The country impact is generally greater in France than in the Netherlands at all quantiles, implying that *BMI* has a greater impact on *wspeed* in the former country.

2.4   A summary of key points

Analogously to quantile regressions, expectiles allow to compute a regression model away from the conditional mean. An asymmetric weighting system modifies the least squares objective function to set the position of the estimated regression away from the center. Both real and artificial data sets are analyzed, and the results are compared with the quantile regression estimates to look at the similarities and at the differences of the two approaches. The main advantage of expectiles is its ease of computation, while the major drawback is its lack of robustness. Next the M-estimators are considered. The latter are robust methods defined to control the impact of outliers when estimating the model at the conditional mean. Their weighting system focuses on curbing extreme values in the data set and provides an excellent diagnostic tool. The simple exclusion of the outliers from the sample, assigning weights equal to zero to the outliers and equal to one elsewhere, provides the trimmed OLS and the trimmed quantile regression estimators. The combination of M-estimators and expectiles yield the M-quantiles. The latter estimator allows to compute a model at various points of the conditional distribution, meanwhile controlling and bounding outliers. Real and artificial data sets are analyzed implementing the M-estimators and the M-quantiles. The comparison of the results shows the advantages and the weaknesses of each one of the above approaches.

References

Andrews, D., Bickel, P., Hampel, R., Huber, P., and Rogers, W. 1972. *Robust estimates of location: Survey and Advances*. Princeton University Press.

Beaton, A., and Tukey, J. 1974. "The fitting of power series, meaning polynomials illustrated on band-spectroscopic data." Technometrics 25, 119-163.

Breckling, J., and Chambers, R. 1988. "M-quantiles." Biometrika 75, 761–71.

Chambers, R., and Tzavidis, N. 2006. "M-quantile models for small area estimation." Biometrika 93, 255–268.

Efron, B. 1992. "Poisson overdispersion estimates based on the method of asymmetric maximum likelihood." Journal of the American Statistical Association 87, 98–107.

He, X. 1997. "Quantile Curves without Crossing." The American Statistician 51, 186–192.

Huber, P. 1964. "Robust estimation of a location parameter." Annals of Mathematical Statistics 35, 73–101.

Huber, P. 1981. *Robust Statistics*. Wiley.

Jones, M. 1994. "Expectiles and M-quantiles are quantiles." Statistics and Probability Letters 20, 149–153.

Kemp, G., and Santos Silva, J. 2012. "Regression towards the mode." Journal of econometrics 170, 92–101.

Krasker, W., and Welsh, R. 1982. "Efficient bounded-influence regression estimation." Journal of the American Statistical Association 77, 595–604.

Lee, M. 1993. "Quadratic mode regression." Journal of Econometrics 57, 1–20.

Neykov, M., Cizek, P., and Neytchev, P. 2012. "The least trimmed quantile regression." Computational Statistics and Data Analysis 56, 1757–1770.

Newey, W., and Powell, J. 1987. "Asymmetric Least Squares Estimation and Testing." Econometrica 55, 819–847.

Pratesi, M., Ranalli, G., and Salvati, N. 2009. "Nonparametric M-quantile regression using penalised splines." Journal of Nonparametric Statistics 21, 287–304.

Schnabel, S., and Eilers, P. 2009. "Optimal expectile smoothing." Computational Statistics and Data Analysis 53, 4168–4177.

Sobotka, F., Schnabel, S., and Schulze Waltrup, L. 2013. "R 'expectreg' package." http://cran.r-project.org/web/packages/expectreg/vignettes/expectreg.pdf.

Tzavidis, N., and Brown, J. 2010. "Using M-quantile models as an alternative to random effects to model the contextual value-added of schools in London." DoQSS working paper 10-11, Institute of Education, University of London.

Yang, Y., and Zou, H. 2014. "Nonparametric multiple expectile regression via ER-Boost." Journal of Statistical Computation and Simulation, on line.

Appendix: computer codes

R codes for quantiles and expectiles:	

`library(expectreg)`
`library(quantreg)`
`library(BayesX)`
`library(mboost)`
`library(splines)`
`library(quadprog)`
	load libraries

`dataname <- read.dta("datafile")`	read data set
`qreg7 <- rq(y ~ x, tau=.7, data=name)` `summary(qreg7)`	compute 70th quantile print results
`qreg3 <- rq(y ~ x, tau=c(.30,.80), data=name)` `summary(qreg3)`	compute 30th and 80th quantiles print results

`ex <- expectreg.ls(y ~ x,`
`        data=name,`
`        expectiles=c(.10,.25,.75,.90),`
`        ci=TRUE)`
`summary(ex)`
`fitted(ex)`
	compute expectiles
	.10, .25, .75, .90
	
	print results
	print expectiles fitted values

`eenorm(dataname$y)`
`eenorm(dataname$x)`
`qqnorm(dataname$y)`
`qqnorm(dataname$x)`
	expectile plot of y
	expectile plot of x
	quantile plot of y
	quantile plot of x

Stata codes for quantiles and expectiles:

`qreg y x, q(.7)`	compute the 70th quantile
`qreg y x, q(.3)`	compute the 30th quantile
`gen q1=1`	generate a unit constant
`reg y x`	compute OLS regression
`forvalues i=1(1)10 {`	starts 10 iterations
`  predict res,resid`	save residuals
`  replace q1=.10 if res >=0`	generate asymmetric weights
`  replace q1=.90 if res <0`	for the 10th expectile
`  reg y x [w=q1]`	10th asymmetrically weighted least
`  drop res`	squares
`}`	end of instructions to be iterated
`reg y x [w=q1]`	repeat last estimated regression
`predict ex10`	fitted values 10th expectile
`la var ex10 "10th expectile"`	
`twoway (scatter y x)`	plot the 10th expectile
`        (connect ex10 x)`	

Stata codes for expectiles without iterations:

`gen q1=1`	generate a unit constant
`reg y x`	compute OLS regression
`predict res,resid`	save residuals
`replace q1=.10 if res>=0`	generate asymmetric weights
`replace q1=.90 if res<0`	for the 10th expectile
`reg y x [w=q1]`	10th asymmetrically weighted least squares

R codes for M-estimators:	
`library(MASS)`	load library

`mod.huber < - rlm(y ~x,`
`            data=dataname,`
`            psi=psi.huber)`
	compute Huber M-estimator

`mod.hampel < - rlm(y ~x,`
`            data=dataname,`
`            psi=psi.hampel)`
	compute Hampel
	   M-estimator

`mod.tukey < - rlm(y ~x,`
`            data=dataname,`
`            psi=psi.bisquare)`
	compute Tukey M-estimator

`summary(mod.huber)`	print results
`plot(mod.huber$w,` `     ylab="Huber Weight",xlab="x")`	plot weights
`list(mod.huber$w)`	print Huber weights
`sqrtw < - (mod.huber$w)^.5`	square root weights
`wy < - dataname$y*sqrtw`	weighted y, wy=$y\sqrt{w_H}$ to control outliers
`wx < - dataname$x*sqrtw`	weighted x, wx=$x\sqrt{w_H}$, to control outliers

Stata codes for M-estimators, Huber-Tukey weights:	
`use datafile`	read the data set

`rreg y x, genwt(nameweight)`	compute M-estimators using
	   Huber in the initial
	   iterations and then Tukey
	   weights

`list nameweight`	print Huber-Tukey weights
`twoway (scatter y x) (scatter` `   nameweight x,msymbol(triangle))`	plot weights
`gen wy=sqrt(nameweight)*y`	modify y to control outliers, wy=$y\sqrt{w_{HT}}$

`gen wx=sqrt(nameweight)*x`	modify each explanatory
	   variable to control outliers,
	   wx= $x\sqrt{w_{HT}}$

`qnorm y`	quantile plot of y
`qnorm x`	quantile plot of x
`keep wy wx`	save weighted data
`save weightedname`	save wy and wx in weightedname

Stata codes for trimmed OLS and trimmed quantile regressions:

`reg y x if y<=k`	excludes values of y> k in OLS
`reg y x if x<=k`	excludes values of x> k in OLS
`qreg y x if y<=k`	excludes values of y> k in QR
`qreg y x if x<=k`	excludes values of x> k in QR

R codes for M-quantiles, Huber weights:

`em <- expectreg.ls(wy ~ wx,`
`        data=name,`
`        expectiles=c(.25,.75,.90),`
`        ci=TRUE)`
	compute expectiles using robustly weighted data

`summary(em)`	print results

R codes for iterative M-quantiles, starting from Huber regression:

`modhub <- rlm(y~x,`
`            data=name,`
`            psi=psi.huber,`
`            maxit=10)`	

`summary(modhub)`	robust regression, Huber weights
`rres <- resid(modhub)`	robust residuals
`scale <- median(abs(rres))/0.6745`	estimate the scale (MAD)
`stres <- rres/scale`	standardize robust residuals
`weightRes <- ifelse(stres < 0,` `                0.25, 0.75)`	asymmetric weights

`Weighted <- rlm(y1c~x1,`
`            data=name,`
`            weights=weightRes)`
	M-quantile

`summary(Weighted)`	results of the M-quantile regression
`rres2 <- resid(Weighted)`	keep residuals for iteration
`scale2 <- median(abs(rres2))/0.6745`	recompute MAD
`stres2 <- rres2/scale2`	standardize robust residuals
`weightRes2 <- ifelse(stres2 < 0,` `                0.25, 0.75)`	asymmetric weights

`Weighted2 <- rlm(y1c~x1,`
`            data=name,`
`            weights=weightRes2)`
	M-quantile

`summary(Weighted2)`	results of the M-quantile regression

Stata codes for M-quantiles, Huber-Tukey weights:

```rreg y x```	Huber-Tukey robust regression
```predict rfit5```	fitted robust regression values
```predict res,resid```	save robust residuals
```gen abres=abs(res)```	absolute value of residuals
```su abres,d```	compute the median
```scalar mad=r(p50)/.6745```	median absolute deviation
```gen rres=res/mad```	standardized residuals u_i
```gen q1=.10 if rres >= 0 &``` ``` rres< 1.345```	generate asymmetric weights
```replace q1 = (1.345/abs(rres))*.10``` ``` if rres >1.345```	for the 10th M-quantile
```replace q1 = (1.345/abs(rres))*.90``` ``` if rres <-1.345```	
```replace q1 = .90``` ``` if rres <= 0 & rres>=-1.345```	
```reg y x [w=q1]```	asymmetrically weighted least squares for the 10th M-quantile
```predict mq10```	fitted values 10th M-quantile
```gen q2=.25``` ``` if rres >=0 & rres < 1.345```	asymmetric weights for the 25th M-quantile
```replace q2 = (1.345/abs(rres))*.25``` ``` if rres >1.345```	
```replace q2 = (1.345/abs(rres))*.75``` ``` if rres <-1.345```	
```replace q2 = .75 if rres <= 0```	M-quantile
```reg y x [w=q2]```	asymmetrically weighted least squares for the 25th M-quantile
```predict mq25```	fitted values 25th
```la var mq10 "10th M-quantile"```	
```la var mq25 "25th M-quantile"```	
```la var rfit5 "M-estimator"```	
```twoway (scatter y x)```	plot of the
``` (connect mq10 x)```	10th, 25th, 50th M-quantile
``` (connect mq25 x)```	
``` (connect rfit5 x)```	

Stata codes for small area estimation:

`rreg wspeed bmi`	robust regression
`predict res, resid`	save residuals
`matrix coef1=e(b)`	save coefficients
`scalar b1=el(coef1,1,1)`	slope coefficient
`scalar b0=el(coef1,1,2)`	intercept
`su wspeed if group==j & res!=.`	size of group j
`scalar smalln=r(N)`	
`scalar aj=r(mean)*smalln`	compute $\sum_{j \in s} y_{ij}$
`gen dumm=1 if group==j`	select non regression data $j \notin s$
`replace dumm=0 if group==j & res!=.`	
`gen meanxbj=(b1*bmi) if`	$x_{ij}^T \widehat{\beta}$ for $j \notin s$
`            group==j & dumm==1`	
`su meanxbj if group==j & dumm==1`	
`scalar ndif=r(N)`	
`scalar bj=r(mean)*ndif`	compute $\sum_{j \notin s} x_{ij}^T \widehat{\beta}$
`scalar groupj=(aj+bj)/(ndif+smalln)`	$N_j^{-1}[\sum_{j \in s} y_{ij} + \sum_{j \notin s} x_{ij}^T \widehat{\beta}]$
`scalar list groupj`	print result

3

Resampling, subsampling, and quantile regression

Introduction

The chapter focuses on subsampling and resampling methods to further analyze the characteristics of the quantile regression estimator. The small size of the Anscombe (1973) model allows to consider the elemental set interpretation of this estimator. Next the estimates at the extreme quantiles are considered. The latters focus on the quantiles in the far tails, characterized by nonstandard behavior, where inference is implemented through a resampling approach. In Section 3.4 a brief review of the asymptotic distribution of the non-extreme quantile regression estimator is reported for comparison's sake. Finally treatment effect estimators computed on average and at the various quantiles are discussed. The decomposition of the treatment effect into impact explained by the covariates and unexplained coefficients effect is considered on average, by the Oaxaca-Blinder approach, and at the various quantiles, by heavily relying on resampling methods.

3.1   Elemental sets

This section considers the elemental sets approach, which provides a very intuitive tool to compare OLS and quantile regressions (Farebrother, 1985; Hawkins, 1993; Heitmann and Ord, 1985; Hussain and Sprent, 1983; Subrahmanyam, 1972). The elemental sets are subsets of size p, where p is the number of unknown parameters in the regression. In the simple two- parameter model, $y_i = \beta_0 + \beta_1 x_i + e_i$, there are

Quantile Regression: Estimation and Simulation, Volume 2. Marilena Furno and Domenico Vistocco.
© 2018 John Wiley & Sons Ltd. Published 2018 by John Wiley & Sons Ltd.
Companion website: www.wiley.com/go/furno/Quantileregression

$\binom{n}{p} = \frac{n!}{p!(n-p)!}$ elemental sets each comprising only two data points, where each set has dimension $p = 2$. Two points i and j univocally identify only one line, and its coefficients are exactly computed by $b_0(i,j) = \frac{1}{2}[y_i + y_j - b_1(i,j)(x_i + x_j)]$ for the intercept and $b_1(i,j) = \frac{(y_i - y_j)}{(x_i - x_j)}$ for the slope. The regression coefficients β_0 and β_1 can be expressed in terms of the elemental sets coefficients $b_0(i,j)$ and $b_1(i,j)$. While the OLS estimator is given by the weighted average of all the possible b(i,j), $\beta_{p,OLS} = \sum_{ij} w_{ij} b_p(i,j)$ for $p = 0, 1$, with weights given by $w_{ij} = \frac{(x_i - x_j)^2}{\sum_{ij}(x_i - x_j)^2}$, the median regression summarizes all the $b_p(i,j)$ by selecting their median value (Theil, 1950). Analogously, for the quantiles away from the median, the estimator selects the chosen quantile of the $b_p(i,j)$ series. The elemental sets provide the building block of one of the methods implemented to compute the quantile regression estimator, the simplex approach (Koenker 2000). The results of the Theil (1950) estimator, however, may differ from the ones provided by the Koenker and Bassett (1978) quantile regression estimator since the latter considers elemental sets as the starting point of an iterative process to select the optimal elemental set (see Section 5.3).[1]

A reduced version of the Anscombe data set $[Y_1 \, X_1]$ provides an easy example. In a sample of size $n = 11$ there are a total of $\binom{11}{2} = \frac{11!}{2!9!} = 55$ elemental sets, that is, 55 pairs of observations that identify 55 different lines defined by 55 parameters vectors. In order to reduce the computational burden, only the first six observations of the $[Y_1 \, X_1]$ sample are here considered. This shrinks the number of elemental sets to $\binom{6}{2} = \frac{6!}{2!4!} = 15$, that is, there are 15 distinct lines passing through all the possible pairs of data in a sample of $n = 6$ observations. These lines are computed without errors, since each pair of observations univocally identifies one straight line. The fifth and sixth column of Table 3.1 report the $b_p(i,j)$ values, while the first four columns in the table collect the observations to compute them. The column w_{ij} provides the weights of the OLS estimator based on elemental sets, $w_{ij} = \frac{(x_{1i} - x_{1j})^2}{\sum_{ij=1,6}(x_{1i} - x_{1j})^2}$, while the last two columns of the table, headed $r(b_1(i,j))$ and $r(b_0(i,j))$, report the rank of the ordered $b_1(i,j)$ and $b_0(i,j)$ coefficients needed to find a given quantile in the b(i,j)s series.

The left-hand side graph of Figure 3.1 presents all the 15 lines identified by the elemental sets. The weighted average of the $b_1(i,j)$ and $b_0(i,j)$ exactly computed in the 15 elemental sets provide the OLS estimated coefficients. The results of the standard OLS regression considering only the first six observations of $[Y_1 \, X_1]$ yields the estimates $\hat{\beta}_0 = 5.52$ and $\hat{\beta}_1 = 0.25$, and they coincide with the weighted averages of the coefficients computed in the elemental set: $\hat{\beta}_{0OLS} = \sum_{ij=1,6} w_{ij} b_0(i,j) = 5.52$ and $\hat{\beta}_{1OLS} = \sum_{ij=1,6} w_{ij} b_1(i,j) = 0.25$. Figures 3.2 and 3.3 depict the elemental sets coefficients times the OLS weights and the OLS weights respectively for the slope and the intercept. By definition, w_{ij}, which is given by the ratio of squared terms, assigns a positive value to each observation in X_1. This provides an additional insight

---

[1] When $p > 2$ the OLS estimator is given by $\hat{\beta}_{OLS} = \frac{\sum |X_i|^2 b_i}{\sum |X_i|^2}$ where $|X_i|$ is the determinant of X_i, X_i is a $(p \times p)$ submatrix including p observations of the p explanatory variables and b_i is a p vector of the exactly computed coefficients defining the ith hyperplane. The θ quantile regression estimator selects the θ order statistic in the set of all the hyperplanes passing through exactly p observations.

Table 3.1   Elemental sets for the first six observations of the $[Y_1\ X_1]$ data set.

x_{1i}	x_{1j}	y_{1i}	y_{1j}	$b_1(i,j)$	$b_0(i,j)$	w_{ij}	$r(b_1(i,j))$	$r(b_0(i,j))$
10	8	8.04	6.95	.545	2.59	.0248	13	4
10	13	8.04	7.58	−.153	9.573	.0559	5	11
10	9	8.04	8.81	−.77	15.74	.00621	1	15
10	11	8.04	8.33	.29	5.14	.00621	8	8
10	14	8.04	9.96	.48	3.24	.09937	10	6
8	13	6.95	7.58	.126	5.942	.1553	6	9
8	9	6.95	8.81	1.86	−7.93	.00621	14	2
8	11	6.95	8.33	.46	3.27	.0559	9	7
8	14	6.95	9.96	.502	2.936	.2236	11	5
13	9	7.58	8.81	−.307	11.572	.09937	3	13
13	11	7.58	8.33	−.375	12.455	.0248	2	14
13	14	7.58	9.96	2.38	−23.36	.00621	15	1
9	11	8.81	8.33	−.24	10.97	.0248	4	12
9	14	8.81	9.96	.23	6.74	.1553	7	10
11	14	6.95	9.96	.543	2.353	.0559	12	3

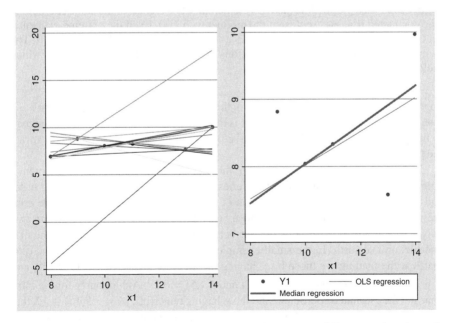

Figure 3.1   The left panel plots all the 15 exactly computed lines passing through each pair of observations. In the right graph is the scatterplot of the six data points together with the OLS and the median regression computed using the 15 elemental sets coefficients. The subsample of size $n = 6$ selects the first six observations of the $[Y_1\ X_1]$ Anscombe data set.

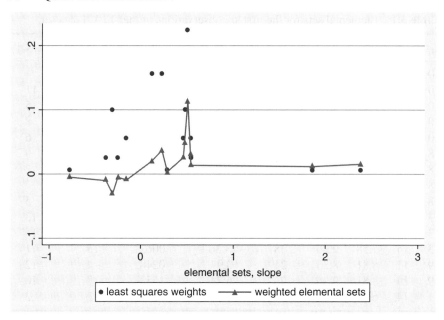

Figure 3.2   The graph plots the OLS elemental sets weights for the slope coefficient. The connected triangles represent the product $w_{ij}b_1(i,j)$. Even the extreme elemental sets receive a nonzero weight in the OLS estimator, causing a lack of robustness. The sample is given by the first six observations of the Anscombe $[Y_1 \ X_1]$ data set, which yield 15 elemental sets.

on the vulnerability of OLS to outliers in the explanatory variables: the farthest is x_{1i} from the other observations, the greater is the numerator of the OLS weight assigned to this observation, w_{ij}, and the greater is the capability of this observation to attract the OLS regression line.

Following Theil (1950), to estimate the median regression coefficients - which coincide with the median of the $b(i,j)$ series - the $b(i,j)$ are sorted in ascending order and are ranked in the $r(b(i,j))$ columns of Table 3.1. The estimates coincide with the median in these series, that is, those $b(i,j)$ that split the $r(b(i,j))$ series into two halves for the median, into one quarter below and three quarters above for the first quartile, and so forth. In the small example of size $n = 6$ here considered, the estimates at the median are those $b(i,j)$ having median rank, $r(b(i,j)) = .5(15) \cong 8$, and at position eight there are $\widehat{\beta}_0(.5) = 5.14$ and $\widehat{\beta}_1(.5) = 0.29$. Analogously for the third quartile, the estimated coefficients are those having rank $r(b(i,j)) = .75(15) \cong 12$, that is, $\widehat{\beta}_0(.75) = 10.97$ and $\widehat{\beta}_1(.75) = 0.54$. The median regression estimates, together with the OLS estimated line, are depicted in the right graph of Figure 3.1.

The ordered elemental sets coefficients of Table 3.1 allow to easily compute confidence intervals by finding in the $r(b(i,j))$ series those values occupying the positions α and $1 - \alpha$. For instance, the 95% confidence interval for the median, $\theta = .5$, can be found by looking at the coefficient at position 0.025 for the lower

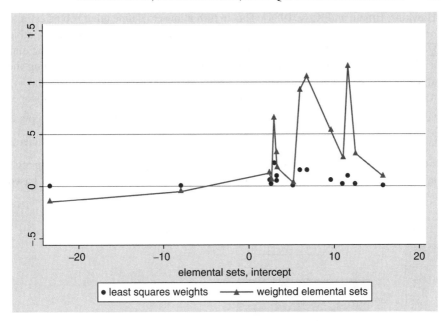

Figure 3.3   The graph plots the OLS elemental sets weights for the intercept. The connected triangles represent the product $w_{ij} b_0(i, j)$. Even the extreme elemental sets receive a nonzero weight in the OLS estimator, thus causing lack of robustness. The sample is given by the first six observations of the Anscombe data set $[Y_1 \; X_1]$, which yield 15 elemental sets.

bound and at position 0.95 for the upper bound of the confidence interval. In the example, this defines the confidence intervals $ci(\beta_0(.5)) = [-23.3 \quad 15.7]$ for the intercept and $ci(\beta_1(.5)) = [-0.77 \quad 2.38]$ for the slope. The above approach is the direct percentile method to compute confidence intervals.

A measure of dispersion of the b(i,j) is one of the possible ways to measure the precision of the quantile regression estimates, that is, the coefficients standard deviations (Furno, 1997). The dispersion of the $b_0(i, j)$ yields the standard deviation $se(b_0(i, j)) = 9.47$, while the dispersion of the $b_1(i, j)$ provides the estimated standard deviation of the slope, $se(b_1(i, j)) = 0.819$. As usual, the ratio between the estimated coefficient and the estimated standard deviation is distributed as a Student-t with $(n - p)$ degrees of freedom under the null hypothesis $H_0 : \beta(.5) = 0$. In this very small example with only $n = 6$ data points, at the median it is $t(\hat{\beta}_1(.5)) = \frac{0.29}{0.819} = 0.354$ and $t(\hat{\beta}_0(.5)) = \frac{5.14}{9.47} = 0.54$, and the null is not rejected in both tests.

The drawback of the elemental sets analysis is in the very quick growth of their number so that they can be implemented only in very small data sets. The advantage lies in providing an intuitive explanation of quantile regression, in granting an estimator for confidence intervals and variance of the coefficients, and last but not least, in supplying an excellent diagnostic tool (Hawkins, 1993). The b(i,j) are indeed useful to locate outliers, since extreme observations generate extreme b(i,j) values

(Hawkins, Bradu, and Kass, 1984). This in turn allows to define other robust estimators, for instance by trimming the extreme values of the b(i,j) series before computing $\sum_{ij} w_{ij} b(i,j)$ (Wu, 1986). The box plot of the 15 elemental sets estimated coefficients in Figure 3.4 allows to easily spot the farthest b(i,j). In detail, after excluding the largest $b_0(i,j) = -23.36$ and $b_1(i,j) = 2.38$ and after modifying accordingly the w_{ij}, in order to keep their sum equal to 1, the weighted sum of the remaining 14 slopes yields $\widehat{\beta}_{1OLS} = \sum_{ij} w_{ij} b_1(i,j) = 0.24$, while the weighted sum of the 14 intercepts is $\widehat{\beta}_{0OLS} = \sum_{ij} w_{ij} b_0(i,j) = 5.71$. The estimates of the median regression coefficients computed after trimming the b(i,j)s are $\widehat{\beta}_1(.5) = 0.26$ and $\widehat{\beta}_0(.5) = 5.54$. The top rows of Table 3.2 report the estimated coefficients for the $[Y_1\ X_1]$ subset of size $n = 6$, based

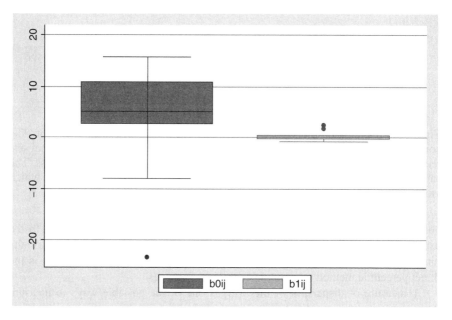

Figure 3.4   Box plot of the 15 elemental set estimated coefficients $b_0(i,j)$ and $b_1(i,j)$ provided by the first six observations of the $[Y_1\ X_1]$ Anscombe data set.

Table 3.2   Median and OLS estimates based on elemental sets.

$[Y_1\ X_1]$	OLS	median	trimmed OLS	trimmed median
slope	0.25	0.29	0.24	0.26
intercept	5.52	5.14	5.71	5.54
$[Y_1^*\ X_1]$	OLS	median	trimmed OLS	trimmed median
slope	−0.17	0.29	−0.096	0.37
intercept	11.14	5.14	10.28	4.20

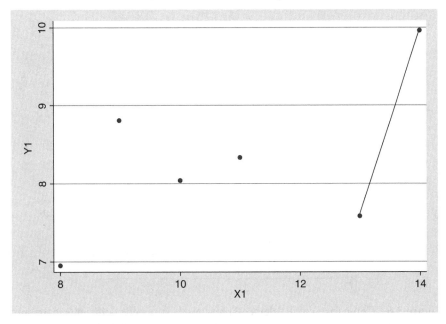

Figure 3.5 Scatter plot of the first six observations in the sample $[Y_1 \ X_1]$. The two observations to the right are the ones determining the farthest elemental set coefficients, $b_0(i,j) = -23.36$ and $b_1(i,j) = 2.38$. These values are dropped when computing the trimmed OLS and trimmed median estimates.

on the elemental sets approach. These results do not differ much from one another since this data set does not comprise outliers. Figure 3.5 presents the scatterplot of the data, and the farthest elemental set coefficients are the ones connecting the last two data points to the right of the plot. Figure 3.6 reports the elemental sets estimated lines.

Next consider the first six observations of the data set $[Y_1^* \ X_1]$ of Table 1.1, which includes one anomalous value. OLS and median regression results are in the bottom rows of Table 3.2. As can be seen in Figures 3.7 and 3.8, the elemental sets signal the presence of outliers, and the farthest elemental set coefficients are:

$b_0(i,j) = 77.64$ and $b_1(i,j) = -6.96$, which are the coefficients of the line passing through the first and the fourth point, $(8.04 \ 10)_1$ and $(15 \ 9)_4$;

$b_0(i,j) = -57.45$ and $b_1(i,j) = 8.05$, which define the line passing through the fourth and the second observation, $(15 \ 9)_4$ and $(6.95 \ 8)_2$;

$b_0(i,j) = 45.015$ and $b_1(i,j) = -3.335$, passing through the fourth and the fifth point, $(15 \ 9)_4$ and $(8.33 \ 11)_5$.

The fourth observation with $Y_1^* = 15$ is clearly under scrutiny since it yields all the three extreme values of the $b(i,j)$s. The trimmed OLS and the trimmed median

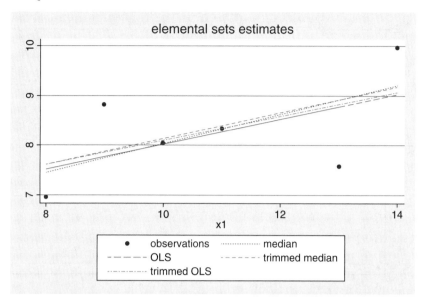

Figure 3.6 Comparison of the elemental set-based estimated coefficients in the subset of $[Y_1\ X_1]$. The trimming excludes the values $b_1(i,j) = 2.38$ and $b_0(i,j) = -23.36$, defined by the observations $(7.58\ 13)_3\ (9.96\ 14)_6$ of Table 1.1. The line connecting the third and the sixth observations is excluded when computing the trimmed estimators. However, these observations keep a role in computing the final estimates through the other elemental sets, that is, through the lines connecting the third and in turn the sixth observation to all the other observations of the sample.

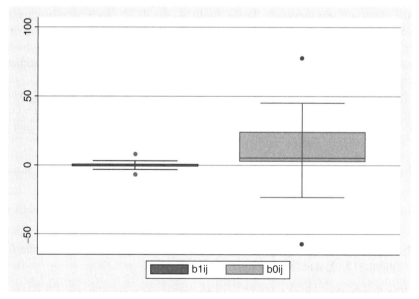

Figure 3.7 Box plot of the 15 elemental sets estimated coefficients $b_0(i,j)$ and $b_1(i,j)$ provided by the first six observations of the $[Y_1^*\ X_1]$ modified Anscombe data set. Here the elemental set coefficients are more dispersed when compared to the box plot of Figure 3.4 depicting the elemental sets coefficients for the clean $[Y_1\ X_1]$ data.

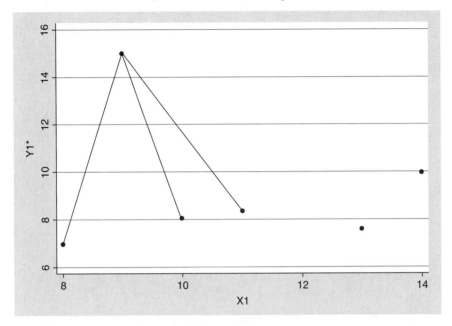

Figure 3.8   Scatterplot of the first six observations in the sample $[Y_1^* \ X_1]$, comprising one outlier in the dependent variable. The lines to the left are defined by the extreme elemental set coefficients, and all of them include the outlying value in the dependent variable $(Y_1^*)_4 = 15$. The coefficients of these three lines are excluded when computing the trimmed OLS and trimmed median regression coefficients.

regression are computed by excluding the above three elemental set values. The last two columns in the bottom rows of Table 3.2 report these results. In this data set with one outlier OLS and trimmed OLS estimates yield a negative slope, while median and trimmed median results are equal or close to the estimates in the top rows, computed in the clean data set $[Y_1 \ X_1]$.

3.2   Bootstrap and elemental sets

When the number of elemental sets is too large, it is possible to combine the bootstrap and the elemental sets analysis. Consider the $[Y_1 \ X_1]$ sample of size $n = 11$. The design matrix bootstrap[2] draws with replacement $M = 100$ samples of size $n = 11$, yielding a series of 100 estimated coefficients for the slope and for the intercept, as computed by OLS and by the median regression. The graphs in Figure 3.9 present the empirical distribution of these estimates. To measure the dispersion of the estimated coefficients, the design matrix bootstrap considers as center of location the mean of the series of the estimated values (Efron, 1979;

---

[2] Details on the design matrix bootstrap can be found in volume 1 (Davino et al. 2014), Section 4.3.2.1.

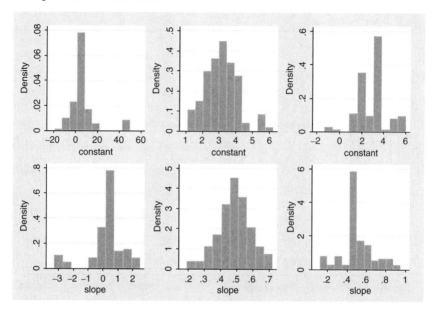

Figure 3.9   Histograms of the elemental sets bootstrap computed in samples of size $n = 2$ in the left graphs, histograms of the design matrix bootstrap OLS estimates in the middle graphs, and histograms of the design matrix bootstrap median regression in the right graphs. Top histograms for the intercept and bottom histograms for the slope, M = 100 replicates, Anscombe data set $[Y_1 \, X_1]$ of size $n = 11$.

Buchinsky, 1995), $\overline{\widehat{\beta}}_0(.50) = \frac{1}{M}\sum_m \widehat{\beta}_0(.50)_m$ and $\overline{\widehat{\beta}}_1(.50) = \frac{1}{M}\sum_m \widehat{\beta}_1(.50)_m$, where $m = 1, \ldots, M$ is the bootstrap replicate indicator and refers to the mth bootstrap estimate. The average value of the bootstrap estimated coefficients, reported in Table 3.3, is equal to $\overline{\widehat{\beta}}_0(.50) = \frac{1}{M}\sum_m \widehat{\beta}_0(.50)_m = 2.86$ for the intercept and $\overline{\widehat{\beta}}_1(.50) = \frac{1}{M}\sum_m \widehat{\beta}_1(.50)_m = 0.526$ for the slope.

The variance of the regression coefficients at the median is then estimated by

$$\mathrm{var}(\widehat{\beta}(.50)) = \mathrm{E}\left[\widehat{\beta}(.50)_m - \overline{\widehat{\beta}}(.50)\right]^2 = \frac{1}{M\text{-}p}\sum_m\left[\widehat{\beta}(.50)_m - \overline{\widehat{\beta}}(.50)\right]^2$$

For M $= 100$ replicates and $p = 2$ estimated parameters, intercept and slope, the dispersion of the estimated coefficients around $\widehat{\beta}(.50)$ is $\widehat{\mathrm{var}}(\widehat{\beta}_0(.50)) = 2.05$ and $\widehat{\mathrm{var}}(\widehat{\beta}_1(.50)) = 0.024$, and their square root values are reported in the middle rows of Table 3.3.

The value of $\overline{\widehat{\beta}}(.50)$ allows to measure the bias of the bootstrap estimator, which is computed as $\overline{\widehat{\beta}}(\theta) - \widehat{\beta}(\theta)$, where $\widehat{\beta}(\theta)$ is the coefficient estimated in the original sample, without bootstrap. The $[Y_1 \, X_1]$ sample of size $n = 11$ provides, at the median regression, the values of $\widehat{\beta}_0(.50) = 3.24$ and $\widehat{\beta}_1(.50) = 0.48$, as reported in Table 1.2,

Table 3.3 Summary statistics of the estimated coefficients, design matrix bootstrap with samples of size $n = 11$, and elemental sets bootstrap with samples of size $n = 2$; M = 100 replicates, Anscombe data set $[Y_1\ X_1]$.

| *mean* | design matrix | | | | |
	$\theta = .25$	$\theta = .50$	$\theta = .75$	OLS	elem. sets
intercept	2.651	2.860	3.448	3.126	6.432
slope	0.464	0.526	0.536	0.486	0.231

| *s.d.* | design matrix | | | | |
	$\theta = .25$	$\theta = .50$	$\theta = .75$	OLS	elem. sets
intercept	1.524	1.434	1.686	0.981	12.72
slope	0.167	0.154	0.197	0.104	1.199

| *bias* θ | design matrix | | | | |
	$\theta = .25$	$\theta = .50$	$\theta = .75$	OLS	elem. sets
intercept	0.671	−0.38	0.728	0.126	3.432
slope	−0.105	0.046	−0.140	−.0141	−0.268

and the bias associated with the bootstrap estimates is $\overline{\hat{\beta}_0}(.50) - \hat{\beta}_0(.50) = 2.86 - 3.24 = -0.38$ for the intercept and $\overline{\hat{\beta}_1}(.50) - \hat{\beta}_1(.50) = 0.526 - 0.48 = 0.046$ for the slope as shown in the bottom rows of Table 3.3. Looking at the histograms of the M = 100 estimates for the slope and for the intercept as computed at the median regression, presented to the right of Figure 3.9, at the median the slope is centered at 0.52 and the intercept at 2.86. These values do not exactly coincide with the values of the parameters estimated in the original sample, and the difference is the bias. The first four columns of Table 3.3 summarize mean, variance, and bias for the coefficients estimated at the three quartile regressions and by OLS in the design matrix bootstrap. The last column reports these results for the elemental sets bootstrap. Instead of resampling with replacement n observations, the elemental set bootstrap resamples p observations so that the regression coefficients are exactly computed, without error. This column presents the largest biases and variances. The left graphs in Figure 3.9 show the histograms of the bootstrapped elemental set coefficients in M = 100 iterations.

Consider next the French data set of Section 1.2.1, relating walking speed and body mass index. In a sample of size $n = 393$ there are $\binom{393}{2} = 77028$ elemental sets, which is a quite large number. In the Dutch case of Section 1.2.2, the number of elemental sets is $\binom{307}{2} = 46971$, too large to be applicable. Instead of computing all of them, a smaller number of elemental sets can be analyzed (Hawkins, 1993; Hall and Mayo, 2008). The last column of Table 3.4 reports the summary statistics of the elemental set bootstrap approach for the French data, selecting M = 100 replicates

Table 3.4   Bootstrapping the elemental sets and the design matrix, OLS, and median regression results, French data set of size $n = 393$, M $= 100$ replicates.

mean	OLS	design matrix median regression	elem. sets
intercept	0.556	0.640	3.575
slope	0.003	−0.001	−0.012

s.d.	OLS	design matrix median regression	elem. sets
intercept	0.105	0.089	19.43
slope	0.004	0.003	0.793

bias	OLS	design matrix median regression	elem. sets
intercept	0.005	0.016	3.025
slope	−.0001	−.0009	−0.124

each having size $p = 2$, while the first two columns report the statistics for OLS and median regression results in the design matrix bootstrap of size $n = 393$. The elemental sets distributions present larger bias and variance when compared to the results of the standard bootstrap approach, and the standard bootstrap approach is decidedly preferable.

Table 3.5 considers the elemental sets bootstrap results in samples of size $p = 2$, the OLS and the median regression estimates in the design matrix bootstrap, for the Dutch example replicates, each having sample size equal to $n = 307$. Figure 3.10 for the French example and Figure 3.11 for the Dutch case compare the histograms of intercept and slope computed in the M $= 100$ replicates bootstrapping elemental sets in the left histograms, the bootstrap OLS estimates in the middle histograms, and the bootstrap quantile regression estimates in the right graphs. The greater dispersion in the estimates of the elemental sets bootstrap is due to the size of each subsample, which is a key element for the accuracy of the results: the smaller the size, the lower the precision (Camponovo et al., 2012). Therefore, while the analysis of the entire group of elemental sets grants robustness, bootstrapping the elemental sets causes a greater dispersion in the distribution of the estimates.

Besides estimation, subsampling can be of use in quantile regression inference as well. Chernozhukov and Fernandez-Val (2005, 2011) use bootstrap to implement inference in extremal quantiles, which is the issue analyzed in the next section, while Escaciano and Goh (2014) define a bootstrap-based test to verify the correct specification of quantile regression models.

Table 3.5 Bootstrapping the elemental sets and the design matrix, OLS, and median regression results, Dutch sample of size $n = 307$, M $= 100$ replicates.

mean	OLS	design matrix median regression	elem. sets
intercept	0.556	0.640	27.016
slope	0.003	−0.001	−1.116

s.d.	OLS	design matrix median regression	elem. sets
intercept	0.088	0.109	41.39
slope	0.003	0.004	1.52

bias	OLS	design matrix median regression	elem. sets
intercept	0.005	0.001	3.675
slope	−.0001	−.0001	0.151

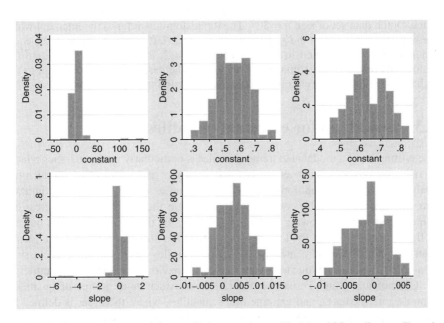

Figure 3.10 Histograms of the coefficients estimated in M $= 100$ replicates, French data set of size $n = 393$. The top histograms depict the intercept while the bottom ones refer to the slope. To the left are the elemental sets results, with coefficients exactly computed in samples of size $n = 2$; in the middle graphs are the results of bootstrapping the OLS regression; to the right are those of the bootstrap median regression.

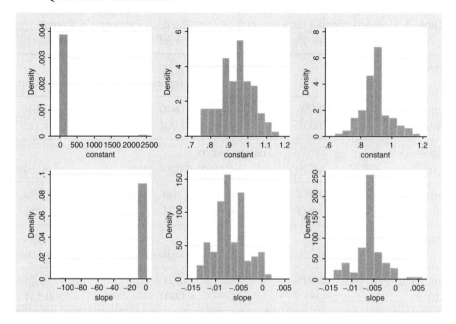

Figure 3.11   Histograms of the bootstrap regression coefficients in M = 100 replicates, Dutch data set of size n = 307. The top histograms refer to the intercept while the bottom graphs report the slope estimates. To the left are the elemental sets coefficients, exactly computed in samples of size n = 2; in the middle graphs are the histograms of the OLS estimated coefficients in the design matrix bootstrap; to the right are the median regression estimates in the design matrix bootstrap.

3.3   Bootstrap for extremal quantiles

The estimation of a model at extreme quantiles is particularly useful in issues related to risk management, such as stress tests, portfolio performance, or productivity analysis in adverse conditions, as well as in studies on unemployment duration and wage inequality. The conventional asymptotic theory of quantile regression, summarized in the following Section 3.4, applies to the central quantiles but not to those quantiles far in the tails. As a consequence, standard tests and confidence intervals cannot be implemented in the extreme quantiles. Even bootstrap and classical resampling methods fail due to the nonstandard behavior of the extremal quantiles.

Chernozhukov and Fernandez-Val (2005) consider three groups of quantiles: central-, intermediate-, and extreme-order quantiles, where the order is defined by $\theta n/p$ with p being the number of regression coefficients, n the sample size, and θ the chosen quantile. The cutoff value, discriminating standard from nonstandard asymptotic distribution, is $\theta < 0.2$ or, alternatively, the value $\theta n/p < 30$ (Chernozhukov and Fernandez-Val, 2011). Above these values, the quantiles are of intermediate or central order, but below this value, the asymptotic behavior of the quantile regression estimators differs. While central and intermediate quantiles abide by the conventional large sample theory, extremal quantiles follow extreme value laws and converge to extreme type variates so that inference in the far tails requires a different approach. Consider the centered statistic

$$Z = A(\widehat{\beta}(\theta) - \beta(\theta))$$

$$A = \frac{\sqrt{k}}{\overline{X}^T(\widehat{\beta}(m\theta) - \widehat{\beta}(\theta))}$$

where A is a normalizing term, the constant $k > 0$ is approximated by θn, m is such that $k(m-1) > p$, and $\overline{X}$ is the vector of sample means of the explanatory variables in X.[3] Chernozhukov and Fernandez-Val (2011) present a resampling approach to compute self-normalized statistics. This avoids the estimation of tail parameters, which are generally difficult to compute in a reliable way. The method exploits the different rates of convergence of extreme and intermediate quantile regressions. The basic assumption is that the dependent variable has Pareto-type tails with tail index ξ.[4] For instance in the Student-t distribution with g degrees of freedom, the tail index is $\xi = 1/g$.

Median-unbiased estimation and inference in the tails can be implemented as follows. The first step considers B subsamples, each of size $b < n$, to compute the statistics

$$\widehat{V}_{i,b,n} = \widehat{A}_b(\widehat{\beta}_i(\theta) - \widehat{\beta}(\theta))$$

$$\widehat{A}_b = \frac{\sqrt{n\theta}}{\overline{X_i}(\widehat{\beta}_i(m\theta_b) - \widehat{\beta}_i(\theta_b))}$$

where $\widehat{\beta}_i(\theta)$ is the θ quantile regression estimated coefficient in the ith subset of size b, $\widehat{\beta}(\theta)$ is the same quantile regression coefficient estimated in the entire sample of size n, $\overline{X_i}$ is the sample mean of the regressors in the ith subsample of size b, m is set to $m = \frac{\theta n + p}{\theta n}$, and $\theta_b = \theta n/b$. The median value of $\widehat{V}_{i,b,n}$ and the $\alpha/2$ and $(1 - \alpha/2)$ quantiles of the $\widehat{V}_{i,b,n}$ series provide the appropriate elements to compute the median-unbiased regression coefficients and the critical values for their confidence intervals, given by:

$$\widehat{\widehat{\beta}}(\theta) = \widehat{\beta}(\theta) - \widehat{V}_{1/2}$$

$$lim_{n \to \infty} P(\widehat{\widehat{\beta}}(\theta) - \widehat{V}_{1-\alpha/2} < \beta(\theta) < \widehat{\widehat{\beta}}(\theta) - \widehat{V}_{\alpha/2}) = 1 - \alpha$$

and $V_{1/2}$ is a bias-correcting factor. The key point in the above approach is the normalizing term $\widehat{A}_b$. The classical subsampling approach would recenter the statistics by $\widehat{\beta}(\theta)$ estimated in the full sample, while this approach considers $\widehat{\beta}_i(\theta_b)$, that is no longer an extreme- but an intermediate-order quantile.

---

[3] In the limit, Z is a function of Γ distributions, not necessarily centered at zero and possibly median biased.

[4] A Pareto-type tail implies a tail decaying by a regularly varying function. The Pareto-type tail definition includes both thick and thin tails distributions.

In the Anscombe data $[Y_1 \ X_1]$ with sample size $n = 11$, the 5th quantile regression is an extreme order quantile since $\theta = .05 < .2$, $\theta n/p = .275 < 30$. The 5th quantile regression provides the estimates $\hat{\beta}_1(.05) = 0.46$ and $\hat{\beta}_0(.05) = 1.6$. These coefficients are both statistically different from zero according to the usual inference approach, with Student-t values $t(\hat{\beta}_1) = 10.84$ and $t(\hat{\beta}_0) = 3.87$ and 95% confidence intervals $ci[\beta_0(.05)] = [0.664 \quad 2.535]$, $ci[\beta_1(.05)] = [0.363 \quad 0.556]$. Next, the extreme-value approach is implemented. The resampling scheme considers $B = 100$ samples, each of size $b = 7$. In each subsample $\hat{\beta}_i(\theta)$, $\hat{\beta}_i(m \ \theta_b)$, $\hat{\beta}_i(\theta_b)$ and $\overline{X}_i$ are computed, and for $\theta_b = \theta n/b = .078 \approx .08.$, $m = \frac{\theta n + p}{\theta n} = 4.6$, the central-intermediate-order quantile regressions for recentering are $\hat{\beta}_i(\theta_b) = \hat{\beta}_i(.08)$ and $\hat{\beta}_i(m \ \theta_b) = \hat{\beta}_i(.37)$. Figure 3.12 depicts the histograms of the slope coefficient $\hat{\beta}_{1i}(.05)$, $\hat{\beta}_{1i}(.08)$ and $\hat{\beta}_{1i}(.37)$ as estimated in the $B = 100$ replicates in samples of size $b = 7$. It can be noticed that the top-left histogram in Figure 3.12 is skewed and longer tailed, while the other two histograms present higher frequencies at the mode and minor skewness. It is such different behavior between the extreme and the central-intermediate-order quantiles, that it is exploited in the normalization of $(\hat{\beta}_i(\theta) - \hat{\beta}(\theta))$. Once computed $\hat{V}$, its median allows to modify the 5th quantile regression estimated coefficients. However, in the $[Y_1 \ X_1]$ data set $\hat{V}_{1/2}$ is practically equal to 0 and does not affect the estimates so that $\hat{\beta}(.05) = \hat{\beta}(.05)$ since

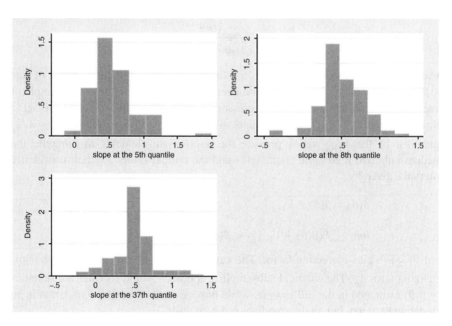

Figure 3.12   Histogram of the bootstrap estimates of the slope coefficient at the 5th, 8th, and 37th quantile regression, $B = 100$ iterations, each with subsamples of size $b = 7$. Anscombe data set $[Y_1 \ X_1]$, sample size $n = 11$.

$\widehat{V}_{1/2}(\widehat{\beta}_0(.05)) = 1e^{-8}$ and $\widehat{V}_{1/2}(\widehat{\beta}_1(.05)) = -5e^{-9}$. The 5th and 95th percentiles of the $\widehat{V}$ series provide the values to compute the $1 - \alpha = 90\%$ confidence intervals in the extreme value approach:

$$ci[\beta_0(.05)] = [\widehat{\beta}_0(.05) - \widehat{V}(.95) \quad \widehat{\beta}_0(.05) - \widehat{V}(.05)]$$

$$= [1.6 - 2.119 \quad 1.6 + 0.512] = [-0.519 \quad 2.11]$$

$$ci[\beta_1(.05)] = [\widehat{\beta}_1(.05) - \widehat{V}(.95) \quad \widehat{\beta}_1(.05) - \widehat{V}(.05)]$$

$$= [0.46 - 0.990 \quad 0.46 + 0.383] = [-0.530 \quad 0.843]$$

The extreme-value confidence intervals are larger than the ones computed relying on the standard asymptotic theory. This mirrors the greater uncertainty characterizing the far tails.

3.3.1   The French data set

In this data set, $n = 393$ and $p = 2$. When $\theta = .05$ is the selected quantile, $\frac{\theta n}{p} = 9.8$ is below the cutoff value of 30, and the extreme-value theory is called for. In the standard approach, the estimated coefficients are $\widehat{\beta}_0(.05) = 0.259$ and $\widehat{\beta}_1(.05) = -0.0005$, with t-values $t(\widehat{\beta}_0(.05)) = 2.06$ and $t(\widehat{\beta}_1(.05)) = -.11$, with 95% confidence intervals $ci(\beta_0(.05)) = [0.0120 \quad 0.507]$ and $ci(\beta_1(.05)) = [-0.0099 \quad 0.0089]$. To implement the extreme value approach, consider $B = 100$ subsamples of size $b = 70$ so that $\theta_b = \theta n/b = .28$, $m = \frac{\theta n + p}{\theta n} = 1.10$, and $m\theta_b = .30$. In each subsample $\widehat{\beta}_i(.05)$, $\widehat{\beta}_i(.30)$, $\widehat{\beta}_i(.28)$, and $\overline{X}_i$ are computed to provide the $\widehat{V}$ series. Figure 3.13 presents the histograms of the slope coefficient estimated in the $B = 100$ subsets of size $b = 70$ at the 5th, the 28th, and the 30th quantile regressions. Analogously to the Anscombe example, the histogram of the 5th quantile regression estimated slope is more dispersed and has lower frequency at the mode than the histogram of the slope estimated at the 28th and the 30th quantiles. Once computed the $\widehat{V}$ series, the median values are $\widehat{V}_{1/2}(\widehat{\beta}_0(.05)) = 0.0491$ and $\widehat{V}_{1/2}(\widehat{\beta}_1(.05)) = 0.0399$. The extreme-value estimated coefficients and their confidence intervals at $1 - \alpha = 90\%$ are

$$\widehat{\widehat{\beta}}_0(.05) = \widehat{\beta}_0(.05) - \widehat{V}_{1/2}(\widehat{\beta}_0(.05)) = 0.2597 - 0.0491 = 0.2106$$

$$ci[\beta_0(.05)] = [\widehat{\widehat{\beta}}_0(.05) - \widehat{V}(.95) \quad \widehat{\widehat{\beta}}_0(.05) - \widehat{V}(.05)]$$

$$= [0.2106 - 0.9765 \quad 0.2106 + 0.9408] = [-0.7659 \quad 1.151]$$

$$\widehat{\widehat{\beta}}_1(.05) = \widehat{\beta}_1(.05) - \widehat{V}_{1/2}(\widehat{\beta}_1(.05)) = -0.00052 - 0.03989 = -0.04041$$

$$ci[\beta_1(.05)] = [\widehat{\widehat{\beta}}_1(.05) - \widehat{V}(.95) \quad \widehat{\widehat{\beta}}_1(.05) - \widehat{V}(.05)]$$

$$= [-0.04041 - 1.027 \quad -0.04041 + 0.7864]$$

$$= [-1.0674 \quad 0.7460].$$

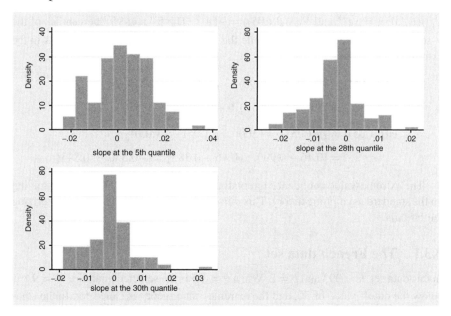

Figure 3.13   Histogram of the bootstrap estimates of the slope coefficient at the 5th, 28th, and 30th quantile regressions, $B = 100$ replicates in subsets of size $b = 70$, French data set of size $n = 393$.

Thus, the confidence intervals computed in the extreme tails are wider than those computed using the standard inference for the central quantiles.

3.3.2   The Dutch data set

In the Dutch data set, with $n = 307$ and $p = 2$ explanatory variables, $B = 100$ subsamples of size $b = 70$ are selected to compute the 5th quantile regression. In the standard quantile regression, the estimated coefficients are $\widehat{\beta}_0(.05) = 0.257$ and $\widehat{\beta}_1(.05) = 0.0024$, with t-values $t(\widehat{\beta}_0(.05)) = 1.85$ and $t(\widehat{\beta}_1(.05)) = 0.51$, with the following 95% confidence intervals $ci(\beta_0(.05)) = [-0.0171 \quad 0.532]$ and $ci(\beta_1(.05)) = [-0.0070 \quad 0.0119]$. The ratio $\frac{\theta n}{p} = 7.6$ is below the cutoff value of 30, and the extreme-value theory is implemented. To compute the $\widehat{V}$ series, set $\theta_b = \theta n/b = .21$, $m = \frac{\theta n + p}{\theta n} = 1.13$, $m\theta_b = .24$ so that $\widehat{\beta}_i(.05)$, $\widehat{\beta}_i(.21)$, $\widehat{\beta}_i(.24)$ and $\overline{X}_i$ are the elements to be computed in the subsets. Figure 3.14 reports the histograms of the slope coefficient as estimated respectively at the 5th, the 21st, and the 24th quantile regression in the $B = 100$ replicates in subsets of size $b = 70$. Once again, the histogram of the slope estimated at the extreme quantile $\theta = .05$ is more

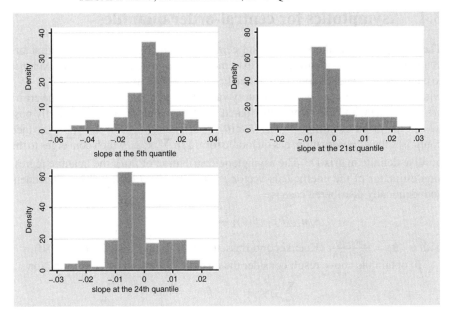

Figure 3.14   Histogram of the bootstrap estimates of the slope coefficient at the 5th, 21st, and 24th quantile regressions, $B = 100$ replicates in subsets of size $b = 70$, Dutch data set of size $n = 307$.

dispersed and has lower frequency at the mode than the other two. The median values of the $\widehat{V}$ series are $\widehat{V}_{1/2}(\widehat{\beta}_0(.05)) = 0.01445$ and $\widehat{V}_{1/2}(\widehat{\beta}_1(.05)) = 0.03966$. The extreme-value estimated coefficients and their confidence intervals for the 5th quantile regression at $1 - \alpha = 90\%$ are the following

$$\widehat{\widehat{\beta}}_0(.05) = \widehat{\beta}_0(.05) - \widehat{V}_{1/2}(\widehat{\beta}_0(.05)) = 0.257 - 0.01445 = 0.2425$$

$$ci[\beta_0(.05)] = [\widehat{\widehat{\beta}}_0(.05) - \widehat{V}(.95) \quad \widehat{\widehat{\beta}}_0(.05) - \widehat{V}(.05)]$$

$$= [0.2425 - 1.4014 \qquad 0.2425 + 0.8622] = [-1.159 \qquad 1.105]$$

$$\widehat{\widehat{\beta}}_1(.05) = \widehat{\beta}_1(.05) - \widehat{V}_{1/2}(\widehat{\beta}_1(.05)) = 0.0024 - 0.03966 = -0.03726$$

$$ci[\beta_1(.05)] = [\widehat{\widehat{\beta}}_1(.05) - \widehat{V}(.95) \quad \widehat{\widehat{\beta}}_1(.05) - \widehat{V}(.05)]$$

$$= [-0.03726 - 1.327 \qquad -0.03726 + 2.0704] = [-1.364 \qquad 2.033].$$

Once again these confidence intervals are wider than those provided by the inference for central quantiles.

3.4   Asymptotics for central-order quantiles

The asymptotics of intermediate- and central-order quantile regression estimators could be obtained by modifying the extreme-value approach. However, it is simpler to consider the standard approach. In the linear regression model $y_i = \beta_0 + \beta_1 x_i + e_i$, with a constant term and one explanatory variable x_i, in a sample of size n, the errors are independent and identically distributed and have common density f strictly positive at the given quantile $F^{-1}(\theta), f(F^{-1}(\theta)) > 0$. The row vector $\mathbf{x}_i = [1 \ x_i]$, which comprises the ith observation, has quadratic form $1/n \sum_i \mathbf{x}_i^T \mathbf{x}_i$, which converges to the positive definite matrix D.[5] The asymptotic distribution of $\hat{\beta}(\theta)$, the quantile regression estimator of the coefficients vector $\beta^T(\theta) = [\beta_0(\theta) \ \beta_1(\theta)]$, in the independent and identically distributed case is

$$\sqrt{n}[\hat{\beta}(\theta) - \beta(\theta)] \to N(0, \omega^2(\theta)D^{-1})$$

with $\omega^2(\theta) = \frac{\theta(1-\theta)}{f(F^{-1}(\theta))^2}$ (Koenker and Bassett, 1978; Ruppert and Carroll, 1980).
To obtain the above result consider the quantile regression objective function

$$\sum_{i=1,...n} \rho(y_i - \beta_0 - \beta_1 x_i) = \min$$

The check function is defined as $\rho(e_i) = [\theta - 1(e_i < 0)]|e_i|$, and the gradient of the slope is

$$\sum_{i=1,...n} \psi(y_i - \beta_0 - \beta_1 x_i)x_i = \sum_{i=1,...n} \psi(e_i)x_i = 0,$$

with $\psi(y_i - \beta_0 - \beta_1 x_i) = \theta - 1(e_i < 0)$.
The ψ function is nondifferentiable when $e_i = 0$. The expected value of the gradient, however, is differentiable in $\hat{\beta}(\theta) - \beta(\theta)$. A stochastic equicontinuity condition on the sample average of $\psi(e_i)$, $\overline{\psi}_n(\beta) = 1/n\sum_{i=1,..n}x_i\psi(y_i - \beta_0 - \beta_1 x_i)$, states that

$$\sqrt{n}\{\overline{\psi}_n(\hat{\beta}) - \overline{\psi}_n(\beta) - E[\overline{\psi}_n(\hat{\beta}) - \overline{\psi}_n(\beta)]\}$$

$$= \sqrt{n}\{[\overline{\psi}_n(\hat{\beta}) - E\overline{\psi}_n(\hat{\beta})] - [\overline{\psi}_n(\beta) - E\overline{\psi}_n(\beta)]\} \to 0$$

for any consistent estimator $\hat{\beta}(\theta) \to \beta(\theta)$, the difference $\sqrt{n}\{[\ \overline{\psi}_n(\hat{\beta}) - E\ \overline{\psi}_n(\hat{\beta})]$ evaluated at the consistent estimator $\hat{\beta}$ is asymptotically equivalent to its value at $\beta = plim\ \hat{\beta}$.
In the above equation, $E(\overline{\psi}_n(\beta)) = 0$ since it is the gradient evaluated at the true parameters. The term $\overline{\psi}_n(\hat{\beta})$ is $o_p(1/\sqrt{n})$, and by Taylor expansion (Hayashi, 2000) the expression becomes

$$E(\overline{\psi}_n(\hat{\beta})) = \overline{\psi}_n(\beta) + \frac{\partial E(\overline{\psi}_n(\beta))}{\partial \beta'} \sqrt{n}[\hat{\beta}(\theta) - \beta(\theta)]$$

---

[5] The case of $k > 2$ explanatory variables, $y = X\beta + e$, comprises the (n, k) matrix $X = [1 \ x_1 \ x_2...x_{k-1}]$, and the $(k, 1)$ vector of coefficients is $\beta^T = [\beta_0 \ \beta_1 \ \beta_2...\beta_{k-1}]$. The ith observation is given by the row vector $\mathbf{x}_i = [1 \ x_{1i} \ x_{2i} \ \cdots \ x_{k-1i}]$.

and

$$\sqrt{n}[\hat{\beta}(\theta) - \beta(\theta)] = \left[\frac{\partial E(\overline{\psi}_n(\beta))}{\partial \beta'}\right]^{-1} \sqrt{n}\overline{\psi}_n(\beta) + o_p(1)$$

By the Lindeberg-Levy central limit theorem $\sqrt{n}[\hat{\beta}(\theta) - \beta(\theta)]$ is asymptotically normal with zero mean and covariance $\omega^2(\theta)D^{-1}$.

In case of non-identically distributed errors, the asymptotic distribution is given by (Koenker and Bassett, 1982):

$$\sqrt{n}[\hat{\beta}(\theta) - \beta(\theta)] \to N(0, \theta(1 - \theta)D_1(\theta)^{-1}DD_1(\theta)^{-1}).$$

and the positive definite matrix $D_1(\theta)$ is the limit of $1/n \sum_i f_i(F^{-1}(\theta)) \, x_i^T x_i$. In case of i.i.d. errors, the covariance simplifies to $\omega^2 D^{-1}$. This occurs because $f_i(F^{-1}(\theta))$ is constant, $f_i(F^{-1}(\theta)) = f(F^{-1}(\theta))$, so it can be pulled out of the $D_1(\theta)$ matrix so that D and $D_1(\theta)$ coincide and cancel out. Thus, the term $D_1(\theta)^{-1}DD_1(\theta)^{-1}$ collapses into D^{-1}.

The case of dependent observations is considered in Weiss (1990). In the median regression model with a first-order serial correlation, $y_i = \beta_0 + \beta_1 x_i + e_i$, $e_i = \delta e_{i-1} + a_i$, the quantile regression estimator is unbiased and asymptotically normal, with covariance matrix $\omega^2 D^{-1} D_2 D^{-1}$, with $D_2 = D + 1/n \sum_i \psi(e_i)$ $\psi(e_{i-1})(x_i^T x_{i-1} + x_{i-1}^T x_i)$. Once again, if the data are independent, the term $1/n \sum_i$ $\psi(e_i) \, \psi(e_{i-1})(x_i^T x_{i-1} + x_{i-1}^T x_i) = 0$, D_2 and D coincide, and the covariance matrix collapses to $\omega^2 D^{-1}$.

3.5 Treatment effect and decomposition

This section analyzes an additional quantile regression approach relying on bootstrap, the quantile treatment effect (QTE) estimator, which is implemented to evaluate the impact of a treatment or a policy. Studies concerning the impact of welfare measures, of education, training programs, medical treatment, and policy intervention are some of the issues that can be assessed by the treatment effect estimators. They generally compare two groups of observations, one referring to the treated and the other to the untreated, the control set. For instance when assessing the effect of a training program, the trained/treated are compared with the non-trained/control group.[6] In the linear regression model $Y = X\beta + e$ the outcome variable Y assumes values Y_1 for the treated, where the treatment variable is $Z = 1$, and Y_0 in the control group, where treatment does not take place and $Z = 0$. The values $Y_0 = X_0\beta_0 + e_0$ and $Y_1 = X_1\beta_1 + e_1$ are the observed outcomes within each group. From one group to the other, both covariates and coefficients may vary and it is relevant to assess which one drives the change. Two independent regressions are implemented, one in each group, and the two vectors of coefficients are separately computed.

---

[6] The simple two-group analysis can be generalized to many groups to model different treatment levels, such as different length of training programs, for instance.

Then the average difference in outcome between treated and untreated, that is, the average treatment effect, can be measured as

$$E(Y_1 - Y_0) = E(X_1\beta_1 - X_0\beta_0) = \beta_1 E(X_1) - \beta_0 E(X_0)$$

In the above equation, one can add and subtract the term $Y_{1/0} = X_1\beta_0$ without affecting the results. $X_1\beta_0$ evaluates the group 1 covariates at the group 0 coefficients and measures how the dependent variable would be if the treated had the same coefficients of the control group. This yields the counterfactual, which is obtained by multiplying the covariates of the treated by the coefficients of the control group. For instance, in a comparison of gender differences in wages, the $Y_{1/0}$ counterfactual considers women's covariates X_1 evaluated at male coefficients β_0: what would be the wage if women's characteristics had the same coefficients/remuneration of the male group. The introduction of $Y_{1/0}$ allows to decompose the difference between the two groups into difference in the covariates and difference in the coefficients as follows (Oaxaca, 1973; Blinder, 1973)

$$E(Y_1 - Y_{1/0} + Y_{1/0} - Y_0) = E(X_1\beta_1 - X_1\beta_0 + X_1\beta_0 - X_0\beta_0)$$
$$= E[X_1(\beta_1 - \beta_0) + \beta_0(X_1 - X_0)]$$
$$= (\beta_1 - \beta_0)EX_1 + \beta_0(EX_1 - EX_0)$$

The last equality shows how the difference between the two groups can be split in two factors, where the term $\beta_0(EX_1 - EX_0)$ provides the average difference in covariates keeping β_0 constant at the control group values, while the term $(\beta_1 - \beta_0)EX_1$ yields the difference in the coefficients of the two regressions keeping the covariates at their average treatment values. While a difference in the covariates would fully explain the difference in the outcomes of the two groups, the difference in the coefficients is instead unexplained by the model and can be ascribed to the treatment, the policy, or the training program implemented. Alternatively, depending on the issue analyzed, the unexplained difference previously defined as treatment effect can also be interpreted as a discrimination effect. This is the case, for instance, of a difference in wages between men and women, or between immigrants versus non-immigrants, or unionized versus non-unionized workers. If the difference depends on the coefficients and cannot be ascribed to a difference in the characteristics of the two groups, then it can be interpreted as discrimination.

Finally, depending on the problem under study, the counterfactual can be defined as $Y_{0/1} = X_0\beta_1$, which provides the group 0 covariates evaluated at group 1 coefficients. $Y_{0/1}$ tells how the outcome would be when changing the coefficients while keeping the covariates unchanged. In the example on wage comparison between men and women, the $Y_{0/1}$ counterfactual considers male covariates X_0 evaluated at women's coefficients β_1: what would be the wage if men's characteristics had the same coefficients/remuneration of women's group. This would change the terms in the decomposition as follows

$$E(Y_1 - Y_0) = (\beta_1 - \beta_0)EX_0 + \beta_1(EX_1 - EX_0).$$

where the change in coefficients is evaluated at group 0 averages of covariates while the average change in covariates is measured at group 1 coefficients. Sometimes an additional term is included, to model interaction between coefficients and covariates, $(\beta_1 - \beta_0)(EX_1 - EX_0)$, yielding

$$E(Y_1 - Y_0) = (\beta_1 - \beta_0)EX_0 + \beta_1(EX_1 - EX_0) + (\beta_1 - \beta_0)(EX_1 - EX_0).$$

As an example, consider the data set on walking speed and *BMI* in Europe analyzed in the previous sections (SHARE, year 2004). The sample can be split according to gender, and the treatment variable Z assumes value Z = 0 in the male group and Z = 1 for women. The outcome Y is the walking speed within each group while the explanatory variables in X are *BMI* and the constant term. The two groups have respectively sample size $n_0 = 1282$ in the male group and $n_1 = 1677$ in the women's subset. Figures 3.15 and 3.16 show the histograms of the dependent and the explanatory variable, walking speed and *BMI*, in each group. Women are characterized by a lower average speed and by a larger variance in *BMI* with respect to the male group. The anthropometric target is to check the statistical significance of a difference in walking speed between men and women together with its decomposition into covariates and coefficients effects, that is, into gap explained and unexplained by *BMI*.

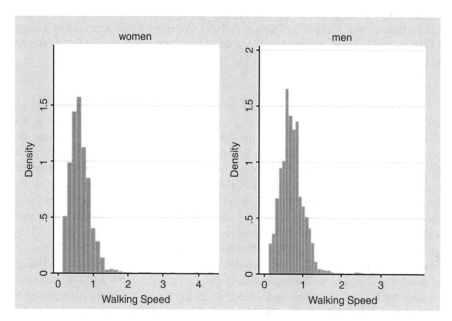

Figure 3.15 Histogram of walking speed in the two groups, the male sample size is $n_0 = 1282$ while women's subset is of size $n_1 = 1677$. Women are characterized by a smaller mean, $\overline{Y}_1 = 0.63$ versus $\overline{Y}_0 = 0.73$ in the male group, while the variances are similar, respectively $var(Y_1) = 0.10$ and $var(Y_0) = 0.09$.

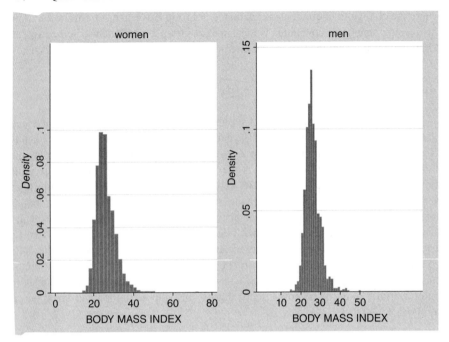

Figure 3.16   Histograms of *BMI* in the two groups, the male sample size is $n_0 = 1782$ while women's subset is of size $n_1 = 1677$. Women are characterized by a larger variance, $\text{var}(X_1) = 22$ versus $\text{var}(X_0) = 14$ in the male group, while the means are similar, respectively $\overline{X}_1 = 26.2$ and $\overline{X}_0 = 26.1$.

The Oaxaca-Blinder decomposition would look at the difference between men's and women's walking speed on average. If the between-groups difference is due to the independent variable, body mass index, then the model does explain the average difference between the two groups. Vice versa, if the difference between the independent variables of the two groups is statistically irrelevant, then the average-groups difference is related to the coefficients and is unexplained by the model, thus providing the treatment/gender effect. The average difference in walking speed is equal to $E(Y_1 - Y_0) = EY_1 - EY_0 = 0.637 - 0.731 = -0.094$, with men walking faster than women. This value, representing the total difference in walking speed, coincides with the OLS coefficient of the dummy variable Z in Table 3.6, where Z assumes unit value in the women's subset. The latter computes the impact of gender on *wspeed* in the regression *wspeed* $= \alpha + \beta \; BMI + \delta \; Z + e$, estimated pooling the two subsets, men and women together. These estimates are reported in the last two rows of Table 3.6. The decomposition allows to split the -0.094 total difference into covariates and coefficients effects. Focusing on the slope, to compute $E[\beta_0(X_1 - X_0)] = \beta_0(EX_1 - EX_0)$, the group 0 slope coefficient β_0 is replaced by its OLS estimated value, which in the male subset is small and not statistically

Table 3.6   OLS estimates.

group 0 = men	intercept	BMI	
	0.7788	−0.0018	
se	(0.060)	(0.0022)	

group 1 = women	intercept	BMI	
	0.7267	−0.0034	
se	(0.042)	(0.0016)	

0 and 1 together	intercept	BMI	Z
	0.713	−0.0029	−0.0937
se	(0.035)	(0.0013)	(0.0118)

different from zero, $\hat{\beta}_{0,OLS} = -0.0018$ with $se(\hat{\beta}_{0,OLS}) = 0.0022$, as reported in the top rows of Table 3.6. Then the covariate impact evaluated at the group 0 slope can be computed by $\hat{\beta}_{0,OLS}(EX_1 - EX_0)_{slope} = -0.0018(26.20 - 26.16) = -0.00007$, which is quite small. For the coefficients effect, given by $EX_1(\beta_1 - \beta_0)$, the OLS slope coefficient in the women's group has to be computed. The OLS estimate of β_1 is given by $\hat{\beta}_{1,OLS} = -0.0034$ with $se(\hat{\beta}_{1,OLS}) = 0.0016$, as reported in the middle rows of Table 3.6, and it is statistically different from zero. The coefficient effect for the slope evaluated at group 1 covariates is then $EX_1(\hat{\beta}_{1,OLS} - \hat{\beta}_{0,OLS})_{slope} = 26.20(-0.0034 + 0.0018) = -0.042$, negative and decidedly more sizable than the covariate effect. Finally, the interaction term is very small, $(\hat{\beta}_{1,OLS} - \hat{\beta}_{0,OLS})(EX_1 - EX_0)_{slope} = (-0.0034 + 0.0018)(26.20 - 26.16) = -0.00006$. Summarizing, in the male subset, *BMI* is irrelevant in explaining walking speed, while for women this coefficient is statistically significant. The average total difference between the two groups is −0.094, and about half of it is related to a difference in the slopes: men walk faster than women, but this is hardly explained by a difference in *BMI*. Figure 3.17 depicts the scatterplot of the data in each subset, clearly showing the presence of dissimilar patterns in the two subsets. Figure 3.18 depicts the OLS estimated regressions for men and women. These lines differ from one group to the other in both intercept and slope.

However, in the above decomposition inference has not been implemented, and nothing so far can be said about the statistical significance of the total average difference, of the average difference in covariates, and in coefficients. To do so, the estimated variances are needed. The variance of the sample mean of the dependent variable for group 0, $\overline{Y}_0 = \overline{X}_0\hat{\beta}_0$ where $\overline{Y}$ and $\overline{X}$ are the sample means, is approximated by (Jann, 2008)

$$var(\hat{\beta}_{0,OLS}\overline{X}_0) \simeq \overline{X}_0 var(\hat{\beta}_0)\overline{X}_0^T + \hat{\beta}_0 var(\overline{X}_0)\hat{\beta}_0^T$$

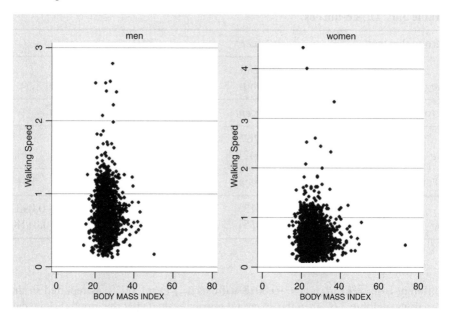

Figure 3.17   Scatterplot of walking speed and *BMI* in each subset, men and women. The two graphs present different patterns.

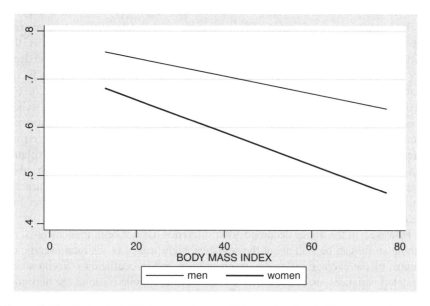

Figure 3.18   Estimated OLS regressions within each subset. The women's group presents smaller intercept and slope so that at large *BMI* values, women's walking speed decreases faster than in the male group.

Under the assumption of independence between groups, the above equation defines the dispersion of each component in the decomposition, that is:

$$var[\hat{\beta}_{0,OLS}(\overline{X}_1 - \overline{X}_0)] \simeq (\overline{X}_1 - \overline{X}_0)var(\hat{\beta}_{0,OLS})(\overline{X}_1 - \overline{X}_0)^T$$

$$+ \hat{\beta}_{0,OLS}var(\overline{X}_1 - \overline{X}_0)\hat{\beta}_{0,OLS}^T$$

$$var[(\hat{\beta}_{1,OLS} - \hat{\beta}_{0,OLS})\overline{X}_1] \simeq \overline{X}_1[var(\hat{\beta}_{1,OLS}) + var(\hat{\beta}_{0,OLS})]\overline{X}_1^T$$

$$+ (\hat{\beta}_{1,OLS} - \hat{\beta}_{0,OLS})var(\overline{X}_1)(\hat{\beta}_{1,OLS} - \hat{\beta}_{0,OLS})^T$$

This allows to establish the statistical relevance of each component. The total difference $E(Y_1 - Y_0) = -0.094$, under the assumption of independence between groups, has standard error $se(\overline{Y}_1 - \overline{Y}_0) = 0.012$, which allows to state that the group difference is significantly different from zero. Next, the variance of the difference in covariates, $var[\hat{\beta}_{0,OLS}(\overline{X}_1 - \overline{X}_0)]$, and of the difference in coefficients, $var[(\hat{\beta}_{1,OLS} - \hat{\beta}_{0,OLS})\overline{X}_1]$, are computed according to the above equations. Table 3.7 reports the results of the Oaxaca-Blinder decomposition for the entire equation, that is, considering both slope and intercept, *BMI* and constant term. The standard errors are respectively $se[\hat{\beta}_{0,OLS}(\overline{X}_1 - \overline{X}_0)] = 0.0003$ for the covariates and $se[(\hat{\beta}_{1,OLS} - \hat{\beta}_{0,OLS})\overline{X}_1] = 0.012$ for the coefficients effect. This allows to conclude that the covariates effect, which for both constant term and *BMI* is given by $\hat{\beta}_{0,OLS}(\overline{X}_1 - \overline{X}_0) = -0.00008$, is not statistically significant. The coefficients effect for slope and intercept, which measures the gap unexplained by the model, is equal to $EX_1(\hat{\beta}_{1,OLS} - \hat{\beta}_{0,OLS}) = -0.093$ and is statistically different from zero. When analyzing intercept and slope together, the coefficients effect accounts for most of the total difference in walking speed between the two groups. The standard error of the interaction term is comparatively large, $se(\beta_1 - \beta_0)(EX_1 - EX_0) = 0.0003$, and this term is not statistically significant. These results confirm the previous conclusion: on average, men walk faster than women, and the difference cannot be explained by the average difference in *BMI* between the two groups.

3.5.1   Quantile treatment effect and decomposition

The Oaxaca-Blinder decomposition so far discussed evaluates the difference between two groups at the mean. The quantile treatment effect (QTE) allows to asses the impact of treatment not only on average but also in the tails of the outcome distribution. It allows to characterize the presence of heterogeneous behaviors of the outcomes at different points of their distributions. In QTE the focus is on the difference between treated and untreated not only at the mean but also in the tails of the

Table 3.7   Oaxaca-Blinder decomposition on average European data on walking speed and *BMI*.

	total	coefficients	covariates	interaction
difference	−0.094	−0.0937	−0.00008	−0.00007
se	(0.012)	(0.012)	(0.0003)	(0.0003)

distributions of the two groups, at various quantiles. Far from being a mere technicality, QTE allows to point out discrepancies between treatment and control changing across quantiles. It may be the case that the discrepancy in the tails sizably differs from the average difference, since treatment can be more effective at the lower (higher) quantiles. For instance, a tax cut may induce an increase in consumption expenditure at lower (higher) incomes and not on average, its impact depending on the income level: at lower incomes its effect may be larger (smaller) than on average or than at the higher income levels. The Oaxaca-Blinder analysis of treatment effect cannot capture these heterogeneities since it relies on average values.

Quantile regressions consider the outcome distribution conditional on the selected covariates. Unlike the analysis on average, the interpretation of the unconditional effects is slightly different from the interpretation of the conditional effects in the quantile framework, due to the definition of the quantile. While the conditional average $E[Y \mid X] = X\beta$ yields the unconditional average $E[Y] = E[X]\beta$, the conditional quantiles $Y(\theta) \mid X = X\beta(\theta)$ do not average up to their unconditional quantiles $Y(\theta)$, since $Y(\theta) \neq Y(\theta) \mid X$ for any quantile θ. Thus, unconditional and conditional quantiles, $Y(\theta)$ and $Y(\theta) \mid X$, do not generally coincide (Frolich and Melly, 2010). Looking at a low quantile, the conditional quantile will summarize the effect for individuals with relatively low outcome given the covariates, even if the level of the outcome is high. The unconditional quantile, on the other hand, summarizes the effect at a low outcome regardless of the covariates so that the outcome level is unquestionably low. For instance, the analysis at low wages considers the unconditional distribution, while the analysis at low wages of highly educated workers looks at the wage distribution conditional to high education. These two wages do not usually coincide since highly educated workers with low wage earn generally more than unschooled workers. While conditional QTE allows to analyze the heterogeneity of the effects with respect to the observables, unconditional QTE aggregates the conditional effects for the entire population. The unconditional quantile functions are one-dimensional functions and do not depend on the explanatory variables in X, whereas the conditional quantile functions are multidimensional. However, conditional and unconditional QTE are trivially the same when the impact of the explanatory variable is not statistically significant.

Different approaches have been proposed to compute QTE, relying in turn on quantile regression, influence function, bootstrap, propensity score, and some of their combinations. The bootstrap-based quantile regression approach in Machado and Mata (2005) computes the unconditional distributions of treated and untreated. Then they decompose the total difference between the two groups into coefficients and covariates differences at various quantiles by introducing the counterfactual distribution.

In a first step, the conditional distributions of the dependent variable are estimated by quantile regression at many quantiles, $\theta = 1, .., m$, separately in the treated and the control group. The estimates of m different quantile regressions within each group yields two sets of size m of estimated coefficients, $\widehat{\beta}_1(\theta)$ for the treated and $\widehat{\beta}_0(\theta)$ for the control group. From the m vectors of estimated coefficients, the corresponding fitted values, $\widehat{Y}_0(\theta) \mid X_0 = X_0\widehat{\beta}_0(\theta)$ and $\widehat{Y}_1(\theta) \mid X_1 = X_1\widehat{\beta}_1(\theta)$ can be computed. The fitted values provide the distribution of the outcome within each group conditional on the covariates X.

Next, the estimated coefficients are separately bootstrapped within each group, yielding $\widetilde{\beta}_0(\theta)$ and $\widetilde{\beta}_1(\theta)$, and the covariates are separately bootstrapped within each group, yielding $\widetilde{X}_0$ and $\widetilde{X}_1$. The unconditional distributions can be computed by $\widetilde{Y}_0(\theta) = \widetilde{X}_0\widetilde{\beta}_0(\theta)$ and $\widetilde{Y}_1(\theta) = \widetilde{X}_1\widetilde{\beta}_1(\theta)$. By bootstrapping both covariates and estimated coefficients, the unconditional distributions of the dependent variable within each group can be computed. The idea behind is the following: for a given X and a random θ uniformly distributed in the $[0;1]$ interval, the term $X\widehat{\beta}(\theta)$ has the same distribution of $Y \mid X$. Instead of keeping X fixed, random samples $\widetilde{X}$ from the population are drawn with replacement, and $\widetilde{X}\widehat{\beta}(\theta)$ has the same distribution as Y (Melly, 2005). This bootstrap procedure is equivalent to integrate the conditional distribution over the X and $\beta(\theta)$ distributions.

Finally, to decompose the treatment effect into impact of covariates and impact of coefficients, the counterfactual is obtained by resampling with replacement the covariates of one group, say $\widetilde{X}_1$, and multiplying them by the bootstrapped coefficients of the other group, $\widetilde{Y}_{1/0} = \widetilde{X}_1\widetilde{\beta}_0(\theta)$, yielding the counterfactual unconditional distribution.

In the example comparing men's and women's walking speed, Table 3.8 reports the quantile regression estimated coefficients in each group for the first nine out

Table 3.8 Quantile regression estimates within each subset in the first nine bootstrap replicates European data on walking speed and *BMI*.

	men, $Z = 0$		women, $Z = 1$	
	slope	constant	slope	constant
$\theta = .87$	−0.0077	1.268	−0.0068	1.116
se	(0.004)	(0.123)	(0.003)	(0.084)
$\theta = .55$	−0.0040	0.8435	−0.0043	0.7404
se	(0.003)	(0.089)	(0.001)	(0.043)
$\theta = .06$	0.0016	0.2718	0.0009	0.2204
se	(0.003)	(0.094)	(0.001)	(0.041)
$\theta = .65$	−0.0038	0.9144	−0.0051	0.8348
se	(0.002)	(0.059)	(0.002)	(0.049)
$\theta = .10$	−0.0001	0.3711	−0.0002	0.3027
se	(0.003)	(0.076)	(0.001)	(0.033)
$\theta = .70$	−0.0040	0.9474	−0.0050	0.8781
se	(0.002)	(0.069)	(0.002)	(0.048)
$\theta = .60$	−0.0033	0.8646	−0.0048	0.7871
se	(0.003)	(0.082)	(0.002)	(0.045)
$\theta = .64$	−0.0034	0.8965	−0.0050	0.8259
se	(0.002)	(0.063)	(0.002)	(0.047)
$\theta = .13$	0.0002	0.3973	0.00001	0.3313
se	(0.004)	(0.099)	(0.001)	(0.037)

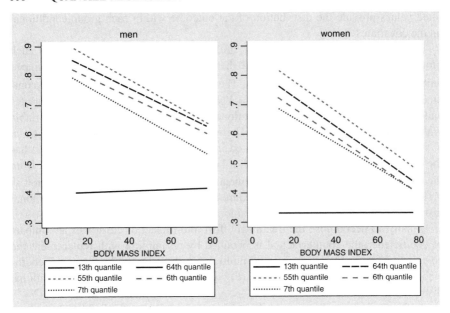

Figure 3.19 Quantile regression estimated lines within each subset for the first five bootstrap replicates reported in Table 3.8. The estimated coefficients in the women's group are smaller than the estimates in the male group.

of $m = 100$ selected quantiles, while Figure 3.19 depicts the first five estimated regressions. The estimates in the women's group are generally smaller. Next, by bootstrapping both covariates and estimated coefficients within each group, the unconditional distributions of Y_0 and Y_1 are obtained, and by bootstrapping covariates of one group and coefficients of the other group, the counterfactual distribution is computed. The top graphs in Figure 3.20 report the unconditional distributions for each group as computed in one of the $m = 100$ replicates implemented. The counterfactual distribution in this same replicate is depicted in the bottom graph of Figure 3.20.

When the number of bootstrap replicates goes to infinity, the Machado and Mata results are numerically identical to the approach proposed by Melly (2006). The latter basically considers the same starting point, quantile regressions to provide the conditional distributions of the outcome in each group, $\hat{Y}_0(\theta) \mid X_0 = X_0\hat{\beta}_0(\theta)$ and $\hat{Y}_1(\theta) \mid X_1 = X_1\hat{\beta}_1(\theta)$. However, instead of implementing bootstrap, Melly (2006) proposes to compute the unconditional distributions analytically, by integrating the conditional distribution $\hat{Y}(\theta) \mid X$ over X and $\beta(\theta)$. This greatly reduces the computational burden. Finally, Chernozhukov et al. (2013) generalize the Machado and Mata (2005) approach by providing a consistent estimator of the variance.

Through QTE, the decomposition can be now computed at various quantiles. Table 3.9 reports the QTE decomposition evaluated at all deciles for the walking speed difference between men and women. The top section of the table reports the

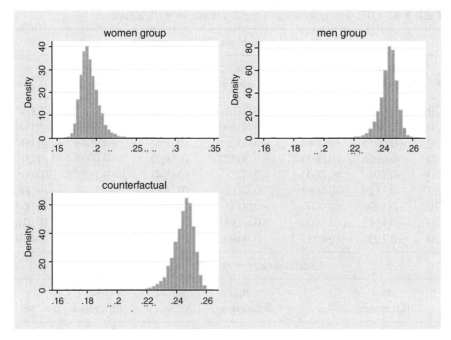

Figure 3.20 Unconditional distributions within each group in the top graphs as computed in one out of 100 replicates of the Machado and Mata approach, $\widetilde{Y}_1 = \widetilde{X}_1 \widetilde{\beta}_1(\theta)$ and $\widetilde{Y}_0 = \widetilde{X}_0 \widetilde{\beta}_0(\theta)$. The bottom graph depicts the counterfactual distribution in the same replicate, $\widetilde{Y}_{1/0} = \widetilde{X}_1 \widetilde{\beta}_0(\theta)$ European data on walking speed and *BMI*.

results provided by the Machado and Mata (2005) approach, while the bottom section collects the results of the Chernozhukov et al. (2013) method. The two approaches yield very similar results. The top section of the table shows that there is a slight increase in the total difference between treated and untreated unconditional distributions, raising from 6% at the 10th decile to 11% above the median. This signals that at high quantiles, the walking speed discrepancy is wider. The same increasing pattern across deciles can be found in the coefficients effect, while the covariates discrepancy is not statistically significant at all quantiles. Thus the total discrepancy cannot be explained by *BMI*, that is, it is not explained by the selected model. Figure 3.21 depicts the behavior of the total, the explained and the unexplained gaps across deciles as computed by the Machado and Mata approach. In the picture, the explained difference is close to zero while the unexplained gap coincides with the total difference almost everywhere.

A different QTE approach is proposed by Frolich and Melly (2010). They compute QTE by implementing a weighted quantile regression, with weights related to the inverse probability of being treated. The quantile regression objective function becomes

$$\sum_{Y < X\beta}[w\theta | Y - X\beta|] + \sum_{Y > X\beta}[w(1 - \theta) | Y - X\beta|]$$

Table 3.9   QTE decomposition of the difference in walking speed.

	Machado and Mata (2005) approach					
θ	total difference	se	covariates difference	se	coefficients difference	se
.10	−0.0600	(0.005)	−0.0026	(0.006)	−0.0627	(0.007)
.20	−0.0850	(0.018)	−0.000001	(0.009)	−0.0850	(0.004)
.30	−0.0975	(0.006)	−0.0002	(0.006)	−0.0977	(0.004)
.40	−0.0966	(0.005)	−0.0022	(0.002)	−0.0988	(0.005)
.50	−0.1040	(0.004)	−0.0017	(0.004)	−0.1057	(0.005)
.60	−0.1120	(0.006)	0.0015	(0.006)	−0.1136	(0.006)
.70	−0.1008	(0.007)	−0.0002	(0.007)	−0.1011	(0.007)
.80	−0.1142	(0.007)	0.0030	(0.016)	−0.1112	(0.016)
.90	−0.1058	(0.011)	0.0080	(0.014)	−0.1138	(0.014)

	Chernozhukov et al. (2013) approach					
θ	total difference	se	covariates difference	se	coefficients difference	se
.10	−0.0711	(0.013)	−0.00007	(0.0008)	−0.0710	(0.013)
.20	−0.0836	(0.012)	−0.0001	(0.0006)	−0.0834	(0.012)
.30	−0.1009	(0.011)	−0.0005	(0.0005)	−0.1004	(0.011)
.40	−0.0994	(0.010)	−0.0004	(0.0006)	−0.0989	(0.010)
.50	−0.1044	(0.013)	−0.0008	(0.0008)	−0.1036	(0.013)
.60	−0.1168	(0.013)	0.0003	(0.0008)	−0.1171	(0.013)
.70	−0.1010	(0.012)	−0.0006	(0.001)	−0.1004	(0.013)
.80	−0.1125	(0.018)	0.0003	(0.001)	−0.1129	(0.018)
.90	−0.1096	(0.023)	0.00009	(0.001)	−0.1097	(0.023)

The weights w are function of the probability of being in group 1 given the covariates, the probability of exposure to treatment conditional on the observed covariates. The w depends upon the probability of each observation of being in the treatment or in the control group. Its definition is related to another popular method to estimate the difference between treatment and control at the mean, the propensity scores (Rosenbaum and Rubin, 1983). Weighting the observations by the inverse of the probability of belonging to a given group provides the potential output: within each group, treated and untreated are weighted by the inverse probability of belonging to that group. This creates a pseudo-population in which there is no confounding factor. Then any difference in the comparison of the potential output of treated and control can be ascribed to the policy/training program under analysis, thus coinciding with the coefficients effect. This yields the Inverse Probability Weight (IPW) estimator of the treatment effect, which is generally implemented on average, measuring the effect

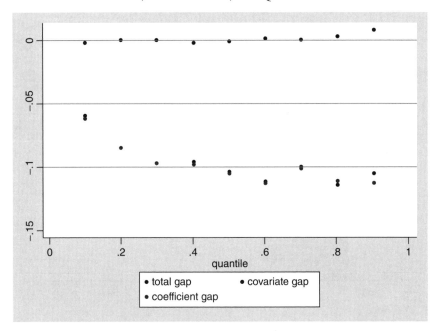

Figure 3.21   Behavior of the total, covariates and coefficients gaps in walking speed between gender across quantiles as computed by the Machado and Mata approach. The total difference coincides almost everywhere with the unexplained/coefficients difference, while the covariates gap is at the top of the graph, close to zero.

of a treatment as a difference between the means: the weighted averages computed within each group reflects the averages in the true population.[7]

In practice, the propensity score is estimated by assuming that $P(Z = 1 \mid X)$ follows a parametric model, that is, a logistic regression model where the probability of treatment is function of the covariates and is estimated by maximum likelihood,

$$P(Z = 1 \mid X) = \frac{\exp(X\beta)}{1 + \exp(X\beta)}$$

The probability of being in group 0 or 1 provides the weights to compute the potential outcome. The IPW estimates of the average treatment effect is given by the difference between the potential output of the two groups

$$n^{-1}\left[\sum \frac{Z_i Y_i}{P(Z = 1 \mid X)} - \sum \frac{(1 - Z_i)Y_i}{1 - P(Z = 1 \mid X)}\right]$$

---

[7] Another way to interpret the IPW estimator is to consider the estimated probability weights as a correction for missing data, where the latter are due to the fact that each subject is observed in only one of the potential outcomes, as belonging to the treated or the untreated group (Cattaneo et al., 2013).

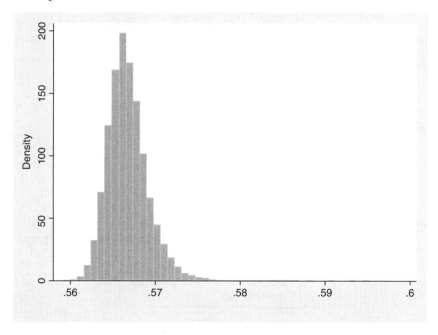

Figure 3.22   Propensity scores estimates, $P(Z = 1 \mid X) = \frac{\exp(X\beta)}{1+\exp(X\beta)}$, where women provide the reference group.

In the data set on walking speed and *BMI* (SHARE, year 2004), the IPW average group difference in walking speed is $n^{-1} \left[\sum \frac{Z_i Y_i}{P(Z=1|BMI)} - \sum \frac{(1-Z_i)Y_i}{1-P(Z=1|BMI)} \right] = -0.094$ with standard error $se = 0.012$. These results agree with the values provided by the Oaxaca-Blinder analysis of Table 3.7, signaling once again a greater walking speed in the male group, which is statistically significant. Since IPW modifies the two groups in order to create only one pseudo-population, any statistically significant difference is due to the coefficients.

To analyze the gender difference at the quantiles, the IPW set of weights are used to implement the weighted quantile regression in each group. Then, treatment and control at their unconditional distributions can be compared, as proposed by Frolich and Melly (2010). In the walking speed example, the histogram of the estimated propensity scores $P(Z = 1 \mid BMI)$ is in Figure 3.22. The histogram of the IPW weights, $w = \frac{Z_i}{P(Z=1|BMI)} + \frac{(1-Z_i)}{1-P(Z=1|BMI)}$, is in Figure 3.23, where women's group 1 receives lower weights since the probability of being in this group is higher. Figure 3.24 depicts the potential output in the two subsets, wY_0 and wY_1. These distributions are centered at $E(wY_1) = 1.12$ in the women's group and at $E(wY_0) = 1.68$ in the men's subset. Table 3.10 reports the QTE estimates of the difference between treated and control at various quantiles as computed by the Frolich and Melly approach. It computes the difference between the quantiles of

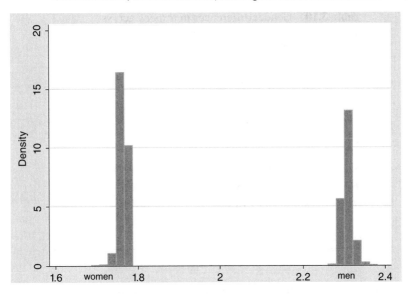

Figure 3.23   Inverse propensity scores weights, $w_i = \frac{Z_i}{P(Z=1|BMI)} + \frac{(1-Z_i)}{1-P(Z=1|BMI)}$. For $Z_i = 1$ the weight is $w_i = \frac{1}{P(Z=1|BMI)}$, while for $Z_i = 0$ it becomes $w_i = \frac{1}{1-P(Z=1|BMI)}$. Women receive a lower weight (around 1.78, strictly less than 2.3 in men's group) since $P(Z = 1 \mid X) > 0.5$.

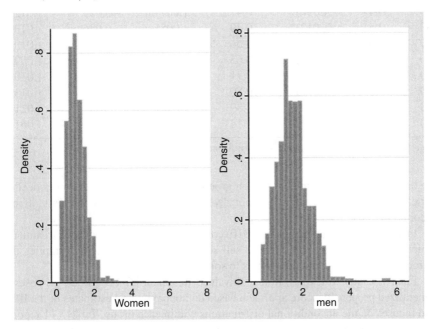

Figure 3.24   Potential output in each subset, with $E(wY_1) = 1.12$ in the women's group and $E(wY_0) = 1.68$ in the male subset.

Table 3.10   Gender differences in *wspeed* across quantiles, Frolich and Melly approach.

	difference in potential output between groups across θ	
θ	difference	*se*
.10	−0.0735	(.014)
.20	−0.0862	(.013)
.30	−0.1037	(.012)
.40	−0.1003	(.012)
.50	−0.1025	(.012)
.60	−0.1175	(.013)
.70	−0.0948	(.014)
.80	−0.1172	(.018)
.90	−0.1105	(.025)

Table 3.11   Potential output across deciles, Cattaneo approach.

	deciles of the potential output distributions					
θ	men	*se*	women	*se*	diff.	*se*
.10	0.34	(.009)	0.31	(.007)	−0.025	(.011)
.20	0.44	(.010)	0.41	(.007)	−0.025	(.012)
.30	0.51	(.012)	0.49	(.005)	*−0.016*	(.013)
.40	0.59	(.008)	0.56	(.007)	−0.036	(.010)
.50	0.65	(.009)	0.63	(.005)	−0.025	(.009)
.60	0.72	(.009)	0.71	(.007)	*−0.014*	(.010)
.70	0.81	(.009)	0.80	(.012)	*−0.011*	(.013)
.80	0.88	(.010)	0.89	(.022)	*0.011*	(.021)
.90	1.05	(.020)	1.14	(.060)	*0.082*	(.061)

Note: In italics are the not statistically significant differences

the unconditional distributions of each group. The coefficients effect across deciles varies from 7% to 11% and does not diverge much from the coefficients effects reported in Table 3.9.

In addition, Frolich and Melly (2010) consider the case of endogenous treatment, which occurs when the treatment is self-selected. This generalization implies the introduction of instrumental variables to solve the endogeneity.

Finally, Cattaneo (2010), in a multivalued treatment effects model, proposes a generalized propensity score approach based on nonparametric estimates of the IPW and focuses on the unconditional distributions within each group. Table 3.11 reports the estimates of the potential outcome distributions across quantiles computed using

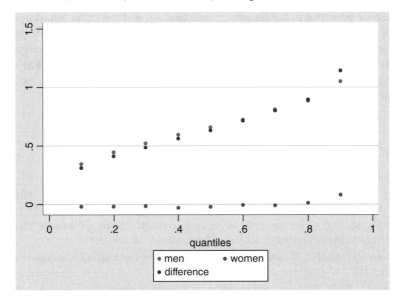

Figure 3.25   Estimated quantiles for each groups, men and women, and for their difference in the Cattaneo treatment effect estimator. With this approach the difference in walking speed between men and women, depicted at the bottom of the graph, is close to zero.

this approach (see also Cattaneo et al., 2013). The treatment effect is provided by the difference between the estimates in the two groups. For instance, at the first decile the treatment effect is given by $0.31-0.34 = -0.03$. But in order to check the statistical relevance of the group difference, standard errors are needed. The last two columns of the table provide the estimated differences together with their standard errors, the latter computed by bootstrap. Compared to the previous results, this approach provides very small estimates of the groups difference, which ranges from -0.02 to -0.03. However, depending on the issue analyzed, it may be relevant to look at each unconditional distribution and not at the sole difference between the two, as provided by the other QTE decomposition approaches. Figure 3.25 depicts the estimated quantiles for each group and for their difference as computed by the Cattaneo treatment effect estimator.

3.6   A summary of key points

Sections 3.1 and 3.2 discuss alternative interpretations of the quantile regression estimators, based respectively on the p-dimensional subsets and on the use of bootstrap when the former approach is unfeasible due to the large size of the sample. Section 3.3 implements bootstrap to the extreme quantile regressions. Indeed, the behavior of the quantile estimator in the far tails is nonstandard, and a specific bootstrap-based approach has been presented by Chernozhukov and Fernandez Val (2011) to perform inference. Section 3.4 summarizes the asymptotic behavior of the quantile regression estimator for central quantiles. The last section is devoted to the estimate of treatment effects at the mean and at various quantiles and to its decomposition into explained and unexplained components.

References

Anscombe, F. 1973. "Graphs in statistical analysis." The American Statistician 27, 17–21.

Blinder, A. 1973. "Wage discrimination: reduced form and structural estimates." Journal of Human Resources 8, 436–455.

Buchinsky, M. 1995. "Estimating the asymptotic covariance matrix for quantile regression models: a Monte Carlo study." Journal of Econometrics 68, 303–338.

Camponovo, L., Scaillet, O., and Trojani, F. 2012. "Robust subsampling." Journal of Econometrics 167, 197–210.

Cattaneo, M. 2010. "Efficient semiparametric estimation of multi-valued treatment effects." Journal of Econometrics 155, 138–154.

Cattaneo, M., Drukker, D., and Holland, A. 2013. "Estimation of multivalued treatment effects under conditional independence." Stata Journal 13, 407–450.

Chernozhukov, V., and Fernandez-Val, I. 2005. "Subsampling inference on quantile regression processes." Sankyha 67, 253–276.

Chernozhukov, V., and Fernandez-Val, I. 2011. "Inference for extremal conditional quantile models with an application to market and birthweight risks." Review of Economic Studies 78, 559–589.

Chernozhukov, V., Fernandez-Val, I., and Melly, B. 2013. "Inference on counterfactual distributions." Econometrica 81, 2205–2268.

Davino, C., Furno, M., and Vistocco, D. 2014. Quantile Regression: Theory and Applications, Wiley.

Efron, B. 1979. "Bootstrap methods: another look at jackknife." Annals of Statistics 7, 1218–1228.

Escaciano, J., and Goh, S. 2014. "Specification analysis of linear quantile models." Journal of Econometrics 178, 495–507.

Farebrother, R. 1985. "Relations among subset estimators: a bibliographical note." Technometrics 27, 85–86.

Frolich, M., and Melly, B. 2010. "Estimation of quantile treatment effects with Stata." The Stata Journal 10, 423–457.

Furno, M. 1997. "Estimating the variance of the LAD regression coefficients." Computational Statistics & Data Analysis 27, 11–26.

Hall, M., and Mayo, M. 2008. "Confidence intervals and coverage probabilities of regression parameter estimates using trimmed elemental estimation." Journal of Modern Applied Statistical Methods 7, 514–525.

Hawkins, D. 1993. "The accuracy of elemental set approximations for regression." Journal of the American Statistical Association 88, 580–589.

Hawkins, D., Bradu, D., and Kass, G. 1984. "Location of several outliers in multiple-regression data using elemental sets." Technometrics 26, 197–208.

Hayashi, F. 2000. Econometrics. Princeton University Press.

Heitmann, G., and Ord, K. 1985. "An interpretation of the least squares regression surface." The American Statistician 39, 120–123.

Hussain, S., and Sprent, P. 1983. "Non-Parametric Regression." Journal of the Royal Statistical Society Series A 146, 182–191.

Jann, B. 2008. "The Blinder-Oaxaca decomposition for linear regression models." The Stata Journal 8, 453–479.

Koenker, R., and Bassett, G. 1978. "Regression quantiles." Econometrica 46, 33–50.

Koenker, R., and Bassett, G. 1982. "Robust tests for heteroskedasticity based on regression quantiles." Econometrica 50, 43–61.

Koenker, R. 2000. "Galton, Edgeworth, Frisch, and prospects for quantile regression in econometrics." Journal of Econometrics 95, 347–374.

Machado, J., and Mata, J. 2005. "Counterfactual decomposition of changes in wage distributions using quantile regression." Journal of Applied Econometrics 20, 445–465.

Melly, B. 2005. "Public-Private sector wage differentials in Germany: Evidence from quantile regression." Empirical Economics 39, 505–520.

Melly, B. 2006. "Estimation of counterfactual distributions using quantile regression." Discussion papers Universitat St. Gallen, 1–50.

Oaxaca, R. 1973. "Male-female wage differentials in urban labor markets." International Economic Review 14, 693–709.

Rosenbaum, P., and Rubin, D. 1983. "The central role of the propensity score in observational studies for causal effects." Biometrika 70, 41–55.

Ruppert, D., and Carroll, R. 1980. "Trimmed least squares estimation in the linear model." Journal of the American Statistical Association 75, 828–838.

Subrahmanyam, M. 1972. "A property of simple least squares estimates." Sankhya B34, 355–356.

Theil, H. 1950. "A rank-invariant method of linear and polynomial regression analysis, parts 1-3." Nederlandsche Akademie voor schap Proceedings Series A 53, 386–392, 521–525, 1397–1412.

Weiss, A. 1990. "Least absolute error estimation in the presence of serial correlation." Journal of Econometrics 44, 127–159.

Wu, C. 1986. "Jacknife, bootstrap and other resampling methods in regression analysis." Annals of Statistics 14, 1261–1295.

Appendix: computer codes

A) elemental sets when p = 2:

```
input x1i x1j y1i y1j          input data as in first
10 8 8.04 6.95                   four columns of Table 3.1
10 13 8.04 7.58
10 9 8.04 8.81
10 11 8.04 8.33
10 14 8.04 9.96
 8 13 6.95 7.58
 8 9 6.95 8.81
 8 11 6.95 8.33
 8 14 6.95 9.96
13 9 7.58 8.81
13 11 7.58 8.33
13 14 7.58 9.96
 9 11 8.81 8.33
 9 14 8.81 9.96
11 14 6.95 9.96
end
gen dx=x1i-x1j
gen dy=y1i-y1j
gen bij=dy/dx                  elemental set slopes
gen yij=y1i+y1j
gen xij=bij*(x1i+x1j)
gen b0ij=(yij-xij)/2           elemental set intercepts
list bij b0ij x1i x1j y1i y1j
su bij,d
scalar bmedian=r(p50)         median slope
su b0ij,d
scalar b0median=r(p50)        median intercept
centile bij, centile(45 50 52)   slope confidence interval
centile b0ij, centile(45 50 52)  intercept confidence interval
gen sumdxsq=sum(dx*dx)
scalar denom=sumdxsq[15]
gen wij=(dx*dx)/denom         OLS weights
gen bo1=wij*bij
gen b0o1=wij*b0ij
gen bols=sum(wij*bij)         OLS slope
gen b0ols=sum(wij*b0ij)       OLS intercept
list wij                      print OLS weights
```

B) trimmed regression:

`reg y x if y<=k`	exclude values of $y \geqslant k$ in OLS
`qreg y x if y<=k`	exclude values of $y \geqslant k$ in quantile regression

C) design matrix bootstrap:

`use file.dat`	input data
`bootstrap _b,` `  reps(100) sa(OLScoef): reg y x`	OLS coefficients
`estat bootstrap`	compute bias and variance

`bootstrap _b,`
`  reps(100) si(2) sa(elem): reg`
`  y x`
	elemental sets coefficients

`bootstrap "qreg y x,q(.5)" _b,` `  reps(100) sa(bootcoeff5)`	median regression coefficients
`matrix varian=e(V)`	save var-cov matrix
`matrix list varian`	print var-cov matrix
`matrix confid=e(ci_percentile)`	percentile confidence intervals
`matrix confidn=e(ci_normal)`	normal confidence intervals
`matrix list confid`	print confidence intervals

* to look at the computed bootstrap median regression coefficients

`clear`	clear to use new data set
`use bootcoeff5`	input median bootstrap coefficients file

*or alternatively to look at the OLS bootstrap coefficients

`use OLScoef`	input OLS bootstrap coefficients file

*or alternatively to look at the elemental set bootstrap coefficients

`use elem`	input elemental set bootstrap coefficients file
`su b_x, d`	show summary statistics of bootstrap slope
`scalar varian=r(Var)`	save variance
`scalar mean=r(mean)`	save mean
`scalar perc1=r(p5)`	keep the 5th percentile
`scalar perc2=r(p95)`	keep the 95th percentile
`histogram b_x`	draw histogram of slope coeff.
`quantile b_x`	draw quantile plot of slope coefficients

D) extremal quantiles:

`clear` `use datafile`	input data

*extreme value
`set seed 88`
`qreg y x, q(.05)`
	compute 5th quantile regression

*bootstrap the $\theta = .05$ quantile regression, $B = 100$ $n = 7$

```
bootstrap "qreg y x,q(.05)" _b, reps(100) si(7)
  sa(coeff05)
```

*bootstrap the $\theta_b = \theta n/b = .05 * 11/7 = .078 \approx .08$ quantile regression

```
bootstrap "qreg y x,q(.08)" _b, reps(100) si(7)
  sa(coeff08)
```

*bootstrap the $m\theta_b = ((\theta n + p)/\theta n) * \theta_b = .37$ quantile regression

```
bootstrap "qreg y x,q(.37)" _b, reps(100)si(7)
  sa(coeff37)
```

*bootstrap the mean of X

```
bootstrap, reps(100) si(7) sa(mediax05): mean x
```

*organize coeff05 mediax05 coeff08 coeff37 side by side

`clear`
`u coeff05`
`gen id=_n`
`sort id`
`save boot05, replace`
`u mediax05, clear`
`ren _b_x medx`
 	
`gen id=_n`
`sort id`
`save xxmedia05, replace`
`u coeff37an,clear`
`ren b_x b137`
	organize the data by subsample
	for the 5th quantile regression
	set the iteration number
	
	for the mean of X in each subset
	assign a name to the series
	   "mean of X"
	set the iteration number
	
	for the $m\theta_b$ quantile regression
	assign a name to the series
	   "37th quantile slope"

`ren b_cons b037`	assign a name to the series "37th quantile intercept"
`gen id=_n`	set the iteration number
`sort id`	
`save boot37, replace`	
`u coeff08,clear`	for the θ_b quantile regression
`ren b_x b108`	assign a name to the series "8th quantile slope"
`ren b_cons b008`	assign a name to the series "8th quantile intercept"
`gen id=_n`	set the iteration number
`sort id`	
`save boot08, replace`	
`u boot05, clear`	merge the bootstrap results by subset
`merge id using boot37`	
`drop _merge`	
`sort id`	
`merge id using xxmedia05`	
`drop _merge`	
`sort id`	
`merge id using boot08`	
`drop _merge`	
`save bboot05, replace`	final data set, data organized by iteration
`clear`	
`u bboot05`	

*numerator of V, differences of each subset from the 5th estimates

```
gen diff=b_x-.46
gen diff0=b_cons-1.6
```

*denominator of A, difference of the $(m\theta_b)$ and (θ_b) estimates

```
gen difdenb0=medx*(b037-b008)
gen difdenb1=medx*(b137-b108)
```

---

*numerator of $A = \sqrt{n\theta} = \sqrt{11 * .05}$

```
gen biga0=.741/difdenb0
gen biga1=.741/difdenb1
gen bigv0=diff0*biga0
gen bigv1=diff*biga1
su bigv0,d
scalar low0=r(p5)
scalar up0=r(p95)
su bigv1,d
scalar low1=r(p5)
scalar up1=r(p95)
scalar list low0 up0 low1 up1
```
5th and 95th quantiles of $\hat{V}$, intercept, slope

E) quantile treatment effect

---

*download at http://www.stata-journal.com/software/sj8-4
*the routine to compute the Oaxaca Blinder decomposition

```
oaxaca y x, by(gender)
```
Oaxaca-Blinder decomposition

*download at http://www.econ.brown.edu/fac/Blaise_Melly/
*to install Melly routine on decomposition

```
net install counterfactual,
    from("http://www.econ.brown.edu/fac/Blaise_Melly/")
cdeco y x, group(gender)
```
Chernozhukov et al. decomposition

*download at http://fmwww.bc.edu/RePEc/bocode/m/mmsel.ado

```
ssc install mmsel
```
to install Machado and Mata routine

```
gen pid=_n
```
identification of individuals

*Machado and Mata decomposition
*(in the tmp folder are saved, for all the replicates, the unconditional and the
*counterfactual distributions)

```
mmsel y x, group(gender) filename(name) reps(#) group1
```
*download at http://www.stata-journal.com/software/sj13-3
*the routine to compute IPW and Frolich and Melly approach

*average treatment effect by IPW

```
teffects ipw (y)(gender x,logit)
```

compute Frolich weights and propensity scores at 10th decile

```
ivqte y (gender), c(x) v generate_p(pp) generate_w(ww)
   quantile(.1)
```

*saves propensity scores in pp and weights in ww

*download at http://www.stata-journal.com/software/sj10-3 the routine
*to compute Cattaneo approach

*compute potential output at 10th 50th 90th deciles

```
poparms (gender y x)(y x), q(.1 .5 .9)
```

*compute potential output at 10th 50th 90th deciles

```
margins gender, pwcompare predict(equation(#2))
```

*estimate difference at second decile with standard errors

```
margins gender, pwcompare predict(equation(#3))
```

4

A not so short introduction to linear programming

Introduction

Quantile regression is a statistical method suitable to model the whole conditional distribution of a response variable in terms of a set of explanatory variables. The beginning of wide dissemination was the formulation in terms of a linear programming problem. Indeed, the availability of sound and efficient methods for solving linear programming problems offered the opportunity to test and apply quantile regression on realistic problems and appreciate its added value.

This chapter proposes a (not so short) journey into the world of linear programming. Mathematics is softened in order to focus on geometric intuition and teaching examples. The general simplex algorithm and its main variants are presented and then applied to generic problems. The aim is to enable the reader to manage the linear programming machinery and to apply it to the quantile regression framework.

4.1 The linear programming problem

The decisional structure of an optimization problem consists of several items:

- feature of the available information, essentially classified in certain or uncertain,
- number of decision makers (one or more),
- number of decision goals (one or more).

Quantile Regression: Estimation and Simulation, Volume 2. Marilena Furno and Domenico Vistocco.
© 2018 John Wiley & Sons Ltd. Published 2018 by John Wiley & Sons Ltd.
Companion website: www.wiley.com/go/furno/Quantileregression

Different combinations of the previous items lead to different decision problems. Mathematical programming deals with the simplest setting characterized by a certain information, a single decision maker, and a unique decision goal. In such a setting, the decision maker formulates the problem in terms of a set of variables (decisional variables), defining an objective function suitable to quantitatively rank the different available choices. The operating principle of the decisional system is described through a set of conditions defined on the decisional variables, conditions resulting in a set of constraints typically formulated using equalities and/or inequalities. Mathematical programming selects the solution that determines the best value of the objective function (optimal solution) among the set of all possible solutions (feasible solutions).

In formal terms, denoting by $f(.)$ the objective function and with $\mathbf{x}$ the n–dimensional vector of decisional variables, a mathematical programming problem is defined as:

$$\text{minimize} \quad z = f(\mathbf{x})$$

$$\mathbf{x} \in \mathcal{X}. \tag{4.1}$$

The set $\mathcal{X} \subseteq \mathbb{R}^n$ of the solutions satisfying all the constraints is the feasible region. The solution of the problem (4.1), in case it exists, is the feasible solution $\mathbf{x}^*$ such that $f(\mathbf{x}^*) \le f(\mathbf{x})$ for each $\mathbf{x} \in \mathcal{X}$. The value $z^* = f(\mathbf{x}^*)$ is the optimal value. The problem is known as unfeasible in case a feasible solution does not exist ($\mathcal{X} = \varnothing$), whereas it is known as unbounded when the feasible region exists but is not bounded. It is worth noting that the formulation in terms of a minimization problem is not restrictive, since any maximization problem, $\max f(\mathbf{x}) : \mathbf{x} \in \mathcal{X}$, can be easily expressed through the equivalent minimization problem, $\min[-f(\mathbf{x})] : \mathbf{x} \in \mathcal{X}$. Therefore, a mathematical programming problem consists in the search of a constrained minimum (maximum), which can be found by resorting to the classical mathematical tools grounded on the first and second derivative.

In the class of mathematical programming problems, the linear programming one is a special case where the objective function and all the constraints are linear, and the decisional variables are continuous and non-negative. Such a problem can be formulated as follows:

$$\text{minimize} \quad z = \mathbf{c}^\top \mathbf{x} = \sum_{i=1}^{n} c_i x_i$$

$$\text{subject to} \quad \mathbf{A}\mathbf{x} \le \mathbf{b}$$

$$\mathbf{x} \ge \mathbf{0} \tag{4.2}$$

where $\mathbf{c}$ denotes the n–dimensional vector of the objective function coefficients (costs in case of a minimization problem or returns in case of a maximization problem), $\mathbf{A}$ is the $m \times n$ matrix expressing the m constraints on the n decisional variables, and $\mathbf{b}$ is the m–dimensional vector of the known coefficients associated with the constraints.

The exclusive use of less–than inequalities in the previous formulation is not restrictive as it is easy to switch from greater–than inequalities to less–than inequalities and turn equalities in two equivalent inequalities. The conversion exploits simple manipulations, as the one previously shown for the equivalence between a minimization and a maximization problem. In fact, a greater–than–or–equal inequality constraint:

$$a_{i1}x_{i1} + \ldots + a_{ij}x_{ij} + \ldots + a_{in}x_{in} \geq b_i$$

is transformed in less–than–or–equal form when multiplying it for -1:

$$-a_{i1}x_{i1} - \ldots - a_{ij}x_{ij} - \ldots - a_{in}x_{in} \leq -b_i.$$

Likewise, an equality constraint:

$$a_{i1}x_{i1} + \ldots + a_{ij}x_{ij} + \ldots + a_{in}x_{in} = b_i$$

is equivalent to the simultaneous occurrence of the two following inequality constraints:

$$a_{i1}x_{i1} + \ldots + a_{ij}x_{ij} + \ldots + a_{in}x_{in} \leq b_i$$

$$a_{i1}x_{i1} + \ldots + a_{ij}x_{ij} + \ldots + a_{in}x_{in} \geq b_i$$

and therefore to the two less–than–or–equal inequalities:

$$a_{i1}x_{i1} + \ldots + a_{ij}x_{ij} + \ldots + a_{in}x_{in} \leq b_i$$

$$-a_{i1}x_{i1} - \ldots - a_{ij}x_{ij} - \ldots - a_{in}x_{in} \leq -b_i.$$

In summary, a linear programming problem consists of a linear objective function, a set of m constraints expressed through linear inequalities representing the conditions of the problem, and a set of n constraints related to the non–negativity of the decisional variables. It is worth highlighting that the non–negative conditions restrict the solution set to $\mathbb{R}_+^n$.

4.1.1   The standard form of a linear programming problem

It is useful to introduce an additional representation for all the linear programming (*LP*) problems in order to standardize the required steps in a solving algorithm. This representation is labeled as standard form (Hiller and Lieberman, 2015). It consists in a minimization problem whose structural constraints, referring to a set of non–negative decisional variables, are expressed in equational form:

$$\text{minimize} \quad z = \mathbf{c}^\top \mathbf{x}$$

$$\text{subject to} \quad \mathbf{Ax} = \mathbf{b} \qquad (4.3)$$

$$\mathbf{x} \geq \mathbf{0}.$$

The standard form encompasses all possible LP problems, since each LP problem can be easily transformed in a way to comply with this formulation. The use of equalities for the constraints permits to exploit a basic property of linear systems pertaining to the transformation of linear equations:

> any (linear) transformation of a system of linear equations does not modify the solution set (Strang 2005, 2009).

This property is the core of the simplex algorithm, as shown in section 4.2. In order to obtain the standard form, a simple trick permits to switch from an inequality to an equality. The following inequality:

$$a_{i1}x_{i1} + \dots + a_{ij}x_{ij} + \dots + a_{in}x_{in} \leq b_i$$

is indeed equivalent to:

$$a_{i1}x_{i1} + \dots + a_{ij}x_{ij} + \dots + a_{in}x_{in} + s = b_i$$
$$s \geq 0$$

which is obtained introducing the artificial non–negative variable s, also called slack variable. Likewise, a greater–than–or–equal inequality:

$$a_{i1}x_{i1} + \dots + a_{ij}x_{ij} + \dots + a_{in}x_{in} \geq b_i$$

is equivalent to the following equality:

$$a_{i1}x_{i1} + \dots + a_{ij}x_{ij} + \dots + a_{in}x_{in} - s = b_i$$
$$s \geq 0$$

in which the artificial variable s, or surplus variable, is subtracted. In the following, we refer to the s variable as slack variable in both cases, as commonly done in the LP literature.

A further point that deserves attention is the non–negativity restriction of the standard form on the decisional variables, since real applications could involve variables that are negative or unrestricted in sign. By now, it should be obvious that a negative decisional variable:

$$x_i \leq 0$$

can be expressed as:

$$-x_i \geq 0.$$

A variable x_i, unrestricted in sign, can always be expressed as difference of two non–negative numbers:

$$x_i = x_i' - x_i''$$
$$x_i' \geq 0$$
$$x_i'' \geq 0.$$

Thus, the transformation of a generic LP problem into its standard form involves the introduction of further variables (the artificial variables) resulting in an increased dimensionality. In the following, we continue to denote by $\mathbf{x}$ the vector of the variables: whereas the initial formulation of the problem involves inequalities, the vector will be then composed both of the natural decisional variables and the artificial variables useful to achieve the standard form.

Finally, using the standard form formulation, the optimal solution $\mathbf{x}^*$ is the vector minimizing z such that:

$$\mathbf{A}\mathbf{x}^* = \mathbf{b}$$
$$\mathbf{x}^* \geq \mathbf{0}$$
$$\mathbf{c}^\mathsf{T}\mathbf{x}^* \leq \mathbf{c}^\mathsf{T}\mathbf{x}$$

for each $\mathbf{x} \in \mathcal{R}_+$. The solution vector $\mathbf{x}^*$ contains the values for both the natural variables and the necessary artificial variables.

4.1.2   Assumptions of a linear programming problem

Before introducing the geometry of LP, it is important to stress the effects of the linearity of the objective function and of the structural constraints. Essentially, a function is linear if a unit increase (decrease) in a variable causes a constant increase (decrease) in the value of the function itself. Such linearity engenders some assumptions on the LP problem (Hiller and Lieberman, 2015). The assumptions are related to the decisional variables and to their effects on the problem. They can be briefly summarized in the following three points:

Divisibility The components of the vector $\mathbf{x}$ can assume any value meeting both the structural and the non–negativity constraints. This hypothesis is also referred to as continuity assumption. In many real problems, the decisional variables are restricted to assume only integer values, but divisibility holds for the applications of the linear programming theory to the quantile regression setting. Therefore, we focus on the case of divisible variables, referring the interested reader to the integer programming theory (Matousek and Gartner, 2005; Vanderbei, 2014) where the assumption may be unrealistic;

Proportionality The contribution of each decisional variable x_i to the value of the objective function and to each structural constraint is proportional to the value of the variable. This is evident from the expression of the objective function, $z = \mathbf{c}\mathbf{x} = \sum_{i=1}^{n} c_i x_i$, where the variable x_i contributes with weight equal to $c_i x_i$. The same for the generic constraint $\sum_{i=1}^{n} a_{ij} x_{ij}$, with $j = 1, \ldots, m$, where $a_{ij} x_{ij}$ is the contribution of the variable x_i. In both cases, the proportionality of the contribution of x_i is obvious. In practical terms, the contribution of x_i is constant across the different values of the variable and it is not affected by (dis)economies of scale;

Additivity The different decisional variables x_i, with $i = 1, \ldots, n$, do not interfere with each other in determining the value of the objective function and of the single constraints. In this case too, the assumption is strictly related to the underlying linearity of the problem: the objective function is indeed the sum of the contributions of each decisional variable, and the same holds for each constraint. As c_i and a_{ij} are the contributions of the variable x_i to the objective function and to the j–th constraint, respectively, the total contribution of the

n decisional variables is $\sum_{i=1}^{n} c_i x_i$ for the former and $\sum_{i=1}^{n} a_{ij} x_{ij}$ for the latter. From a practical point of view, the additivity assumption excludes all the synergies between the decisional variables.

4.1.3   The geometry of linear programming

The use of two decisional variables allows us to visually introduce the geometry of linear programming, and the basic idea of its solution methods through geometric figures in a plane. The problem can be represented in the 3–D space in case of three decisional variables. Starting from these simple cases, we can move to more general and realistic problems, involving n decisional variables.

The 2–D problem

Let us consider the following LP problem with two decisional variables:

$$\begin{aligned}
\text{maximize} \quad & z = 5x_1 + 3x_2 \\
\text{subject to} \quad & x_1 + x_2 \leq 600 \quad (C_1) \\
& 2x_1 + x_2 \leq 800 \quad (C_2) \\
& x_1 \geq 0 \quad (C_3) \\
& x_2 \geq 0 \quad (C_4).
\end{aligned} \qquad (4.4)$$

The 2–D problem can be represented and solved using a graphical approach. Each linear inequality associated with a constraint divides the plane into two half–planes: one in which the inequality is satisfied and one in which it is not. The same for the two non–negativity constraints, which restrict the solutions to $\mathbb{R}_+^2$. Therefore, considering the intersection of all the half–planes associated with the technical constraints and with the non–negativity constraints on the same Cartesian plane, we obtain the region of feasible solutions, known as feasible set or feasible region. In particular, in Figure 4.1, the first constraint $x_1 + x_2 \leq 600$ corresponds to the half–plane below the line C_1, while the half–plane below the line C_2 is the geometrical equivalent of the second constraint $2x_1 + x_2 \leq 800$. The two non–negativity constraints $x_1 \geq 0$ and $x_2 \geq 0$ limit the solution to the first quadrant, the former expressing the half–plane above the horizontal axis (C_3) and the latter, the half–plane to the right of the vertical axis (C_4). The polygon $\mathcal{OAED}$, intersection of the four half–planes, is the geometric set of all the points that simultaneously meet the four constraints, and therefore is the feasible set. The optimal solution z^* is the value belonging to the polygon $\mathcal{OAED}$ that maximizes the objective function $z = 5x_1 + 3x_2$.

In order to complete the geometrical interpretation of the problem (4.4), we need to represent the two–variable function z in the same Cartesian plane. To this end, it is useful to resort to the level lines. In fact, the objective function is linear and hence it corresponds to the plane z represented in Figure 4.2. Let us now consider a set of planes parallel to the planes $\mathcal{O}x_1x_2$: in particular, the planes $\alpha : z = 0, \beta : z = 2, \delta : z = 4, \gamma : z = 6, \epsilon : z = 8$ are shown in Figure 4.2. The intersections of z with the set of planes $\alpha, \beta, \gamma, \delta, \epsilon$, provide the section lines a, b, c, d, e. The projections

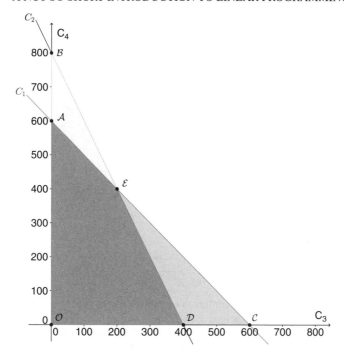

Figure 4.1   Geometrical representation of the constraints in the LP problem (4.4). The intersection of the four constraints C_1, C_2, C_3 and C_4 determines the feasible region corresponding to the polygon $\mathcal{OAED}$.

of such section lines on the plane $\mathcal{O}x_1x_2$ define the level lines a, b_0, c_0, d_0, e_0 of the objective function z: the level lines of z are then a sheaf of parallel lines, orthogonal projections of the points of the plane z sharing the same value (in the example $z = 0$, $z = 2$, $z = 4$, $z = 6$, and $z = 8$). The level lines are parallel to the line $a|z = 3x + 5y = 0$, which is known as the line generating the sheaf. More specifically, in the case of a linear objective function, the level lines are orthogonal to the vector $v(5, 3)$, which is determined by the two coefficients of the decisional variables x_1 and x_2 in z. It denotes the direction to follow in order to move along the level lines.

Figure 4.3 depicts some level lines for the 2D–example (4.4), starting with the case of $z = 0$ up to $z = 2200$, which corresponds to the optimal solution, as shown below. It is obvious that the solution $\mathbf{x}^*$ has to lie on the boundary of the feasible set: in fact, given a generic interior point $\mathcal{P}$, the value of the objective function z can be increased while remaining in the feasible region. We can obtain a higher value of z moving the point $\mathcal{P}$ towards a higher level line, and this until $\mathcal{P}$ is moved on the boundary of the feasible region. In the LP example (4.4), the solution is the corner $\mathcal{E}$ where the level line $z = 2200$ intersects the feasible region $\mathcal{OAED}$. The above line of reasoning entails a correspondence among corners and optimal solution: the search for the solution can be limited to the boundary of the feasible region and, more specifically, to its corners. Therefore, the graphical approach simply requires computing the coordinates of the four corners, and then evaluating the objective function in each of

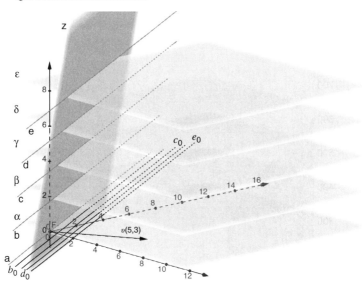

Figure 4.2 Geometrical representation of the objective function z in the LP problem (4.4). The objective function z defines a plane; the intersections of z with the set of planes $(\alpha, \beta, \gamma, \delta, \epsilon)$ corresponding to different values of z, provide a set of lines (a, b, c, d, e), whose projections on the plane $\mathcal{O}x_1 x_2$ define the level lines (a, b_0, c_0, d_0, e_0) of z. Such level lines are orthogonal to the vector $v(5, 3)$, determined by the two coefficients of the decisional variables in z.

them. More in detail, the coordinates of each corner are the solutions of a system of equations, each comprising two equations, since a corner is the intersection of two lines defined by the problem constraints. The four systems of equations along with their solutions are:

$$\mathcal{O}(0,0) : \begin{cases} x_1 = 0 \\ x_2 = 0 \end{cases} \qquad \mathcal{A}(0,600) : \begin{cases} x_1 + x_2 = 600 \\ x_2 = 0 \end{cases}$$

$$\mathcal{E}(200,400) : \begin{cases} x_1 + x_2 = 600 \\ 2x_1 + x_2 = 800 \end{cases} \qquad \mathcal{D}(400,0) : \begin{cases} 2x_1 + x_2 = 800 \\ x_1 = 0. \end{cases}$$

From Figure 4.3, it is immediately clear that the corner $\mathcal{E}$, for instance, corresponds to the intersection of the constraints C_1 and C_2 of the LP problem. When evaluating the objective function z in each corner, the optimal solution z^* is achieved in $\mathcal{E}$:

$$z_{\mathcal{O}} = 0 \qquad\qquad z_{\mathcal{A}} = 2000$$

$$\boxed{z_{\mathcal{E}} = 2200 = z^*} \qquad z_{\mathcal{D}} = 1200.$$

The basic idea underlying the simplex algorithm emerges from the above example: to find the optimal solution, it is sufficient to inspect the corners of the feasible region and choose the one associated with the highest value of z.

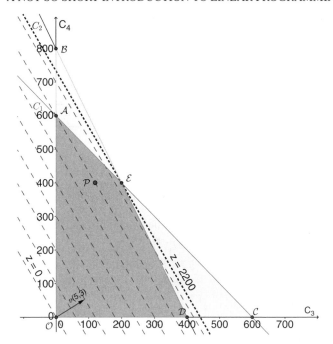

Figure 4.3   Geometrical representation of the LP programming problem (4.4). The optimal solution has to lie on the boundary of the polygon $\mathcal{OAED}$. The choice of any point $\mathcal{P}$ interior to the feasible region is not optimal, since z can be improved considering a point on a higher level line. The optimal solution corresponds to the corner $\mathcal{E}$ where the feasible region intersects the level line $z = 2200$. The level lines are orthogonal to the vector $v(5, 3)$, where 5 and 3 are the two coefficients of the decisional variables x_1 and x_2 in the objective function z.

Many LP problems admit a single solution as in the above example. There exist problems admitting infinite solutions. The approach exploring the corners of the feasible region is valid also when the vector determining the direction of the level lines is orthogonal to one of the edges of the polygon defining the feasible region: in this case, there exist infinite equivalent solutions, since the highest admissible level line overlaps an edge of the polygon, that is, it passes through two corners. An example is provided by the following LP problem:

$$\text{maximize} \quad z = 5x_1 + 5x_2$$
$$\text{subject to} \quad x_1 + x_2 \leq 600 \quad (C_1)$$
$$2x_1 + x_2 \leq 800 \quad (C_2) \quad \quad (4.5)$$
$$x_1 \geq 0 \quad (C_3)$$
$$x_2 \geq 0 \quad (C_4).$$

The problem is the same as in (4.4), except for the coefficient c_2 of the variable x_2, here equal to 5. Such a variant only causes a change in the slope of the level lines, which are now parallel to the line C_1, with no effects on the feasible region. From Figure 4.4 it is evident that any point belonging to the edge $\overline{A\mathcal{E}}$ is an optimal solution. Hence, by focusing only on the corners A and $\mathcal{E}$, the problem could be solved.

A simple check only requires to evaluate the new objective function $z = 5x_1 + 5x_2$ in the four corners, whose coordinates do not change with respect to the previous case:

$$z_\mathcal{O} = 0 \qquad \boxed{z_A = 3000 = z^*}$$

$$\boxed{z_\mathcal{E} = 3000 = z^*} \qquad z_D = 1200.$$

Every linear combination of the two points A and $\mathcal{E}$:

$$\mathcal{P}^* = \alpha A + (1 - \alpha)\mathcal{E}, \text{with } \alpha \in [0, 1]$$

defines a point lying on the edge that joins the two points, and then ensures the same optimal value of the objective function $z_{\mathcal{P}*} = 3000$.

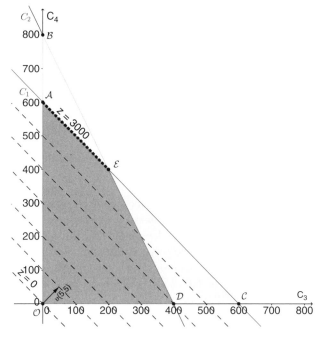

Figure 4.4 Geometrical representation of the LP programming problem (4.5). The optimal solution is any point belonging to the edge $\overline{A\mathcal{E}}$: since the vector $v(5, 5)$ is orthogonal to this edge, the level lines are parallel to $\overline{A\mathcal{E}}$. Moving again on the corners of the feasible region, both the solutions z_A and $z_\mathcal{E}$ provide the same optimal value $z^* = 3000$, as well as any point lying on the edge joining A and $\mathcal{E}$.

Next it is worth mentioning the case in which the feasible region is unbounded. Consider the following maximization problem:

$$\text{maximize} \quad z = 4x_1 - 3x_2$$
$$\text{subject to} \quad -2x_1 + x_2 \le 4 \quad (C_1)$$
$$x_1 - 2x_2 \le 8 \quad (C_2)$$
$$x_1 - x_2 \le 11 \quad (C_3)$$
$$x_1 \ge 0 \quad (C_4)$$
$$x_2 \ge 0 \quad (C_5).$$

(4.6)

Figure 4.5 depicts the corresponding feasible region and the level lines defined by z: here, for any level line, it is always possible to improve the solution because a lower level line always intersects the feasible region. The optimal solution is therefore unbounded, since the two variables, x_1 and x_2, can indefinitely increase, and the objective function z increases with them.

Finally, it is possible also to face inconsistency when the feasible region is empty. In such a case, there is no solution that meets all the problem constraints.

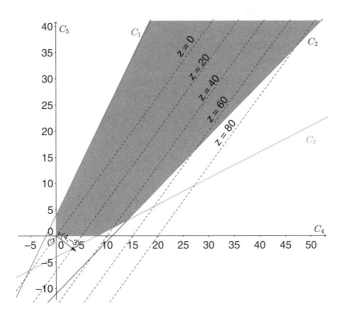

Figure 4.5 Geometrical representation of the LP programming problem (4.6). The feasible region is unbounded and therefore there does not exist any finite optimal solution: considering any level line, it is always possible to consider a lower line still belonging to the feasible region.

From examples (4.4), (4.5), and (4.6), the close link between the corners of the feasible region and the optimal solutions (Hiller and Lieberman, 2015) can be described as follows:

- If a LP problem is solvable with a bounded feasible region, there exists then one corner and at least one optimal solution.

- If the problem admits exactly one optimal solution, it must be a corner.

- If the problem has multiple optimal solutions, at least two of them must be corners.

This key result also works in higher-dimensional spaces and is indeed the foundation for solving the LP problems, as will be shown in section 4.2.

The 3–D problem

By adding a further variable, it is still possible to graphically represent the problem moving to a 3D–space. With three decisional variables, each constraint defines a half–space (no longer a half–plane). The intersection of all the half–spaces corresponding to the technical constraints and non–negativity constraints provides a polyhedron.

As an example, let us consider the following problem:

$$\text{maximize} \quad z = 300x_1 + 200x_2 + 350x_3$$

$$\begin{aligned}
\text{subject to} \quad & x_1 + x_2 + x_3 \leq 40 & (C_1) \\
& 4x_1 + 2x_2 + 3x_3 \leq 120 & (C_2) \\
& x_3 \leq 20 & (C_3) \\
& x_1 \geq 0 & (C_4) \\
& x_2 \geq 0 & (C_5) \\
& x_3 \geq 0 & (C_6).
\end{aligned}$$

$$(4.7)$$

The feasible region, identified by the intersection of the six half–spaces defined by the C_i constraints, $i = 1, \dots, 6$, is the polyhedron depicted in Figure 4.6. The objective function $z = 300x_1 + 200x_2 + 350x_3$ is a hyperplane in $\mathbb{R}_+^4$. As in the case of two decisional variables, it is possible to project the intersections of z with the set of

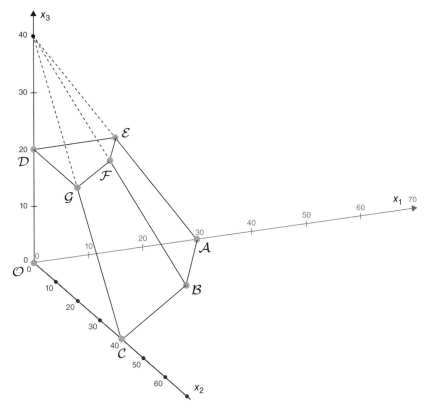

Figure 4.6   Geometrical representation of the LP programming problem (4.7). Each constraint C_i, $i = 1, \ldots, 6$, defines a half–space in the 3D–space; the feasible region is the polyhedron determined by the intersection of the six half–spaces.

hyperplanes corresponding to different values of z, so to obtain a set of planes. As in the previous case, such planes are orthogonal to the vector $v(300, 200, 350)$, which is defined by the coefficients of the decisional variables x_1, x_2 and x_3 in the objective function z. The planes for $z = 0$ (plane generating the sheaf) and for $z = 12000$ (optimal solution) are depicted in Figure 4.7. The optimal solution is obtained in the corner $\mathcal{F}$ of the feasible region. Also in this case, it is possible to exploit a graphical method to achieve the solution, reasoning only on the corners of the region. The eight corners of the feasible region are indeed the solutions of as many systems of three equations considering three constraints at the same time. In fact, each corner is the intersection of three planes associated with the problem constraints. In particular, the eight corners of the polyhedron $\mathcal{OABCDEFG}$ are the solutions of the following

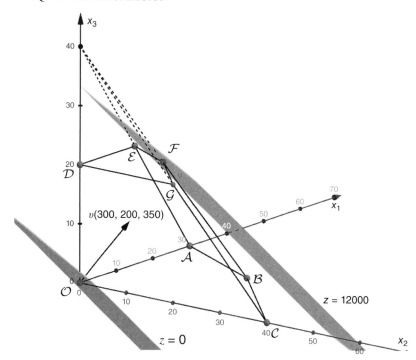

Figure 4.7 Geometrical representation of the LP programming problem (4.7). The objective function z is a hyperplane in $\mathbb{R}_+^4$. The intersections of z with the set of hyperplanes corresponding to different values of z provide a set of planes. Such level planes are orthogonal to the vector $v(300, 200, 350)$. The two planes for $z = 0$ (plane generating the sheaf) and $z = 12000$ (optimal solution) are here depicted.

systems of three linear equations:

$$\mathcal{O}(0,0,0) : \begin{cases} x_1 = 0 \\ x_2 = 0 \\ x_3 = 0 \end{cases} \qquad \mathcal{A}(30,0,0) : \begin{cases} 4x_1 + 2x_2 + 3x_3 = 120 \\ x_2 = 0 \\ x_3 = 0 \end{cases}$$

$$\mathcal{B}(20,20,0) : \begin{cases} x_1 + x_2 + x_3 = 0 \\ 4x_1 + 2x_2 + 3x_3 = 0 \\ x_3 = 0 \end{cases} \qquad \mathcal{C}(0,40,0) : \begin{cases} x_1 + x_2 + x_3 = 40 \\ x_1 = 0 \\ x_3 = 0 \end{cases}$$

$$\mathcal{D}(0,0,20) : \begin{cases} x_1 = 0 \\ x_2 = 0 \\ x_3 = 20 \end{cases} \qquad \mathcal{E}(15,0,20) : \begin{cases} 4x_1 + 2x_2 + 3x_3 = 120 \\ x_2 = 0 \\ x_3 = 20 \end{cases}$$

$$\mathcal{F}(10,10,20) : \begin{cases} x_1 + x_2 + x_3 = 40 \\ 4x_1 + 2x_2 + 3x_3 = 0 \\ x_3 = 20 \end{cases} \qquad \mathcal{G}(0,20,20) : \begin{cases} x_1 + x_2 + x_3 = 40 \\ x_1 = 0 \\ x_3 = 20. \end{cases}$$

The optimal solution z^* is obtained by computing the objective function z in the eight corners of the polyhedron:

$$z_\mathcal{O} = 0 \qquad z_A = 9000 \qquad z_B = 10000 \qquad z_C = 8000$$

$$z_D = 7000 \qquad z_\mathcal{E} = 11500 \qquad \boxed{z_\mathcal{F} = 12000 = z^*} \qquad z_\mathcal{G} = 11000$$

and it takes on the maximum value in corner $\mathcal{F}$, as shown in Figure 4.7.

The multidimensional problem

Starting from the 2–D and 3–D cases, the geometric interpretation of the general problem (4.2), involving n decisional variables and m constraints, naturally follows (Strang, 2005). In particular, each linear inequality divides the n–dimensional space into two half–hyperspaces, one where the inequality is satisfied and the other where it is not. The feasible region is now the intersection of the m half–hyperspaces defined by the problem constraints, and of the m half–hyperspaces corresponding to the non–negativity constraints. Such a region is a hyper–polyhedron in the n–dimensional space, and the optimal solution corresponds to a corner of this polytope. That holds when the problem admits a unique solution. In case of more optimal solutions, these solutions correspond to all the n–dimensional points lying on a facet. Although it is not possible to provide a graphical representation of the multidimensional case, and to exploit the graphical approach to obtain the solution, the leading idea of exploring the corners of the feasible region is still valid: the method that is mostly used for solving the LP problem, detailed in the following section, arises from this insight.

4.2   The simplex algorithm

Before plunging into the simplex algorithm, it is useful to introduce the definition and the meaning of *basic solution*, concept at the core of the simplex method. It is also convenient to define a further formulation of a LP problem, the canonical form, which will be exploited in the solving procedure.

4.2.1   Basic solutions

Let us refer to the generic formulation of a LP problem in standard form introduced in equation (4.3) and focus on the system constraints:

$$\mathbf{Ax} = \mathbf{b}$$

$$\mathbf{x} \geq \mathbf{0}$$

where the vector $\mathbf{x}$ is composed of the natural variables as well as of the artificial variables introduced to obtain the standard form.

Such a system is characterized by m equations (the constraints) and n unknowns (the decisional variables and the slack variables). It is clear that starting from a system of m equations in n unknowns, with $n > m$, we can set $n - m$ unknowns at will, and determine the other m variables, consequently. A basic solution is defined by setting

$n - m$ unknowns to 0. In particular, the variables set to 0, labeled $\mathbf{x}_N$, are the nonbasic variables, and are clearly related to the degrees of freedom of the problem. The remaining variables greater than 0, denoted by $\mathbf{x}_B$, are the basic variables. Therefore, the number of basic variables is equal in number to the functional constraints of the problem, whereas the number of nonbasic variables equals the number of variables minus the number of functional constraints. According to this notation, the vector $\mathbf{x}$ can be partitioned as follows:

$$\mathbf{x} = \left[\begin{array}{c} \mathbf{x}_N \\ \hline \mathbf{x}_B \end{array} \right]$$

and the matrix $\mathbf{A}$ consequently as:

$$\mathbf{A} = \left[\, \mathbf{N} \vdots \mathbf{B} \, \right]$$

where the submatrix $\mathbf{N}$ contains the nonbasic variables and the submatrix $\mathbf{B}$ to the basic variables. To highlight nonbasic and basic variables, the system constraints can be formulated as follows:

$$\mathbf{N}\mathbf{x}_N + \mathbf{B}\mathbf{x}_B = \mathbf{b}$$

$$\mathbf{x}_N = \mathbf{0}$$

$$\mathbf{x}_B \geq \mathbf{0}.$$

Thus, the system constraints are reduced to:

$$\mathbf{N} \underbrace{\mathbf{x}_N}_{=0} + \mathbf{B}\mathbf{x}_B = \mathbf{b}$$

and the values of the basic variables are easily obtained solving the system $\mathbf{B}\mathbf{x}_B = \mathbf{b}$. To this end, the matrix $\mathbf{B}$ has to be non–singular, its inverse exists and $\mathbf{x}_B = \mathbf{B}^{-1}\mathbf{b}$,

The notation highlighting the role of basic variables is extended to the vector $\mathbf{c}$ of the objective function coefficients:

$$\mathbf{c} = \left[\begin{array}{c} \mathbf{c}_N \\ \hline \mathbf{c}_B \end{array} \right]$$

where $\mathbf{c}_N$ is the set of coefficients of the nonbasic variables $\mathbf{x}_N$, and $\mathbf{c}_B$ are the coefficients of the basic variables $\mathbf{x}_B$.

Therefore, the objective function in terms of nonbasic and basic variables becomes:

$$z = \mathbf{c}^\top \mathbf{x} = \mathbf{c}_N^\top \underbrace{\mathbf{x}_N}_{=0} + \mathbf{c}_B^\top \mathbf{x}_B$$

and the value of the objective function for the basic solution is:

$$z = \mathbf{c}_B^\top \mathbf{x}_B = \mathbf{c}_B^\top \mathbf{B}^{-1}\mathbf{b}.$$

The number of possible basic solutions equals the number of ways the m variables $\mathbf{x}_B$ can be chosen among the n variables $\mathbf{x}$, which equals the number of ways that the $n - m$ variables $\mathbf{x}_N$ can be chosen among the n variables $\mathbf{x}$. More formally, they are the number of m–element subsets, or m–combinations, (or, equivalently, the number of $n - m$ elemental subsets) of an n–element set:

$$\binom{n}{m} = \binom{n}{n - m} = \frac{n!}{m!(n - m)!} \; .$$

Since a basic solution is obtained by setting $n - m$ variables to 0 to derive the others m variables, this binomial coefficient is an upper bound for the number of feasible basic solutions. Such an upper bound can be actually achieved only when all the possible sets of m columns chosen from the matrix $\mathbf{A}$ are linearly independent.[1]

Let us refer again to the 2D problem (4.4). Its corresponding standard form provides the following system of $m = 2$ equations (I and II) in $n = 4$ unknowns (x_1, x_2, x_3, x_4), obtained by adding the two artificial variables x_3 and x_4 to the two original variables x_1 and x_2:

$$
\begin{aligned}
\text{minimize} \quad & \tilde{z} = -z = -5x_1 - 3x_2 + 0x_3 + 0x_4 = -5x_1 - 3x_2 \\
\text{subject to} \quad & 1x_1 + 1x_2 + 1x_3 + 0x_4 = 600 && (I) \\
& 2x_1 + 1x_2 + 0x_3 + 1x_4 = 800 && (II) \\
& x_1 \geq 0 && (4.8) \\
& x_2 \geq 0 \\
& x_3 \geq 0 \\
& x_4 \geq 0.
\end{aligned}
$$

In particular, the slack variable x_3 is added to transform the original constraint (C_1) into equation (I), while the slack variable x_4 is added to the original constraint (C_2) to obtain equation (II). The two slack variables enter with coefficients equal to 0 in the objective function and do not modify it.

A first basic solution is easily obtained by setting the two slack variables as basic variables. The system constraints according to the above notation are:

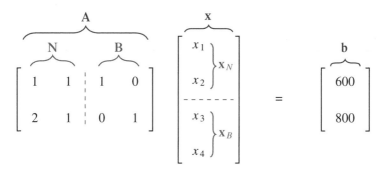

---

and the objective function is:

$$
\begin{bmatrix} \tilde{z} \end{bmatrix} = \begin{bmatrix} \overbrace{\underbrace{-5 \quad -3}_{\mathbf{c}_N^\top} \quad \vdots \quad \underbrace{0 \quad 0}_{\mathbf{c}_B^\top}}^{\mathbf{c}^\top} \end{bmatrix} \overbrace{\begin{bmatrix} \left.\begin{matrix} x_1 \\ x_2 \end{matrix}\right\}\mathbf{x}_N \\ \text{------} \\ \left.\begin{matrix} x_3 \\ x_4 \end{matrix}\right\}\mathbf{x}_B \end{bmatrix}}^{\mathbf{x}}
$$

Setting x_3 and x_4 as basic variables permits to immediately obtain the solution, as matrix $\mathbf{B}$ is the identity matrix:

$$
\mathbf{x}_B = \begin{bmatrix} x_3 \\ x_4 \end{bmatrix} = \mathbf{B}^{-1}\mathbf{b} = \mathbf{I}_2\mathbf{b} = \mathbf{b} = \begin{bmatrix} 600 \\ 800 \end{bmatrix}
$$

The corresponding value of the objective function is:

$$
\tilde{z} = \mathbf{c}_b^\top\mathbf{x}_B = \begin{bmatrix} 0 & 0 \end{bmatrix}\begin{bmatrix} 600 \\ 800 \end{bmatrix} = 0.
$$

The solution $\mathbf{x}_B = \begin{bmatrix} x_3 = 600 \\ x_4 = 800 \end{bmatrix}$ is only one of the possible basic solutions. In order to achieve a geometrical interpretation, it is enlightening to inspect all the possible $\binom{n=4}{m=2} = \binom{n=4}{n-m=2} = 6$ solutions, listed in Table 4.1. The different solutions are obtained by setting $m = 2$ of the $n = 4$ variables equal to 0. Each row of the table presents the values of the four decisional variables and, in case of a feasible solution, the value of the objective function $\tilde{z}$. The last column associates each solution with a corner of the constraint region, depicted in Figure 4.8. The graphical representation of the basic solutions has been obtained by slightly modifying Figure 4.1. Each of

Table 4.1 The six basic solutions for the LP problem (4.4). The last column associates the solutions with the corners of the feasible region depicted in Figure 4.8

#	x_1	x_2	x_3	x_4	feasible	z	corner
1	0	0	600	800	yes	$z_\mathcal{O} = 0$	$\mathcal{O}$
2	0	600	0	200	yes	$z_\mathcal{A} = 2000$	$\mathcal{A}$
3	0	800	-200	0	no	–	$\mathcal{B}$
4	600	0	0	-400	no	–	$\mathcal{C}$
5	400	0	200	0	yes	$z_D = 1200$	D
6	200	400	0	0	yes	$z_\mathcal{E} = 0$	$\mathcal{E}$

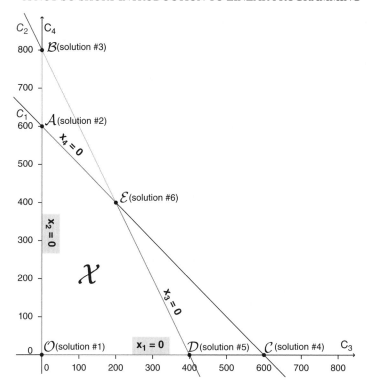

Figure 4.8 Geometrical representation of the constraints in the LP problem (4.4). This figure is a slight modification of Figure 4.1, which highlights the feasible basic solutions ($\mathcal{O}$, $\mathcal{A}$, $\mathcal{E}$, and $\mathcal{D}$) and their association with the corners of the feasible region $\mathcal{X}$ (polygon $\mathcal{O}\mathcal{A}\mathcal{E}\mathcal{D}$). The corners $\mathcal{B}$ and $\mathcal{C}$ are not feasible, as they do not belong to $\mathcal{X}$. Each of the four lines limiting the feasible region is obtained by setting to 0 a decisional variable, so defining the boundary of the half–plane associated with a problem constraint. Two adjacent corners share the same boundary.

the four lines in the graph is the set of points for which a decisional variable assumes the value 0, in particular:

- the horizontal axis is defined by the solutions for which $x_1 = 0$,

- the solutions for which $x_2 = 0$ define the vertical axis,

- the set of points for which $x_3 = 0$ lies on the line that define the boundary of constraint C_2,

- the set of points for which $x_4 = 0$ defines the boundary of constraint C_1.

In the graph a basic solution corresponds to the intersection of two of the four lines, because it is obtained by setting $m = 2$ variables equal to 0. Furthermore, Table 4.1 and Figure 4.8 also pinpoint that two solutions define two adjacent corners if they share $m - 1 = 1$ constraint boundary, that is, if they lie on the same line. This is verified for corners $\mathcal{O}$ and $\mathcal{A}$, which share the constraint C_4, for corners $\mathcal{A}$ and $\mathcal{E}$, sharing the constraint C_1, for corners $\mathcal{E}$ and $\mathcal{D}$, both satisfying constraint C_2, and for corners $\mathcal{O}$ and $\mathcal{D}$, which lie on the horizontal axis, boundary of constraint C_3. Furthermore, only four of the six basic solutions are feasible: solution #3 (corner $\mathcal{B}$) and solution #4 (corner C) are outside the feasible region $\mathcal{X}$, as the first corresponds to a negative value of $x_3 = -200$, and the second is obtained when $x_4 = -400$. Such two points do not satisfy constraints C_2 and C_1, respectively.

It emerges that the number of feasible basic solutions is at most equal to the number of basic solutions: in fact, the set of feasible basic solutions is a subset of the set of basic solutions, and both sets consist of a finite number of solutions. A special case occurs when more than $n - m$ variables are equal to 0 in a basic solution: in such a case the feasible basic solution is said to be degenerate. A degenerate solution is due to the presence of a redundant constraint, its elimination would not change the feasible region $\mathcal{X}$. See section 4.4 for further details about this.

The importance of basic solutions for solving a LP problem is to be attributed to the relationship between the corners of the feasible region and the optimal solutions introduced in the previous section. In fact, a problem with infinite solutions (the points of the feasible set $\mathcal{X}$) can be solved by examining only a finite number of them (the feasible basic solutions, i.e., the corners of $\mathcal{X}$). This is the core of the simplex method and exploits the following fundamental theorem of LP (Vanderbei 2014; Hiller and Lieberman 2015):

> Consider a LP problem in standard form.
>
> • If there exists a feasible solution, then there exists a basic feasible solution.
>
> • If there exists an optimal finite solution, then there exists an optimal finite basic solution.

The 2D–example suggests the method to distinguish a feasible basic solution from a not feasible basic solution. Each intersection between two lines associated with two constraints defines a point, and each point can be a corner of the feasible set (corners $\mathcal{O}$, $\mathcal{A}$, $\mathcal{D}$, and $\mathcal{E}$) or can be external to the feasible set (corners $\mathcal{B}$ and C). Starting from the LP problem expressed in standard form, the following steps can be carried out in the 2D–case to detect the corners of the feasible region:

• two of the variables are set to 0,

• the values of the remaining variables are then computed:

- if such values are not negative, the values of the variables x_1 and x_2 are the coordinates of a corner of the feasible region (feasible basic solution),

- if any of the variables is negative, the values of x_1 and x_2 define a point external to the feasible region (not feasible basic solution).

The previous method exploits the formulation of a point in terms of equivalence between two equations. The same idea can be used in a space with more than three dimensions, since a point in $\mathbb{R}^n$ is the result of a system of n equations. Therefore, the previous method used to detect the corners can be extended to the case of n dimensions, whereas the LP problem $\mathbf{Ax} = \mathbf{b}$ is expressed in standard form:

- $n - m$ of the n variables are set to 0 (geometrically, this is equivalent to intersect the $n - m$ hyperplanes associated with the corresponding constraints),

- the values of the m remaining variables are then computed. If such values are not negative, the values of the variables $x_1, \ldots, x_n$ are the coordinates of a corner of the feasible region (feasible basic solution). If any of the variables is negative, the values of $x_1, \ldots, x_n$ define a point external to the feasible region (not feasible basic solution).

Through such procedure, the number of solutions to explore is restricted to a finite number, that is, the $\binom{n}{m}$ basic solutions. Although the number of feasible basic solutions is finite and less–or–equal than the above upperbound, it exponentially increases with n and m. Therefore it is not possible to search for the optimal solution through the exhaustive enumeration of all the feasible basic solutions. The optimality condition introduced in the next subsection allows us to obviate this problem. It provides a rule to check if the current solution is optimal: in case it is not, the rule makes it possible to move to a new feasible basic solution which ensures a value of the objective function at least equivalent.

4.2.2   Optimality test

A feasible basic solution:

$$\mathbf{x}^* = \begin{bmatrix} \mathbf{x}_B \\ \mathbf{x}_N \end{bmatrix} = \begin{bmatrix} \mathbf{B}^{-1}\mathbf{b} \\ \mathbf{0} \end{bmatrix}$$

is optimal if and only if $\mathbf{c}^\mathsf{T}\mathbf{x}^* \le \mathbf{c}^\mathsf{T}\mathbf{x}, \forall \mathbf{x} \in \mathcal{X}$. Starting from the original system formulated in terms of basic and non–basic variables, $\mathbf{Ax} = \mathbf{Bx}_B + \mathbf{Nx}_N = \mathbf{b}$, the basic variables can be expressed as:

$$\mathbf{x}_B = \mathbf{B}^{-1}\mathbf{b} - \mathbf{B}^{-1}\mathbf{Nx}_N. \tag{4.9}$$

Such expression can then be replaced in the objective function as follows:

$$\begin{aligned}
z = \mathbf{c}^\mathsf{T}\mathbf{x} &= \mathbf{c}_B^\mathsf{T}\mathbf{x}_B + \mathbf{c}_N^\mathsf{T}\mathbf{x}_N \\
&= \mathbf{c}_B^\mathsf{T}\underbrace{(\mathbf{B}^{-1}\mathbf{b} - \mathbf{B}^{-1}\mathbf{Nx}_N)}_{\mathbf{x}_B} + \mathbf{c}_N^\mathsf{T}\mathbf{x}_N \\
&= \mathbf{c}_B^\mathsf{T}\mathbf{B}^{-1}\mathbf{b} - \mathbf{c}_B^\mathsf{T}\mathbf{B}^{-1}\mathbf{Nx}_N + \mathbf{c}_N^\mathsf{T}\mathbf{x}_N \\
&= \mathbf{c}_B^\mathsf{T}\mathbf{B}^{-1}\mathbf{b} + (\mathbf{c}_N^\mathsf{T} - \mathbf{c}_B^\mathsf{T}\mathbf{B}^{-1}\mathbf{N})\mathbf{x}_N \\
&= \underbrace{\mathbf{c}_B^\mathsf{T}\underbrace{\mathbf{B}^{-1}\mathbf{b}}_{\mathbf{x}_B} + \mathbf{c}_N^\mathsf{T}\underbrace{\mathbf{x}_N}_{\mathbf{0}}}_{\mathbf{c}^\mathsf{T}\mathbf{x}^*} + (\mathbf{c}_N^\mathsf{T} - \mathbf{c}_B^\mathsf{T}\mathbf{B}^{-1}\mathbf{N})\mathbf{x}_N + \underbrace{(\mathbf{c}_B^\mathsf{T} - \mathbf{c}_B^\mathsf{T}\underbrace{\mathbf{B}^{-1}\mathbf{B}}_{\mathbf{I}})}_{\mathbf{0}}\mathbf{x}_B
\end{aligned}$$

$$= \mathbf{c}^{\top}\mathbf{x}^* + \underbrace{\mathbf{c}_B^{\top}\mathbf{x}_B + \mathbf{c}_N^{\top}\mathbf{x}_N}_{\mathbf{c}^{\top}\mathbf{x}} - \mathbf{c}_B^{\top}\mathbf{B}^{-1}\underbrace{(\mathbf{B}\mathbf{x}_B + \mathbf{N}\mathbf{x}_N)}_{\mathbf{A}\mathbf{x}}$$

$$= \mathbf{c}^{\top}\mathbf{x}^* + (\mathbf{c}^{\top} - \mathbf{c}_B^{\top}\mathbf{B}^{-1}\mathbf{A})\mathbf{x}$$

$$= \mathbf{c}^{\top}\mathbf{x}^* + \bar{\mathbf{c}}^{\top}\mathbf{x}.$$

The vector:

$$\bar{\mathbf{c}}^{\top} = (\mathbf{c}^{\top} - \mathbf{c}_B^{\top}\mathbf{B}^{-1}\mathbf{A})$$

$$= \begin{bmatrix} \bar{\mathbf{c}}_B & \bar{\mathbf{c}}_N \end{bmatrix}$$

$$= \begin{bmatrix} \mathbf{c}_B^{\top} - \mathbf{c}_B^{\top}\mathbf{B}^{-1}\mathbf{B} & \mathbf{c}_N^{\top} - \mathbf{c}_B^{\top}\mathbf{B}^{-1}\mathbf{N} \end{bmatrix}$$

$$= \begin{bmatrix} \mathbf{0} & \mathbf{c}_N^{\top} - \mathbf{c}_B^{\top}\mathbf{B}^{-1}\mathbf{N} \end{bmatrix}$$

introduced in the last equation, is called vector of reduced costs, because the original costs $\mathbf{c}$ are decreased by the amount $\mathbf{c}_B^{\top}\mathbf{B}^{-1}\mathbf{A}$. From the above algebra, it is easy to derive the optimality test: since $\mathbf{c}^{\top}\mathbf{x}^*$ has to be less than or at most equal to $\mathbf{c}^{\top}\mathbf{x}$ in order for $\mathbf{x}^*$ to be an optimal solution, and since $\mathbf{c}^{\top}\mathbf{x} = \mathbf{c}^{\top}\mathbf{x}^* + \bar{\mathbf{c}}^{\top}\mathbf{x}$, then $\bar{\mathbf{c}}^{\top}\mathbf{x}$ has to be non–negative. From the non–negativity conditions, it follows the optimality condition $\bar{\mathbf{c}} \geq \mathbf{0}$[2]. It is worth highlighting that such a condition is sufficient but not necessary. For further details see Matousek and Gartner (2005) and Vanderbei (2014).

4.2.3   Change of the basis: entering variable and leaving variable

The optimality condition derived above is checked on the current basic solution $\mathbf{x}^{(i)}$. If the condition does not hold, then there exists at least one nonbasic variable $x_h \in \mathbf{x}_N$ for which $\bar{c}_h < 0$. It is therefore possible to obtain a new basic solution $\mathbf{x}^{(i+1)}$ by entering x_h in basis, in place of another variable x_k, which leaves the basis. Hence, the change of the basis from $\mathbf{x}^{(i)}$ to $\mathbf{x}^{(i+1)}$ consists in determining the variable x_h that enters the basis and the variable x_k leaving the basis. To this end, let us consider the objective function computed in the new basis $\mathbf{x}^{(i+1)}$, obtained when x_h enters the basis:

$$z = \mathbf{c}^{\top}\mathbf{x}^{(i+1)} = \mathbf{c}^{\top}\mathbf{x}^{(i)} + \bar{\mathbf{c}}\mathbf{x}_N$$

$$= \mathbf{c}^{\top}\mathbf{x}^{(i)} + \begin{bmatrix} \bar{c}_1, & \dots & ,\bar{c}_{h-1},\bar{c}_h,\bar{c}_{h+1}, & \dots & ,\bar{c}_{n-m} \end{bmatrix} \begin{bmatrix} 0 \\ \vdots \\ 0 \\ x_h \\ 0 \\ \vdots \\ 0 \end{bmatrix}$$

$$= \mathbf{c}^{\top}\mathbf{x}^{(i)} + \bar{c}_h x_h.$$

---

[2] The optimality condition requires that $\bar{\mathbf{c}}$ is non-positive in case of a maximization problem if the standard form introduced in subsection 4.1.1 is not used.

Hence, when x_h enters the basis, the objective function decreases by the amount $\bar{c}_h x_h < 0$ with respect to its value in correspondence of the initial basis $\mathbf{x}^{(i)}$. The improvement in z is therefore proportional to the increase in x_h, through a proportionality factor $\bar{c}_h$. Thus, when there are more nonbasic variables with negative reduced cost, that is, not satisfying the optimality condition, the variable chosen to enter the basis is the variable x_h whose reduced cost $\bar{c}_h$ is minimum, that is, is greater in module[3]. The leading idea is simple: a smaller value of $\bar{c}_h$ leads to a greater improvement in the objective function, and the choice of x_h appears as the most efficient way to minimize the objective function to achieve the optimal solution of the problem. However, it is worth highlighting that this heuristic does not imply a smaller number of iterations, as the efficiency of choosing the variable x_h with the smaller c_h as entering variable is true only locally: therefore there is no guarantee about this choice from a global point of view. Notwithstanding, the use of a local criterion is still rational, in that a global viewpoint is not available.

Once the variable x_h entering the basis has been chosen, the next step consists in detecting the variable x_k leaving the basis. Such step is based on the computation of the maximum value that x_h can assume. In fact, when x_h enters the basis, it moves from the initial value of 0 (it was initially a nonbasic variable) to a positive value. Such a change involves a change in the other basic variables, which can be formulated (according to equation 4.9) as follows:

$$\mathbf{x}_B = \mathbf{B}^{-1}\mathbf{b} - \mathbf{B}^{-1}\mathbf{N}\mathbf{x}_N$$

$$= \mathbf{B}^{-1}\mathbf{b} - \mathbf{B}^{-1} \left[\mathbf{A}_{m+1} \vdots \mathbf{A}_{m+2} \vdots \ldots \vdots \mathbf{A}_h \vdots \ldots \vdots \mathbf{A}_m \right] \begin{bmatrix} 0 \\ \vdots \\ 0 \\ x_h \\ 0 \\ \vdots \\ 0 \end{bmatrix}$$

$$= \mathbf{B}^{-1}\mathbf{b} - \underbrace{\mathbf{B}^{-1}\mathbf{A}_h}_{=\bar{\mathbf{A}}_h} x_h$$

where $\mathbf{A}_h$ denotes the column of matrix $\mathbf{A}$ corresponding to the entering variable x_h. Since $\mathbf{x}_B \geq 0$ for the non-negativity condition, the largest value x_h can assume depends on $\overline{\mathbf{A}}_h = \mathbf{B}^{-1}\mathbf{A}_h$. When all the elements of $\overline{\mathbf{A}}_h$ are negative or equal to 0, any value $x_h \geq 0$ satisfies the condition $\mathbf{x}_B \geq 0$: in such a case the problem is unbounded, as x_h can always increase leading to an improvement in the objective function. If some elements of $\overline{\mathbf{A}}_h$ are greater than 0, the largest value of x_h that still ensures the

---

[3] The choice is arbitrary in case more than one variable take the same minimum negative cost. See section 4.4 for further details.

non negativity condition is the minimum value of $\frac{\mathbf{B}^{-1}\mathbf{b}}{\overline{\mathbf{A}}_h}$, for which:

$$\mathbf{x}_B = \mathbf{B}^{-1}\mathbf{b} - \overline{\mathbf{A}}_h x_h$$

$$= \mathbf{B}^{-1}\mathbf{b} - \overline{\mathbf{A}}_h \min_{\overline{a}_h > 0} \frac{\mathbf{B}^{-1}\mathbf{b}}{\overline{\mathbf{A}}_h}$$

$$= \left[x_1, \ \dots \ , x_{k-1}, 0, x_{k+1}, \ \dots \ , x_m \right]$$

In other words, the first variable x_k that assumes the value 0 when the entering variable x_h increases is selected as leaving variable. The value of x_h is indeed fixed in order to keep the non-negativity condition. In summary, the leaving variable x_k is determined through the following rule:

$$k \ : \ \operatorname*{argmin}_{\overline{a}_h > 0} \frac{\mathbf{B}^{-1}\mathbf{b}}{\overline{\mathbf{A}}_h} \ .$$

Also in this case, as for the choice of the entering variable, there may be more indexes that provide the same minimum. See section 4.4 for a rule of thumb to be applied in this case.

4.2.4   The canonical form of a linear programming problem

It is helpful to introduce a particular standard form, the *canonical form*. It will be indeed very useful for the computation of the basic solutions, allowing us to avoid the inversion of matrix **B**. Recalling the example in subsection 4.2.1, a proper setting of the basic variables speeds up the computation of the corresponding basic solution: in the example, the solution for the corner $\mathcal{O}$ has been immediately computed since the partition **B** of the constraint matrix **A** is the identity matrix $\mathbf{I}_m$. In detail, a LP problem in standard form:

$$\begin{aligned}
\text{minimize} \quad & z = \mathbf{c}_N^{\top} \mathbf{x}_N + \mathbf{c}_B^{\top} \mathbf{x}_B \\
\text{subject to} \quad & \mathbf{N}\mathbf{x}_N + \mathbf{B}\mathbf{x}_B = \mathbf{b} \\
& \mathbf{x}_N = \mathbf{0} \\
& \mathbf{x}_B \geq \mathbf{0}
\end{aligned} \tag{4.10}$$

is also in canonical form if, once a basic solution $\mathbf{x}_B$ is defined, at least the first two out of the following conditions are fulfilled:

(i) $\mathbf{B} \equiv \mathbf{I}_m$

(ii) $\mathbf{c}_B = \mathbf{0}$

(iii) $\mathbf{b} \geq \mathbf{0}$.

A weak canonical form is defined when (i) and (ii) are fulfilled, and a strong canonical form is defined when also (iii) holds.

Therefore, a canonical form is referred to the set of basic variables $\mathbf{x}_B$: starting from a canonical form, it is straightforward to determine a feasible basic solution, which corresponds to the values of the vector $\mathbf{b}$. In fact, each equation contains only one basic variable different from 0, that is, the variable corresponding to the value 1 of the canonical form: x_3 for the first equation and x_4 for the second equation.

Whereas the system $\mathbf{Ax} = \mathbf{b}$ is not in canonical form, simple algebraic manipulations easily transform it. The use of the standard form, that is, the use of equations both for the objective function and for the constraints, permits to exploit a fundamental property of the systems of linear equations (Strang 2005, 2009):

Any transformation of a system of linear equations, obtained multiplying an equation by any real number different from zero or linearly combining any two equations, does not affect the solution.

The simplex algorithm exploits this type of transformations for exploring the set of feasible basic solutions and choosing the optimal one. Starting from a corner of $\mathcal{X}$ (a feasible basic solution), the movement toward another corner (a different feasible basic solution) is obtained through manipulations of the equations suitable to express them in canonical form with respect to the new basis.

Back to the LP problem (4.8) expressed in standard form, focusing only on the equations:

$$\text{①} \quad -5x_1 - 3x_2 + \boxed{0}\,x_3 + \boxed{0}\,x_4 = \tilde{z} = -z$$

$$\text{②} \quad +1x_1 + 1x_2 + \boxed{1}\,x_3 + \boxed{0}\,x_4 = \boxed{600}$$

$$\text{③} \quad +2x_1 + 1x_2 + \boxed{0}\,x_3 + \boxed{1}\,x_4 = \boxed{800}$$

It is evident that the problem is in canonical form with respect to the basis sequence (x_3, x_4), since the conditions (i), (ii), and (iii) hold. Now suppose that we are interested in expressing the problem in canonical form with respect to the basis sequence (x_2, x_4). Variable x_3 must exit the basis and move to the vector x_N, and variable x_2 from the vector $\mathbf{x}_B$ has to replace x_3 in the basis. The system has to be transformed so to obtain:

- a value 0 for the coefficient c_2 in the first equation, in order to move x_2 to the vector $\mathbf{x}_B$,

- a value 1 for the coefficient $a_{1,2}$ in the second equation, as x_2 replaces x_3 in the basis,

- a value 0 for the remaining coefficient $a_{2,2}$ of x_2, in the third equation.

Variable x_2 enters the basis with value $a_{1,2}$. This element is known as *pivot element* and the operation used for changing the basis of the canonical form is known as *pivoting*, since it is carried out pivoting on an element of matrix $\mathbf{A}$. The pivoting operation is essentially a linear combination of equations: once the variable to move to the new basis is chosen, the pivot equation is divided by the pivot element, so to obtain a value 1 in correspondence of the pivot element. The other equations are linearly combined with the pivot equation in order to have all values equal to 0 for the coefficients corresponding to the variable entering the basis. In the example, since the pivot $a_{1,2} = 1$, the second equation does not change, the third equation is transformed subtracting the second from it, and the first equation is modified adding the second equation multiplied by 3. In summary, the pivoting operation along with the new system are the following:

$$① + 3 \times ② \rightarrow -2x_1 - \boxed{0}\, x_2 + 3x_3 + \boxed{0}\, x_4 = \bar{z} + 1800$$

$$② \rightarrow +1x_1 + \boxed{1}\, x_2 + 1x_3 + \boxed{0}\, x_4 = \boxed{600}$$

$$③ - ② \rightarrow +1x_1 + \boxed{0}\, x_2 - 1x_3 + \boxed{1}\, x_4 = \boxed{200}\,.$$

The problem is now in canonical form with respect to the basis (x_2, x_4).

Finally, the change of the basis can also be expressed in matrix form. To express the system in canonical form with respect to the set $\mathbf{x}_B$, consider the matrix $\mathbf{B}$ composed of the initial columns of $\mathbf{A}$ corresponding to the variables $\mathbf{x}_B$. Premultiplying the constraint equation in the LP problem (4.10) by $\mathbf{B}^{-1}$, we obtain:

$$\mathbf{B}^{-1}\mathbf{N}\mathbf{x}_N + \underbrace{\mathbf{B}^{-1}\mathbf{B}}_{=\mathbf{I}_m}\, \mathbf{x}_B = \mathbf{B}^{-1}\mathbf{b}$$

which is a canonical form with respect to $\mathbf{x}_B$. Starting from such a canonical form, and setting $\mathbf{x}_N = 0$, the values for the basic variables are then $\mathbf{x}_B = \mathbf{B}^{-1}\mathbf{b}$. Therefore, to express a system $\mathbf{A}\mathbf{x} = \mathbf{b}$ in canonical form with respect to a set of variables $\mathbf{x_B}$ and determine a basic solution, it is sufficient to premultiply the system by $\mathbf{B}^{-1}$. Such a matrix, inverse of $\mathbf{B}$, is composed of the column of $\mathbf{A}$ corresponding to the relevant variables to express the system in canonical form.

In this example, we have:

$$\begin{array}{cc} x_2 & x_4 \end{array}$$
$$\mathbf{B} = \begin{bmatrix} 1 & 0 \\ 1 & 1 \end{bmatrix} \Rightarrow \mathbf{B}^{-1} = \begin{bmatrix} 1 & 0 \\ -1 & 1 \end{bmatrix}$$

The new constraint matrix is then obtained premultiplying matrix $\mathbf{A}$ by $\mathbf{B}^{-1}$:

$$\mathbf{B}^{-1}\mathbf{A} = \begin{bmatrix} 1 & 0 \\ -1 & 1 \end{bmatrix} \begin{array}{cccc} x_1 & x_2 & x_3 & x_4 \end{array} \begin{bmatrix} 1 & 1 & \boxed{1} & \boxed{0} \\ 2 & 1 & \boxed{0} & \boxed{1} \end{bmatrix} = \begin{array}{cccc} x_1 & x_2 & x_3 & x_4 \end{array} \begin{bmatrix} 1 & \boxed{1} & 1 & \boxed{0} \\ 1 & \boxed{0} & -1 & \boxed{1} \end{bmatrix}$$

and it is now in canonical form with respect to the basis (x_2, x_4). The values for the new basic variables are:

$$\mathbf{x}_B = \begin{bmatrix} x_2 \\ x_4 \end{bmatrix} = \mathbf{B}^{-1}\mathbf{b} = \begin{bmatrix} 1 & 0 \\ -1 & 1 \end{bmatrix} \begin{bmatrix} 600 \\ 800 \end{bmatrix} = \begin{bmatrix} 600 \\ 200 \end{bmatrix}$$

Such solutions are tantamount to those obtained manipulating the three equations through the pivoting operation.

The canonical form and its role for achieving a feasible basic solution are now defined. Next section details the simplex algorithm.

4.2.5   The simplex algorithm

The simplex algorithm consists of the following three steps:

- **initialization**: it consists in the detection of a first feasible basic solution. The system of equations corresponding to the objective function and to the constraints is expressed in canonical form with respect to a first set of basic variables. The presence of a canonical form guarantees an immediate computation of the values of the basic variables;

- **test**: in this phase, the optimality of the detected solution is tested. In case of optimality, the algorithm stops;

- **iteration**: it consists in moving to a better different feasible basic solution, i.e. to an adjacent corner that is feasible and that improves the solution. In this step, the algorithm essentially examines the corners adjacent to the current basic solution and chooses a new corner that differs from the previous one for only one basic variable. This leads to determine a variable entering the basis and a variable exiting the basis.

Figure 4.9 shows the corresponding flowchart. The algorithm starts from a feasible basic solution and moves toward a new feasible basic solution if the value of the objective function is improved. The transition from a given solution to a new one is carried out through a pivoting operation so to move between two different canonical forms. The last two steps are iterated until the optimal solution is achieved. According to the fundamental theorem of LP (see subsection 4.2.1), if there exists an optimal solution, then there exists an optimal finite basic solution. Since the number of basic solutions is finite, the algorithm converges to the optimal solution in a finite number of iterations, exploring solutions that improve gradually the value of the objective function. In most cases the algorithm converges to the optimal solution without visiting all the possible basic solutions. To this end, the optimality test plays a crucial role for limiting the number of iterations.

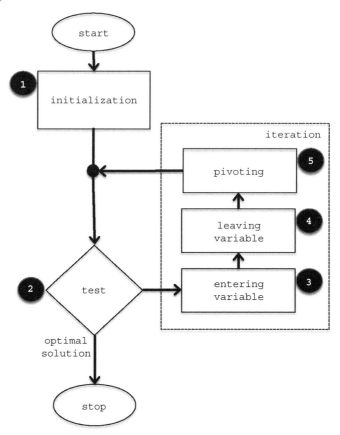

Figure 4.9   The general structure of the simplex algorithm.

Using the notation introduced to distinguish basic and nonbasic variables, it is now possible to detail the simplex algorithm:

1: **B** ← initial basis   ▷ assume B is an initial feasible basic solution
2: *optimal* ← **false**   ▷ flag variable for checking optimality
3: *unbound* ← **false**   ▷ flag variable for checking unboundness
4: **while** *optimal* = **false and** *unbound* = **false do**
5:     $\mathbf{x}_B \leftarrow \mathbf{B}^{-1}\mathbf{b}$ ▷ current basic solution
6:     $\bar{\mathbf{c}}^\mathsf{T} \leftarrow (\mathbf{c}^\mathsf{T} - \mathbf{c}_B^\mathsf{T}\mathbf{B}^{-1}\mathbf{A})$ ▷ vector of reduced costs
7:     **if** $\bar{\mathbf{c}}^\mathsf{T} \geq \mathbf{0}$ **then**   ▷ optimality test
8:         *optimal* ← **true**
9:     **else**   ▷ change of the basis

10:        $h : \text{argmin}_{\bar{c}_i < 0}\ \bar{\mathbf{c}}$, for a non–basic variable ▷ index of the entering variable
11:        $\bar{\mathbf{A}}_h \leftarrow \mathbf{B}^{-1}\mathbf{A}_h$
12:        **if** $(\bar{\mathbf{A}}_h)_i \leq 0\ \forall i = 1, \dots, m$ **then**
13:            $unbound \leftarrow$ **true**
14:        **else**
15:            $k : \text{argmin}_{\bar{a}_h > 0}\ \dfrac{\mathbf{B}^{-1}\mathbf{b}}{\bar{\mathbf{A}}_h}$ ▷ index of the leaving variable
16:            $\mathbf{B} \leftarrow \mathbf{B} \cup \{\mathbf{A}_h\} \setminus \{\mathbf{A}_k\}$ ▷ new basis switching the role of x_h and x_k
17:            $x_h \leftarrow \dfrac{(\mathbf{B}^{-1}\mathbf{b})_k}{(\bar{\mathbf{A}}_h)_k}$ ▷ update the value for the entering variable
18:            $x_k \leftarrow 0$ ▷ set to 0 the leaving variable
19:        **end if**
20:    **end if**
21: **end while**

The algorithm is shown in action on the 2D LP problem (4.4).

❶ **Initialization**

Starting from the standard form (4.8) of the LP problem (4.4), the basis (x_3, x_4) composed of the two slack variables, is a convenient starting point as the system is already expressed in strong canonical form:

(i) $\mathbf{B} = \begin{bmatrix} \mathbf{A}_3 \vdots \mathbf{A}_4 \end{bmatrix} = \begin{matrix} x_3 & x_4 \\ \begin{bmatrix} 1 & 0 \\ 0 & 1 \end{bmatrix} \end{matrix} = \mathbf{I}_2$

(ii) $\mathbf{c}_B^{\mathsf{T}} = \begin{matrix} x_3 & x_4 \\ \begin{bmatrix} 0 & 0 \end{bmatrix} \end{matrix} = \mathbf{0}$

(iii) $\mathbf{b} = \begin{bmatrix} 600 \\ 800 \end{bmatrix} \geq \mathbf{0}.$

The first basic solution is then easily determined:

$$\mathbf{x}_B = \begin{bmatrix} x_3 \\ x_4 \end{bmatrix} = \mathbf{B}^{-1}\mathbf{b} = \begin{bmatrix} 1 & 0 \\ 0 & 1 \end{bmatrix} \begin{bmatrix} 600 \\ 800 \end{bmatrix} = \begin{bmatrix} 600 \\ 800 \end{bmatrix}$$

Such solution is feasible because $\mathbf{x}_B \geq \mathbf{0}$ and it corresponds to the corner $\mathcal{O}$, origin of the axes in Figure 4.8. The two nonbasic variables x_1 and x_2 are indeed equal to 0. This is always the case when a strong canonical form holds, that is, when $\mathbf{b} \geq \mathbf{0}$ and the problem contains only less–than inequalities constraints.

In case of a problem with equality or greater–than constraints and/or with negative known coefficients, a preliminary transformation of the system is required to obtain an initial feasible basic solution. This case will be detailed in subsection 4.3 and will be exploited in the next chapter to solve the quantile regression problem.

❷ Optimality test - iteration 1

To test the optimality of the current feasible solution $\mathbf{x}_B = \begin{bmatrix} x_3 \\ x_4 \end{bmatrix}$, we have to verify if the optimality condition $\overline{\mathbf{c}} \geq \mathbf{0}$ is satisfied. The vector of reduced costs is:

$$\overline{\mathbf{c}}^{\mathsf{T}} = (\mathbf{c}^{\mathsf{T}} - \mathbf{c}_B^{\mathsf{T}} \mathbf{B}^{-1} \mathbf{A})$$

$$= \begin{bmatrix} -5, -3, 0, 0 \end{bmatrix} - \begin{bmatrix} 0, 0 \end{bmatrix} \begin{bmatrix} 1 & 0 \\ 0 & 1 \end{bmatrix} \begin{bmatrix} 1 & 0 & 1 & 0 \\ 2 & 1 & 0 & 1 \end{bmatrix}$$

$$= \begin{bmatrix} -5, -3, 0, 0 \end{bmatrix}$$

Therefore the optimal condition is not satisfied.

❸ Entering variable - iteration 1

Both the nonbasic variables x_1 and x_2 are eligible for entering the basis, as their associated reduced costs are negative: $\overline{c}_1 = -5$ and $\overline{c}_2 = -3$. Using the local criterion introduced above, x_1 is the variable chosen to enter the basis, as the corresponding improvement in z is the greatest (this is a minimization problem and $\overline{c}_1 = -5$ is the smallest).

Once the entering variable is selected, the condition $(\overline{\mathbf{A}}_h)_i \leq 0, \forall i = 1, \ldots, m$ is used to check if the problem is unbounded:

$$\overline{\mathbf{A}}_h = \mathbf{B}^{-1} \mathbf{A}_h = \begin{bmatrix} 1 & 0 \\ 0 & 1 \end{bmatrix} \begin{bmatrix} 1 \\ 2 \end{bmatrix} = \begin{bmatrix} 1 \\ 2 \end{bmatrix}$$

The condition does not hold, and the algorithm proceeds.

❹ Leaving variable - iteration 1

To determine the leaving variable, the ratio $\dfrac{\mathbf{B}^{-1}\mathbf{b}}{\overline{\mathbf{A}}_h}$ is computed only considering the cells $\overline{\mathbf{a}}_h$ of the vector $\overline{\mathbf{A}}_h$ which are greater than zero:

$$\frac{\mathbf{B}^{-1}\mathbf{b}}{\overline{\mathbf{A}}_h} = \frac{\begin{bmatrix} 1 & 0 \\ 0 & 1 \end{bmatrix} \begin{bmatrix} 600 \\ 800 \end{bmatrix}}{\begin{bmatrix} 1 \\ 2 \end{bmatrix}} = \begin{bmatrix} 600 \\ 400 \end{bmatrix}$$

The entry 400 is the minimum, and it corresponds to the second variable in the basis. Therefore, x_4 is selected to leave the basis. The value 400 is the value assigned to x_1 when it enters the basis. This is indeed the maximum value that x_1 can assume preserving the non-negativity condition. This can be simply verified by expressing the basic variables in terms of the nonbasic variables, starting from the two equations associated to the constraints:

$$\begin{cases} x_3 = 600 - x_1 - x_2 \\ x_4 = 800 - 2x_1 - x_2 \end{cases}$$

To ensure the two non-negativity conditions, $x_3 \geq 0$ and $x_4 \geq 0$, we have:

$$\begin{cases} x_3 = 600 - x_1 \geq 0 \\ x_4 = 800 - 2x_1 \geq 0 \end{cases}$$

In fact, x_1 has to be chosen to enter the basis, whereas $x_2 = 0$ in that it is a nonbasic variable. From the two inequalities it follows that $x_1 \leq 400$.

❺ Pivoting - iteration 1

At the end of the first iteration, the value of x_1 is set equal to 400 and the value of x_4 is set equal to 0. The objective function in the new basis is equal to:

$$\tilde{z} = \mathbf{c}^T \mathbf{x} + \bar{c}_h x_h = \begin{bmatrix} -5, -3, 0, 0 \end{bmatrix} \begin{bmatrix} 0 \\ 0 \\ 200 \\ 400 \end{bmatrix} + (-5 \times 400) = -2000.$$

Matrix $\mathbf{B}$ is updated accordingly using column $A_{h=1}$ instead of column $A_{k=4}$. Therefore, the following configuration holds:

$$\begin{array}{cc} A_3 \quad A_1 & \qquad A_3 \quad A_1 \end{array}$$
$$\mathbf{B} = \begin{bmatrix} 1 & 1 \\ 0 & 2 \end{bmatrix}, \quad \mathbf{B}^{-1} = \begin{bmatrix} 1 & -0.5 \\ 0 & 0.5 \end{bmatrix},$$

$$\mathbf{x}_B = \begin{bmatrix} x_3 \\ x_1 \end{bmatrix} = \begin{bmatrix} 1 & -0.5 \\ 0 & 0.5 \end{bmatrix} \begin{bmatrix} 600 \\ 800 \end{bmatrix} = \begin{bmatrix} 200 \\ 400 \end{bmatrix}, \quad \mathbf{x}_N = \begin{bmatrix} x_4 \\ x_2 \end{bmatrix} = \begin{bmatrix} 0 \\ 0 \end{bmatrix}$$

The constraint matrix in canonical form with respect to the new basis (x_3, x_1) is then:

$$\mathbf{B}^{-1}\mathbf{A} = \begin{bmatrix} 1 & -0.5 \\ 0 & 0.5 \end{bmatrix} \cdot \begin{bmatrix} x_1 & x_2 & x_3 & x_4 \\ 1 & 1 & 1 & 0 \\ 2 & 1 & 0 & 1 \end{bmatrix} = \begin{bmatrix} x_1 & x_2 & x_3 & x_4 \\ 0 & 0.5 & 1 & -0.5 \\ 1 & 0.5 & 0 & 0.5 \end{bmatrix}.$$

The new basic solution corresponds to the corner $\mathcal{D}$ in Figure 4.8. Therefore, starting from the initial feasible solution $\mathcal{O}$, the algorithm inspects the adjacent corners $\mathcal{A}$ and $\mathcal{D}$, choosing to move toward the one that ensures the greater improvement in the objective function z. Whereas in step ❸ x_2 would be selected to enter the basis instead of x_1, the new feasible solution would be $\mathcal{A}$.

❷ Optimality test - iteration 2

The vector of reduced costs corresponding to the new basis is:

$$\bar{\mathbf{c}}^T = \begin{bmatrix} -5, -3, 0, 0 \end{bmatrix} - \begin{bmatrix} 0, -5 \end{bmatrix} \begin{bmatrix} 1 & -0.5 \\ 0 & 0.5 \end{bmatrix} \begin{bmatrix} 1 & 1 & 1 & 0 \\ 2 & 1 & 0 & 1 \end{bmatrix}$$

$$= \begin{bmatrix} -5, -3, 0, 0 \end{bmatrix} - \begin{bmatrix} -5, -2.5, 0, -2.5 \end{bmatrix} = \begin{bmatrix} 0, -0.5, 0, 2.5 \end{bmatrix}.$$

Once again the optimality condition is not satisfied, and therefore the algorithm moves to a better different feasible solution. If we refer to Figure 4.8, starting from the current feasible solution $\mathcal{D}$, the only adjacent corner still to explore is $\mathcal{E}$, as we started from $\mathcal{O}$ and the basic solution $\mathcal{C}$ is not feasible.

❸ Entering variable - iteration 2

There is only one negative reduced cost, $\overline{c}_2 = -0.5$, associated with x_2, which is therefore selected to enter the basis.

The unboundedness condition does not hold:

$$\overline{\mathbf{A}}_h = \mathbf{B}^{-1}\mathbf{A}_h = \begin{bmatrix} 1 & -0.5 \\ 0 & 0.5 \end{bmatrix} \begin{bmatrix} 0.5 \\ 0.5 \end{bmatrix} = \begin{bmatrix} 0.25 \\ 0.25 \end{bmatrix}.$$

❹ Leaving variable - iteration 2

The ratio $\dfrac{\mathbf{B}^{-1}\mathbf{b}}{\overline{\mathbf{A}}_h}$ is then computed considering both the cells of $\overline{\mathbf{A}}_h$, as they are positive:

$$\frac{\mathbf{B}^{-1}\mathbf{b}}{\overline{\mathbf{A}}_h} = \frac{\begin{bmatrix} 1 & -0.5 \\ 0 & 0.5 \end{bmatrix} \begin{bmatrix} 200 \\ 400 \end{bmatrix}}{\begin{bmatrix} 0.25 \\ 0.25 \end{bmatrix}} = \begin{bmatrix} 800 \\ 1600 \end{bmatrix}.$$

The leaving variable, corresponding to index $k = 1$, is the first variable in basis, that is, x_3. The corresponding ratio $\dfrac{\mathbf{B}^{-1}\mathbf{b}}{(\overline{\mathbf{A}}_h)_k} = 800$ is the value assigned to x_2 for entering the basis.

❺ Pivoting - iteration 2

The objective function in the new basis is equal to:

$$z = \mathbf{c}^\top \mathbf{x} + \overline{c}_h x_h = \begin{bmatrix} -5, -3, 0, 0 \end{bmatrix} \begin{bmatrix} 400 \\ 0 \\ 200 \\ 0 \end{bmatrix} + (-0.25 \times 800) = -2200.$$

The new basis matrix and its inverse at the end of the second iteration are:

$$\begin{array}{cc} A_2 \quad A_1 \end{array}$$
$$\mathbf{B} = \begin{bmatrix} 1 & 1 \\ 1 & 2 \end{bmatrix}, \qquad \begin{array}{cc} A_3 \quad A_1 \end{array}$$
$$\mathbf{B}^{-1} = \begin{bmatrix} 2 & -1 \\ -1 & 1 \end{bmatrix},$$

$$\mathbf{x}_B = \begin{bmatrix} x_2 \\ x_1 \end{bmatrix} = \begin{bmatrix} 2 & -1 \\ -1 & 1 \end{bmatrix} \begin{bmatrix} 600 \\ 800 \end{bmatrix} = \begin{bmatrix} 400 \\ 200 \end{bmatrix}, \qquad \mathbf{x}_N = \begin{bmatrix} x_4 \\ x_3 \end{bmatrix} = \begin{bmatrix} 0 \\ 0 \end{bmatrix},$$

Therefore, the constraint matrix in canonical form with respect to the new basis (x_2, x_1) is:

$$
\mathbf{B}^{-1}\mathbf{A} = \begin{bmatrix} 2 & -1 \\ -1 & 1 \end{bmatrix}
\begin{array}{c}
\begin{array}{cccc} x_1 & x_2 & x_3 & x_4 \end{array} \\
\begin{bmatrix} 1 & 1 & \boxed{1} & \boxed{0} \\ 2 & 1 & \boxed{0} & \boxed{1} \end{bmatrix}
\end{array}
= \begin{array}{c}
\begin{array}{cccc} x_1 & x_2 & x_3 & x_4 \end{array} \\
\begin{bmatrix} \boxed{0} & \boxed{1} & 2 & -1 \\ \boxed{1} & \boxed{0} & -1 & 1 \end{bmatrix}
\end{array}.
$$

The new basic solution corresponds to the corner $\mathcal{E} = (200,400)$ in Figure 4.8.

❷ Optimality test - iteration 3

Finally, the new vector of reduced costs corresponding to the basis (x_2, x_1) is computed:

$$
\bar{\mathbf{c}}^\mathsf{T} = \begin{bmatrix} -5, -3, 0, 0 \end{bmatrix} - \begin{bmatrix} -3, -5 \end{bmatrix} \begin{bmatrix} 2 & -1 \\ -1 & 1 \end{bmatrix} \begin{bmatrix} 1 & 1 & 1 & 0 \\ 2 & 1 & 0 & 1 \end{bmatrix}
$$

$$
= \begin{bmatrix} -5, -3, 0, 0 \end{bmatrix} - \begin{bmatrix} -5, -3, -1, -2 \end{bmatrix} = \begin{bmatrix} 0, 0, 1, 2 \end{bmatrix}
$$

The optimality condition is now fulfilled, and then the algorithm stops. Therefore the optimal solution is the corner $\mathcal{E}$ with an associated value of the objective function $z = -2200$, as already shown in subsection 4.1.3 using the graphical approach.

4.2.6   The tableau version of the simplex algorithm

The above simplex algorithm can be formulated through a particular coding of the same data, named *tableau*. This is a table composed of $m + 1$ rows and $n + 1$ columns, where only the numeric coefficients are included. The rows refer to the m constraints along with the objective function, while the columns refer to the n decisional variables along with the known coefficients associated with the constraints and contained in the vector $\mathbf{b}$. The general structure of the tableau object is the following:

$$
\begin{array}{cc}
 & \begin{array}{ccc} b & x_1 & \cdots & x_n \end{array} \\
\begin{array}{c} z \\ C_1 \\ \vdots \\ C_m \end{array} &
\begin{array}{|c|ccc|}
\hline
-d & & \mathbf{c}^\mathsf{T} & \\
\hline
 & & & \\
\mathbf{b} & & \mathbf{A} & \\
 & & & \\
\hline
\end{array}
\end{array}
\quad
\begin{array}{l}
\longleftarrow \quad \text{row } 0 \\
\longleftarrow \quad \text{row } 1 \\
\qquad \vdots \\
\longleftarrow \quad \text{row } m
\end{array}
$$

The first row (row 0) is related to the objective function z. A known coefficient d is inserted in the equation $z = \mathbf{cx} + d$ pertaining to the objective function, so to treat this equation in the same way as the equations related to the constraints. This coefficient is inserted in the tableau on the left side of row 0 with an opposite sign, in order to have the equation in the form $z - d = \mathbf{cx}$. The values c_j related to the variables x_j, $j = 1, \ldots, m$ are on the right part of row 0. The other rows (row 1 to m) refer to the problem constraints $\mathbf{Ax} = \mathbf{b}$: the known coefficients $\mathbf{b}$ are in the first column, whereas the a_{ij} coefficients of the n variables in the m equations are on the right part.

The tableau is useful for representing and updating all the variables involved in the problem, since it avoids the wordy repeat of the same symbols in all the equations by relegating them to the row and column labels. Using the notation introduced above for highlighting basic and nonbasic variables, the tableau can also be written as follows:

	b	x_1	$\cdots$	x_m	x_{m+1}	$\cdots$	x_n
z	0	$\mathbf{c}_B^\top$			$\mathbf{c}_N^\top$		
x_1 $\vdots$ x_m	$\mathbf{b}$	$\mathbf{B}$			$\mathbf{N}$		

where we assume $d = 0$, as it does not affect the optimization process, and we refer to the m variables in basis with the labels $x_1, \ldots, x_m$.

The same operation to transform the system in canonical form with respect to a given basis $\mathbf{B}$ can be applied to the tableau, so to obtain:

	b	x_1	$\cdots$	x_m	x_{m+1}	$\cdots$	x_n
z	$-\mathbf{c}_B^\top \mathbf{B}^{-1} \mathbf{b}$	0	$\cdots$	0	$\bar{\mathbf{c}}_N^\top$		
x_1 $\vdots$ x_m	$\mathbf{B}^{-1}\mathbf{b}$	$\mathbf{I}_m$			$\mathbf{B}^{-1}\mathbf{N}$		

The tableau in canonical form shows an identity matrix in correspondence of the variables in basis. For these variables, the reduced costs are null. The value of row 0 in the first column is the opposite of the objective function in the current basic solution. The remaining values refer to the basic variables $\mathbf{x}_B$. The entering variable x_h is determined considering the reduced cost coefficients in row 0. The value the entering variable assumes is determined using the first column along with column $\mathbf{A}_h$, which is associated with the entering variable. Such column is one of the columns of the block $\mathbf{B}^{-1}\mathbf{N}$, pertaining the nonbasic variables.

Therefore, referring again to the LP problem (4.4), its corresponding formulation in standard form, and the initial tableau are:

minimize $\quad \tilde{z} = -z = -5x_1 - 3x_2$

subject to $\quad x_1 + x_2 + x_3 = 600$

$\qquad\qquad 2x_1 + x_2 + x_4 = 800$

$\qquad\qquad x_i \geq 0, \quad \forall i = 1, \ldots, 4$

	b	x_1	x_2	x_3	x_4
$\tilde{z}$	0	-5	-3	0	0
x_3	600	1	1	1	0
x_4	800	2	1	0	1

The first row refers to the objective function z, while the other rows refer to the non-basic variables, x_3 and x_4 in the initial setting, whose values are contained in the first column. The boxed elements highlight that the system is expressed in canonical form with respect to the two slack variables x_3 and x_4. The presence in row 0 of two negative coefficients -5 and -3, associated with x_1 and x_2, respectively, signals that the current solution is not optimal. Using the empirical rule based on the greater improvement in $\tilde{z}$, x_1 is chosen as entering variable. The entering variable is hence marked using an arrow ($\downarrow$) above the corresponding column label. The value to be assigned to the entering variable is determined using the ratios of the values on the first column and the values of the column of the entering variable itself, only for the positive entries:

$$\min\left\{\frac{600}{1}, \frac{800}{2}\right\} = \min\{600, 400\}$$

The value 400 is hence assigned to x_1 for entering the basis. As variable x_4 is designed to leave the basis, its row label is marked with an arrow ($\leftarrow$). The value 2, corresponding to the leaving variable row and the entering variable column, is the pivot element and is highlighted in the tableau. The pivot value is used to switch the role of variables x_1 and x_4, the former entering the basis and the latter leaving it. Following the tableau for the first iteration, along with a summary of the iteration in terms of involved variables/values:

	b	x_1	x_2	x_3	x_4
z	0	-5	-3	0	0
x_3	600	1	1	1	0
$\leftarrow x_4$	800	2	1	0	1

iteration	: 1
entering variable	: x_1
leaving variable	: x_4
pivot	: 2
current solution	: $x_3 = 600, x_4 = 800$
objective function	: 0

Next operation expresses the tableau with respect to the new basis (x_3, x_1). To this end the pivoting operation is carried out. The pivot row is updated by dividing it for the pivot element, so to obtain a value 1 for the corresponding entry:

$$\frac{(800 \mid 2, 1, 0, 1)}{2} = \left(400 \;\middle|\; 1, \frac{1}{2}, 0, \frac{1}{2}\right).$$

The resulting row is multiplied by the value in the pivot column of the other row (1 in this example) and subtracted from the other row. The goal is to obtain a 0 on the other element of the pivot column:

$$(0 \mid 1, 1, 1, 0) - 1 \times \left(400 \;\middle|\; 1, \frac{1}{2}, 0, \frac{1}{2}\right) = \left(-2000 \;\middle|\; 0, \frac{1}{2}, 1, -\frac{1}{2}\right).$$

The same operation is carried out on row 0 in order to have a null reduced cost for the entering variable:

$$(-5, -3, 0, 0) - 5 \times \left(1, \frac{1}{2}, 0, \frac{1}{2}\right) = \left(0, -\frac{1}{2}, 0, \frac{5}{2}\right).$$

At the end of the pivoting operation, x_4 and x_1 switch their role. The new tableau is:

	b	x_1	x_2	x_3	x_4
$\tilde{z}$	-2000	0	$-1/2$	0	$5/2$
x_3	200	0	$1/2$	1	$-1/2$
x_1	400	1	$1/2$	0	$1/2$

This tableau is now in canonical form with respect to the new basis (x_3, x_1), as highlighted by the boxed elements.

The new tableau is used for the second iteration, carrying out the same steps: the optimal condition is not satisfied and x_2 is the only eligible variable to enter the basis, because its reduced cost is the only one having a negative value. The leaving variable is now x_3, whose ratio $\frac{200}{1/2}$ is minimum. Therefore the pivot element is $\frac{1}{2}$. The tableau becomes:

	b	x_1	x_2	x_3	x_4
$\tilde{z}$	-2000	0	$-1/2$	0	$5/2$
$\leftarrow x_3$	200	0	$1/2$	1	$-1/2$
x_1	400	1	$1/2$	0	$1/2$

iteration   : 2
entering variable   : x_2
leaving variable   : x_3
pivot   : $1/2$
current solution   : $x_3 = 200, x_1 = 400$
objective function : -2000

The pivoting operation is carried out accordingly:

- the pivot row (first row) is divided by the pivot element $(1/2)$,

- the new pivot row is multiplied by $\frac{1}{2}$ and then subtracted from the second row,

- the new pivot row is multiplied by $-\frac{1}{2}$ and subtracted from the row 0.

The resulting vector of reduced costs satisfies the optimality condition, as all the costs are non-negative, and the algorithm stops:

	b	x_1	x_2	x_3	x_4
$\tilde{z}$	-2200	0	0	1	2
x_2	400	0	1	2	-1
x_1	200	1	0	-1	1

iteration   : 3
entering variable   : —
leaving variable   : —
pivot   : —
current solution   : $x_2 = 400, x_1 = 200$
objective function : -2200

Using the matrix formulation of the simplex algorithm, the solution $\mathbf{x}_B = \begin{bmatrix} x_2 \\ x_1 \end{bmatrix} = \begin{bmatrix} 400 \\ 200 \end{bmatrix}$, corresponding to corner $\mathcal{E}$ in Figure 4.8, provides the optimal value $z = 2200$ for the objective function.

Using the above notation, it is now possible to detail the simplex algorithm in its tableau formulation:

1: $\mathbf{B} \leftarrow$ initial basis   ▷ assume $\mathbf{B}$ is an initial feasible basic solution
2: $T \leftarrow$ initial tableau   ▷ corresponding tableau
3: *optimal* $\leftarrow$ **false**   ▷ flag variable for checking optimality
4: *unbound* $\leftarrow$ **false**   ▷ flag variable for checking unboundedness
5: **while** *optimal* = **false and** *unbound* = **false do**
6:     **if** $T[0, j] \geq 0, \forall j \in 1, \dots, n$ **then**
7:         *optimal* $\leftarrow$ **true**
8:     **else**
9:         $h : \mathrm{argmin}_{T[0,j]<0} T[0, j], \forall j$ related to nonbasic variables   ▷ index of the entering variable

```
10:        if T[i, h] ≤ 0, ∀i = 1, … , m then
11:            unbound ← true
12:        else
13:            k : argmin_{T[i,h]>0} T[i,0]/T[i,h], ∀i ∈ (1, … , m)          ▷ index of the leaving variable
14:            PIVOTING(T, k, h)                                              ▷ procedure for updating the tableau
15:        end if
16:    end if
17: end while
18: procedure PIVOTING(T, k, h)                                ▷ procedure for carrying out the pivoting operation
19:    m ← nRows(T)                                            ▷ compute the number of rows of the tableau
20:    n ← nCols(T)                                            ▷ compute the number of columns of the tableau
21:    pivot ← T[k, h]                                         ▷ pivot element
22:    for j = 1 to n do                    ▷ compute the new pivot row dividing it for the pivot element
23:        T[k, j] ← T[k, j]/pivot
24:    end for
25:    for i = 0 to m do                    ▷ for all the rows (row 0 is related to the objective function)
26:        if i ≠ k and T[i, h] ≠ 0 then    ▷ except for the pivot row and when the entry on the pivot
                                                column is already null
27:            entryPivotColumn ← T[i, k]                      ▷ store the element on the pivot column
28:            for j = 1 to n do            ▷ subtract from each row the new pivot row multiplied
                                                for the element on the pivot column
29:                T[i, j] ← T[i, j] − entryPivotColumn * T[k, j]
30:            end for
31:        end if
32:    end for
33: end procedure
```

The tableau algorithm in action on the 3-D example

Consider again the 3–D problem (4.7), depicted in Figures 4.6 and 4.7. Its formulation in standard form follows:

$$\text{minimize} \quad \tilde{z} = -z = -300x_1 - 200x_2 - 350x_3$$

$$\text{subject to} \quad x_1 + x_2 + x_3 + x_4 = 40$$

$$4x_1 + 2x_2 + 3x_3 + x_5 = 120$$

$$x_3 + x_6 = 20$$

$$x_i \geq 0, \quad \forall i = 1, \dots, 6.$$

The associated tableau is in canonical form with respect to the basis (x_4, x_5, x_6) composed of the three slack variables introduced to express the three constraints in terms of equations:

	b	x_1	x_2	x_3	x_4	x_5	x_6
$\tilde{z}$	0	-300	-200	-350	0	0	0
x_4	40	1	1	1	1	0	0
x_5	120	4	2	3	0	1	0
x_6	20	0	0	1	0	0	1

The initial solution $\mathbf{x}_B = \begin{bmatrix} x_4 \\ x_5 \\ x_6 \end{bmatrix} = \begin{bmatrix} 40 \\ 120 \\ 20 \end{bmatrix}$ and $\mathbf{x}_N = \begin{bmatrix} x_1 \\ x_2 \\ x_3 \end{bmatrix} = \begin{bmatrix} 0 \\ 0 \\ 0 \end{bmatrix}$ clearly corresponds to

the corner $\mathcal{O}$ in Figure 4.7. It does not satisfy the optimality condition, because there are three negative coefficients in the row 0 of the tableau. The entering variable at the first iteration is x_3, whose coefficient $c_3 = -350$ is minimum. Considering the ratios of each element in the first column to its corresponding non–negative entry in the column of the entering variable:

$$\min\left\{ \frac{40}{1}, \frac{120}{3}, \frac{20}{1} \right\} = \min\{40, 40, 20\}$$

x_6 is selected as leaving variable. Therefore, the first iteration of the algorithm (and its representation in the tableau) is:

iteration	: 1
entering variable	: x_3
leaving variable	: x_6
pivot	: 1
current solution	: $x_4 = 40, x_5 = 120, x_6 = 20$
objective function	: 0

	b	x_1	x_2	x_3	x_4	x_5	x_6
$\tilde{z}$	0	−300	−200	−350	0	0	0
x_4	40	1	1	1	1	0	0
x_5	120	4	2	3	0	1	0
← x_6	20	0	0	1	0	0	1

Being the pivot element equal to 1, the pivot row does not change, while the other rows are transformed in order to obtain all zeros in the pivot column. In particular, the pivoting operation is

- the pivot row is multiplied by -350 and then subtracted from row 0,

- the pivot row is subtracted from row 1,

- the pivot row is multiplied by 3 and then subtracted from row 2.

The resulting tableau is now in canonical form with respect to the basis (x_4, x_5, x_6) and it is used for the second iteration

iteration	: 2
entering variable	: x_1
leaving variable	: x_5
pivot	: 4
current solution	: $x_4 = 20, x_5 = 60, x_3 = 20$
objective function	: -7000

$\downarrow$

	b	x_1	x_2	x_3	x_4	x_5	x_6
$\tilde{z}$	-7000	-300	-200	0	0	0	350
x_4	20	1	1	0	1	0	-1
$\leftarrow x_5$	60	4	2	0	0	1	-3
x_3	20	0	0	1	0	0	1

The current solution $\mathbf{x}_B = \begin{bmatrix} x_4 \\ x_5 \\ x_3 \end{bmatrix} = \begin{bmatrix} 20 \\ 60 \\ 20 \end{bmatrix}$ and $\mathbf{x}_N = \begin{bmatrix} x_1 \\ x_2 \\ x_6 \end{bmatrix} = \begin{bmatrix} 0 \\ 0 \\ 0 \end{bmatrix}$ corresponds to the cor-

ner $D(x_1 = 0, x_2 = 0, x_3 = 20)$ in Figure 4.7.

The variable x_1, whose coefficient on row 0 is minimum, is selected as entering variable, and it replaces the leaving variable x_5, for which the ratio of the element of the first column to the value in the pivot column, $60/4 = 15$, is minimum. The pivot element is now 4, and the pivoting operation consists of the following operations:

- the pivot row is divided by the pivot element:

$$\frac{(60 \mid 4, 2, 0, 0, 1, -3)}{4} = \left(15 \mid 1, \frac{1}{2}, 0, 0, \frac{1}{4}, -\frac{3}{4} \right)$$

- the new pivot row is multiplied by -300 and then subtracted from row 0,

- the new pivot row is subtracted from row 1.

The resulting tableau is now in canonical form with respect to the basis (x_4, x_1, x_3) and is used in iteration 3:

iteration   : 3
entering variable   : x_2
leaving variable   : x_4
pivot   : $1/2$
current solution   : $x_4 = 5, x_1 = 15, x_3 = 20$
objective function   : -11500

$\downarrow$

	b	x_1	x_2	x_3	x_4	x_5	x_6
$\tilde{z}$	-11500	0	-50	0	0	75	125
$\leftarrow x_4$	5	0	$1/2$	0	1	$-1/4$	$-1/4$
x_1	15	1	$1/2$	0	0	$1/4$	$-3/4$
x_3	20	0	0	1	0	1	1

The current basic solution $\begin{bmatrix} x_4 \\ x_1 \\ x_3 \end{bmatrix} = \begin{bmatrix} 5 \\ 15 \\ 20 \end{bmatrix}$, with the variable x_2 out of the basis and hence equal to 0, corresponds to the corner $\mathcal{E}(x_1 = 15, x_2 = 0, x_3 = 20)$ in Figure 4.7.

The only variable with negative coefficient in row 0 is now x_2 and is selected as entering variable. The ratios of the coefficients in the first column to the non–negative values of the pivot column:

$$\min\left\{\frac{5}{1/2}, \frac{20}{1/2}\right\} = \min\{10, 40\}$$

imply x_4 be the leaving variable. The pivot element is now $1/2$, and the pivoting operation is:

- the pivot row is divided by $1/2$,

$$\frac{(5 \mid 0, 1/2, 0, 1, -1/4, -1/4)}{1/2} = \left(10 \,\middle|\, 0, 1, 2, 0, -\frac{1}{2}, -\frac{1}{2}\right)$$

- the new pivot row is multiplied by -50 and subtracted by row 0,

- the new pivot row is multiplied by $1/2$ and subtracted by row 2.

The new tableau is in canonical form with respect to the basis (x_2, x_1, x_3), and the algorithm enters iteration 4:

iteration	: 4
entering variable	: —
leaving variable	: —
pivot	: —
current solution	: $x_2 = 10, x_1 = 10, x_3 = 20$
objective function	: -12000

	b	x_1	x_2	x_3	x_4	x_5	x_6
$\tilde{z}$	-12000	0	0	0	100	50	100
x_2	10	0	1	0	2	$-1/2$	$-1/2$
x_1	10	1	0	0	-1	$1/2$	$-1/2$
x_3	20	0	0	1	0	0	1

The current basic solution $\begin{bmatrix} x_2 \\ x_1 \\ x_3 \end{bmatrix} = \begin{bmatrix} 10 \\ 10 \\ 20 \end{bmatrix}$, corresponding to the corner $\mathcal{F}$ in Figure 4.7, is optimal, as all the coefficient in row 0 are non–negative. The objective function in the corner $\mathcal{F}$ is $\tilde{z} = -12000$.

A special case: an unbounded objective function

Consider the unbounded objective function of the LP problem (4.6) depicted in Figure 4.5. The standard form requires the introduction of three artificial variables, x_3, x_4, x_5, to transform the three inequalities into equalities:

minimize $\tilde{z} = -z = -4x_1 + 3x_2$

subject to $-2x_1 + x_2 + x_3 = 4$

$x_1 - 2x_2 + x_4 = 11$

$x_1 - x_2 + x_5 = 8$

$x_i \geq 0, \quad \forall i = 1, \dots, 5$

	b	x_1	x_2	x_3	x_4	x_5
z	0	−4	3	0	0	0
x_3	4	−2	1	1	0	0
x_4	11	1	−2	0	1	0
x_5	8	1	−1	0	0	1

The resulting tableau is in canonical form with respect to the three artificial variables, which are the initial solution triggering the simplex algorithm. The initial solution corresponds to the corner $\mathcal{O}$ in Figure 4.5. The first entering variable is x_1, because its cost coefficient in row 0 is minimum. The leaving variable has to be selected between x_4 and x_5, because the value in the pivot column associated with the variable x_3 is negative. The ratio between the coefficients of the first and second column is minimum in correspondence of x_5; hence x_5 is selected to leave the basis. The pivot element is therefore equal to 1, corresponding to the value at the cross of the first column with the last row. The first iteration is summarized through the tableau as follows:

	b	x_1	x_2	x_3	x_4	x_5
z	0	−4	3	0	0	0
x_3	4	−2	1	1	0	0
x_4	11	1	−2	0	1	0
← x_5	8	1	−1	0	0	1

iteration : 1
entering variable : x_1
leaving variable : x_5
pivot : 1
current solution : $x_3 = 4$
: $x_4 = 11$
: $x_5 = 8$
objective function : 0

The pivoting operation is

- the pivot row (last row) does not change as the pivot value is already equal to 1,

- the pivot row is multiplied by −4 and subtracted from row 0,

- the pivot row is multiplied by −2 and subtracted from row 1,

- the pivot row is subtracted from row 2.

The resulting tableau, in canonical form with respect to the basis (x_3, x_4, x_1) follows:

	b	x_1	x_2	x_3	x_4	x_5
z	-32	0	-1	0	0	4
x_3	20	0	-1	1	0	2
x_4	3	0	-1	0	1	-1
x_1	8	1	-1	0	0	1

iteration	: 2
entering variable	: x_2
leaving variable	: —
pivot	: —
current solution	: $x_3 = 20$
	: $x_4 = 3$
	: $x_1 = 8$
objective function	: -32

There is only a negative coefficient in row 0 and hence only one variable, x_2, eligible to enter the basis. The pivot column contains only negative coefficients and therefore the problem is characterized by an unbounded objective function, as already shown through the graphical method (see Figure 4.5).

4.3   The two–phase method

All the above examples used to show the simplex algorithm are of the form:

$$\text{minimize}\{z = \mathbf{c}^T\mathbf{x} \,:\, \mathbf{Ax} \le \mathbf{b}, \text{with } \mathbf{x} \ge \mathbf{0}\}, \mathbf{b} \ge \mathbf{0}.$$

The initial solution to start the simplex algorithm can be easily obtained by transforming the system in standard form:

$$\text{minimize}\{z = \mathbf{c}_N^T\mathbf{x}_N + \mathbf{0}^T\mathbf{x}_B \,:\, \mathbf{Nx}_N + \mathbf{I}_m\mathbf{x}_B \le \mathbf{b}, \text{with } \mathbf{x}_N \ge \mathbf{0}, \mathbf{x}_B \ge \mathbf{0}\}, \mathbf{b} \ge \mathbf{0}$$

where $\mathbf{x}_N$ contains the initial variables and $\mathbf{x}_B$ the m slack variables introduced to transform the inequalities constraints into equalities. The new system is in canonical form with respect to $\mathbf{x}_N$. The initial solution for activating the algorithm is immediately available, and, from a geometrical point of view, it corresponds to the origin: as the m slack variables are in basis, then $\mathbf{x}_N = \mathbf{0}$. This initial basic solution is not available when the condition $\mathbf{b} \ge \mathbf{0}$ does not hold, and/or when one or more constraints contain a greater–than inequality and/or an equality. In the last case, as shown above, the equality is replaced by two opposite inequalities (see section 4.1).

To this end, consider the following simple LP problem with two decisional variables, graphically represented in Figure 4.10:

$$\begin{aligned}
\text{maximize} \quad & z = 5x_1 + 6x_2 \\
\text{subject to} \quad & 0.4x_1 + 0.6x_2 \le 12 \quad (C_1) \\
& x_1 + x_2 \le 24 \quad (C_2) \\
& x_1 + x_2 \ge 6 \quad (C_3) \\
& x_1 \ge 0 \quad (C_4) \\
& x_2 \ge 0 \quad (C_5).
\end{aligned}$$

$$(4.11)$$

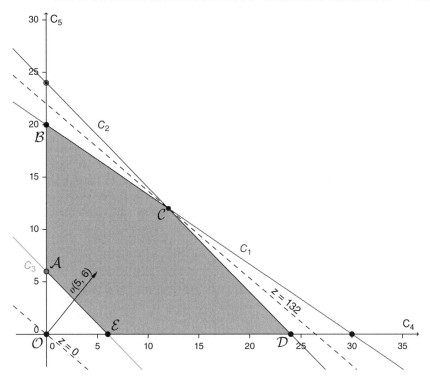

Figure 4.10 Geometrical representation of the LP programming problem (4.11). The presence of a greater–than inequality (C_3) renders that the axis origin $\mathcal{O}$ is unsuitable as starting solution for activating the simplex algorithm, as it is not a feasible point.

The 2–D problem can be easily represented and solved using the graphical approach introduced in section 4.1.3. The coordinates of the five basic solutions are indeed computable by solving the following five systems of two equations, each expressing the intersection of two of the problem constraints and determining the corners of the feasible region:

$$\mathcal{A}(0,6) : \begin{cases} x_1 + x_2 = 6 \\ x_2 = 0 \end{cases} \qquad \mathcal{B}(0,20) : \begin{cases} 0.4x_1 + 0.6x_2 = 12 \\ x_2 = 0 \end{cases}$$

$$\mathcal{C}(12,12) : \begin{cases} 0.4x_1 + 0.6x_2 = 12 \\ x_1 + x_2 = 6 \end{cases} \qquad \mathcal{D}(24,0) : \begin{cases} x_1 = 0 \\ x_1 + x_2 = 24 \end{cases}$$

$$\mathcal{E}(6,0) : \begin{cases} x_1 = 0 \\ x_1 + x_2 = 6 \end{cases}$$

The optimal solution z^* is obtained by computing the objective function z in the five corners of the polygon $\mathcal{ABCDE}$:

$$z_A = 36 \qquad z_B = 120 \qquad \boxed{z_C = 132 = z^*}$$

$$z_D = 120 \qquad z_{\mathcal{E}} = 30$$

and corresponds to the corner C, where the objective function assumes the value 132, as shown in Figure 4.7.

To apply the simplex algorithm to this example, three artificial variables have to be considered to express the problem in standard form. The presence of a greater–than inequality (C_3) implies that the axis origin $\mathcal{O}$ cannot be used as starting solution for activating the simplex algorithm, as it is not a feasible point. This leads to the presence of -1 in the last row, which entails that the corresponding tableau is not in canonical form with respect to the three artificial variables x_3, x_4, x_5:

minimize $\quad \tilde{z} = -z = -5x_1 - 6x_2$

subject to $\quad 0.4x_1 + 0.6x_2 + x_3 = 12$

$\qquad\qquad x_1 + x_2 + x_4 = 24$

$\qquad\qquad x_1 + x_2 - x_5 = 6$

$\qquad\qquad x_i \geq 0, \quad \forall i = 1, \dots, 5$

	b	x_1	x_2	x_3	x_4	x_5
$\tilde{z}$	0	-5	-6	0	0	0
x_3	12	0.4	0.6	1	0	0
x_4	24	1	1	0	1	0
x_5	6	1	1	0	0	-1

To obviate this problem, the two–phase method resorts to a twofold application of the simplex algorithm: the first to detect an initial basis, if it exists, and the second to look for the optimal solution. The method consists in transforming the original problem in the following artificial problem:

$$\text{minimize} \left\{ w = \sum_{i=1}^{m} y_i \; : \; \mathbf{Ax} + \mathbf{I}_m \mathbf{y} = \mathbf{b}, \text{with } \mathbf{x} \geq \mathbf{0}, \mathbf{y} \geq \mathbf{0} \right\},$$

through the addition of the artificial variables y_i. The artificial problem shares the same constraints of the initial problem. In addition to the slack variables used to transform inequalities into equalities, a positive artificial variable with unitary coefficient is inserted for each constraint $\geq$. The same is done for each equality constraint, even in case no transformations are required for this type of constraints (and hence no slack variables have to be inserted). The condition $\mathbf{b} > \mathbf{0}$ is no more impelling, as it is sufficient to multiply the involved inequalities by -1, to invert their directions. It is worth stressing the different role of slack variables and artificial variables. The former do not alter the initial problem, since their only task is to change inequalities into equalities. The latter, whereas they are different from 0, change the problem into a different one.

The addition of the artificial variables entails a basis in the matrix of the coefficients of the slack and of artificial variables. The corresponding system consists of m equations, one for each constraint, and $n_1 + n_2 + n_3 + n_4$ variables, and in particular:

- n_1 original variables,

- n_2 slack variables ($0 \leq n_2 \leq m$) with positive coefficients, one for each constraint of type $\leq$,

- n_3 slack variables ($0 \leq n_3 \leq m$) with negative coefficients, associated with the constraint of type $\geq$,

- n_4 artificial variables ($0 \leq n_4 \leq m$), associated with the constraints of type $=$ or $\geq$.

In the following, we continue to denote by n the total number of variables, including the original variables, the slack variables, and the artificial variables, unless there is ambiguity.

The transformation of the original system determines a system of equations in canonical form with respect to the slack and/or artificial variables. The first basic solution is indeed detected by looking for an identity matrix in the columns of the $n_2 + n_3 + n_4$ slack and artificial variables. The original variables, along with the slack variables with negative sign, are treated as nonbasic variables, while the positive slack variables and the artificial variables are considered as basic variables. The use of the artificial variables extends the feasible set to a new feasible set, called the extended feasible set. As an example, the LP problem (4.11) requires the addition of a single artificial variable in correspondence of the third constraint to obtain a canonical form:

$$\text{minimize} \quad \tilde{z} = -z = -5x_1 - 6x_2$$
$$\text{subject to} \quad 0.4x_1 + 0.6x_2 + x_3 = 12$$
$$x_1 + x_2 + x_4 = 24$$
$$x_1 + x_2 - x_5 + y_1 = 6$$
$$x_i \geq 0, \quad \forall i = 1, \dots, 5$$
$$y_1 \geq 0.$$

The artificial objective function minimized in the first phase coincides in this example with the artificial variable y_1. Starting from the third constraint, the artificial objective function can be easily formulated in terms of the original and slack variables:

$$\text{minimize} \ w = \sum_{i=1}^{m} y_i = y_1 = -x_1 - x_2 + x_5 + 6.$$

Since y_1 has to be non–negative (recall the LP non–negativity constraint), the optimal solution of this artificial problem is $w = 0$, obtained when the artificial variable $y_1 = 0$. With such solution, the problem is again inside the original feasible set.

In order to apply the two–phase method, the tableau is slightly modified with the insertion of two rows 0, one for the artificial problem and one for the original problem. The former is used for the first phase and the latter for the second phase. The pivoting operations are carried out on both rows during the first phase. At the end of the first phase, the objective function of the second phase (second row 0) will be expressed in canonical form and the basic variables will correspond to the optimal solution detected in the first phase. Returning to the example, the introduction of the artificial variable y_1 generates a canonical form with respect to the basis (x_3, x_4, y_1), composed of the two slack variables x_3 and x_4, along with the artificial variable y_1:

	b	x_1	x_2	x_3	x_4	x_5	y_1	
w	-6	-1	-1	0	0	1	0	row 0 (phase 1)
$\tilde{z}$	0	-5	-6	0	0	0	1	row 0 (phase 2)
x_3	12	0.4	0.6	1	0	0	0	
x_4	24	1	1	0	1	0	0	
y_1	6	1	1	0	0	-1	1	

The extended domain is represented in the space (x_1, x_2, y_1) in Figure 4.11. The basic solution $\begin{bmatrix} x_3 \\ x_4 \\ y_1 \end{bmatrix} = \begin{bmatrix} 12 \\ 24 \\ 6 \end{bmatrix}$ corresponds to the corner $F(x_1 = 0, x_2 = 0, y_1 = 6)$ in the extended feasible set, because both x_1 and x_2 are out of the basis and therefore equal to 0. A basic solution in which the artificial variable is out of the basis, and hence null, has to be detected to force the path of the algorithm to the original feasible set. Such a solution is indeed a feasible basic solution also for the original feasible set and it can be exploited as starting point of the second phase. By analyzing row 0 (first phase), the optimal condition is not satisfied as the coefficients of the variables x_1 and x_2 are both negative. Since the two coefficients coincide, it is indifferent which of the two variables is selected to enter the basis. If we choose x_2, the leaving variable is y_1, whose ratio between the elements of the first column and the non–negative coefficients of the pivot column is minimum:

$$\min\left\{\frac{12}{0.6}, \frac{24}{1}, \frac{6}{1}\right\} = \min\{20, 24, 6\}$$

Therefore, by choosing x_2 as entering variable, the first iteration of the first phase is:

iteration   : 1 (first phase)
entering variable   : x_2
leaving variable   : y_1
pivot   : 1
current solution   : $x_3 = 12, x_4 = 24, y_1 = 6$
objective function   : 0

	b	x_1	x_2	x_3	x_4	x_5	y_1
w	-6	-1	-1	0	0	1	0
$\tilde{z}$	0	-5	-6	0	0	0	1
x_3	12	0.4	0.6	1	0	0	0
x_4	24	1	1	0	1	0	0
$\leftarrow y_1$	6	1	1	0	0	-1	1

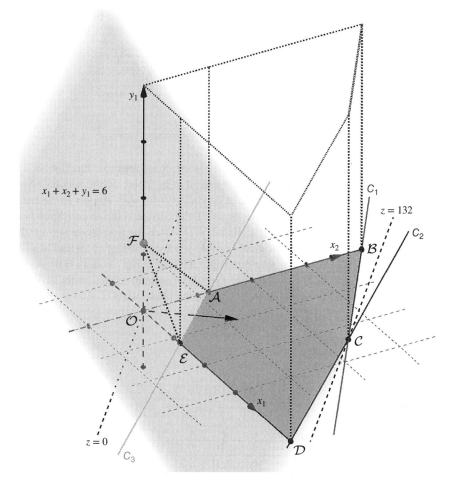

Figure 4.11   Extended feasible set of the LP programming problem (4.11), obtained inserting an artificial variable y_1 in correspondence of the second constraint (C_2), of the type $\geq$. The corner $\mathcal{F}$ is used as starting point of the two–phase method. The two–phase method forces the path of the algorithm to return to the original feasible set.

The pivoting operation consists of the following steps:

- The pivot row does not change as the pivot element is 1,

- The row 0 corresponding to the artificial objective function w is added to the pivot row,

- The pivot row is multiplied by -6 and then subtracted from the row 0 corresponding to the original objective function $\bar{z} = -z$,

- The pivot row is subtracted from the second row.

The resulting tableau satisfies the optimality condition ending the first phase:

	b	x_1	x_2	x_3	x_4	x_5	y_1
w	0	0	0	0	0	0	1
$\tilde{z}$	-36	1	0	0	0	-6	7
x_3	8.4	-0.2	0	1	0	0.6	-0.6
x_4	18	0	0	0	1	1	-1
x_2	6	1	1	0	0	-1	1

The artificial variable y_1 is out of the basis; the artificial objective function $w = 0$ and the current basic solution can be used as starting point of the second phase of the algorithm. The tableau is indeed in canonical form with respect to the basis (x_3, x_4, x_2). Since only x_2 is in the basis ($x_2 = 6$), the current solution corresponds to corner $\mathcal{A}(x_1 = 0, x_2 = 6)$ in Figures 4.10 and 4.11. Once the first phase is over, the algorithm returns to explore the original feasible set. By choosing x_1 instead of x_2 as entering variable, the algorithm would move to the corner $\mathcal{E}(x_1 = 6, x_2 = 0)$. At the end of the first phase, both row 0 (phase 1) and the column related to the artificial variable y_1 are deleted from the tableau. The second phase starts:

iteration	: 1 (second phase)
entering variable	: x_5
leaving variable	: x_3
pivot	: 0.6
current solution	: $x_3 = 8.4, x_4 = 18, x_2 = 6$
objective function	: -36

	b	x_1	x_2	x_3	x_4	x_5
$\tilde{z}$	-36	1	0	0	0	-6
$\leftarrow x_3$	8.4	-0.2	0	1	0	0.6
x_4	18	0	0	0	1	1
x_2	6	1	1	0	0	-1

This solution does not satisfy the optimality condition, and the only eligible variable for entering the basis is x_5, whose coefficient is negative. The variable x_3, whose ratio $8.4/0.6 = 14$ is minimum, leaves the basis and therefore the pivot element is 0.6. Next we switch the role of x_5 and x_3;

- the pivot row is divided by 0.6 in order to have 1 in the pivot position,

$$\frac{(8.4 \mid -0.2, 0, 1, 0, 0.6)}{0.6} = \left(14 \mid -\frac{1}{3}, 0, \frac{5}{3}, 0, 1\right)$$

- the new pivot row is multiplied by -6 and subtracted from row 0,
- the new pivot is subtracted from the second row and added to the third row.

The resulting tableu is:

iteration	: 2 (second phase)
entering variable	: x_1
leaving variable	: x_4
pivot	: 1/3
current solution	: $x_5 = 14, x_4 = 4, x_2 = 20$
objective function	: -120

	b	x_1	x_2	x_3	x_4	x_5
$\tilde{z}$	-120	-1	0	10	0	0
x_5	14	$-1/3$	0	5/3	0	1
$\leftarrow x_4$	4	1/3	0	$-5/3$	1	0
x_2	20	2/3	1	5/3	0	0

This solution, with $x_1 = 0$ and $x_2 = 20$, corresponds to the corner $\mathcal{B}$ in Figures 4.10 and 4.11. The only variable with negative coefficient is x_1, and is then selected to enter the basis. The variable x_4, whose ratio $\frac{4}{1/3} = 12$ is minimum, leaves the basis. The pivot element is $1/3$. Next we move x_1 in the basis:

- the pivot row is multiplied by 3,

- the new pivot row is added to row 0,

- the new pivot row is multiplied by $-1/3$ and subtracted from the first row,

- the new pivot row is multiplied by $2/3$ and subtracted from the last row.

The resulting tableau is optimal, as all the reduced costs are positive:

iteration	: 3 (second phase)
entering variable	: —
leaving variable	: —
pivot	: —
current solution	: $x_5 = 18, x_1 = 12, x_2 = 12$
objective function	: -132

	b	x_1	x_2	x_3	x_4	x_5
$\tilde{z}$	-132	0	0	5	3	0
x_5	18	0	0	0	1	1
x_1	12	1	0	-5	3	0
x_2	12	0	1	5	-2	0

The optimal solution corresponds to corner $\mathcal{C}$, as in the graphical method. In Figure 4.11, the path of the two–phase method is through the corners $\mathcal{F} \rightarrow \mathcal{A} \rightarrow \mathcal{B} \rightarrow \mathcal{C}$. It would have been instead $\mathcal{F} \rightarrow \mathcal{E} \rightarrow \mathcal{D} \rightarrow \mathcal{C}$ if the variable x_1 becomes the entering variable in iteration 1 in the first phase.

To summarize, in case of presence of $\geq$ inequalities and/or equalities, and/or when the condition $\mathbf{b} \geq 0$ is not satisfied, the initial problem is extended with the introduction of an adequate number of artificial variables. The extended problem is in canonical form with respect to a basis in which the artificial variables are included, along with the slack variables associated to the $\leq$ inequalities. The objective function can be expressed in canonical form, deriving the expression of each artificial variable y_i in terms of the variables not belonging to the basis. The first phase of the simplex algorithm minimizes the artificial objective function $w = \sum_{i=1}^{m} y_i$. Let us denote by w^* the optimal value, and by $(\mathbf{x}^*, \mathbf{y}^*)$ the corresponding solution. The following two cases are possible:

$w^* \neq 0$: in this case, there does not exist any solution to the artificial problem for which all the artificial variables are null. Hence the original problem does not admit solution;

$w^* = 0$: in this case $\mathbf{y}^* = 0$, because $\mathbf{y}^* \geq 0$ for the LP non–negativity conditions and $\sum_{i=1}^{m} y_i^* = w^* = 0$. Therefore, $\mathbf{x}^*$ is a feasible basic solution of the initial problem and can be exploited as initial solution for the second phase. Two further possibilities can occur:

- all the artificial variables are nonbasic variables in the detected solution w^*: the artificial variables can be deleted along with the artificial objective function; $\mathbf{x}^*$ can be used as initial feasible solution for the second phase of the algorithm. This is the case of the previous example shown above;

- one or more artificial variables are still in basis, with null value. From a geometrical point of view, in such a case the optimal solution lies in correspondence of a degenerate corner of the extended feasible set. There are more feasible basic solutions for this corner: some solutions present null artificial variables among the basic variables, and some solutions do not. Once the artificial variables are excluded, the detected corner corresponds to a corner of the initial feasible set. To obtain a feasible basic solution for the initial problem, the artificial variables have to leave the basis. To this end, they have to be replaced by the nonbasic variables. From an algebraic point of view, the simplex method can be applied taking care that the artificial variables still in basis do not assume positive values. Therefore, once the entering variable x_h is detected, the leaving variable is determined as follows:

  – if at least one element of the column $\mathbf{a}_h$, corresponding to the artificial basic variable in basis, is still negative, then this variable is forced to leave the basis,

  – if some element of the column $\mathbf{a}_h$, corresponding to an artificial variable, is positive, then this variable will automatically leave the basis.

4.4   Convergence and degeneration of the simplex algorithm

The number of basic solutions in a LP problem is at most equal to $\binom{n}{m}$, as shown in subsection 4.2.1. Although such number could be huge in many real applications, this condition ensures the convergence of the simplex algorithm in a finite number of iterations, except when the algorithm visits the same corner more than one time, triggering cycles and not converging to the optimal solution. If in each iteration the leaving variable is univocally determined, that is, the ratio $\mathbf{B}^{-1}\mathbf{b}/\overline{\mathbf{A}}_h$ is minimum in correspondence of only one entry, then the objective function decreases in a strictly monotone way, ensuring that no corners can be visited more than once. When this condition is not verified, the LP problem, and the corresponding feasible basic solutions, are known as degenerates. In some cases, visiting a degenerate feasible basic solution does not affect the convergence of the algorithm, although it is possible that the algorithm cyclically visits the same basis and does not converge.

As an example of a degenerate solution, consider the following problem:

$$\text{minimize} \quad z = x_1 + 2x_2$$

$$\text{subject to} \quad x_1 + x_2 \leq 5$$

$$- x_1 + x_2 \leq 3$$

$$x_1 + 2.5x_2 \leq 11 \tag{4.12}$$

$$x_1 \geq 0$$

$$x_2 \geq 0.$$

After the addition of three slack variables, one for each inequality, the system is in canonical form with respect to the basis (x_3, x_4, x_5), and its tableau representation is:

	b	x_1	x_2	x_3	x_4	x_5
z	0	-1	-2	0	0	0
x_3	5	1	1	1	0	0
$\leftarrow x_4$	3	-1	1	0	1	0
x_5	11	1	2.5	0	0	1

Tableau 1

The variable x_2 enters the basis, and the variable x_4 leaves the basis. The tableau resulting from the pivot operation is:

	b	x_1	x_2	x_3	x_4	x_5
z	6	-3	0	0	2	0
$\leftarrow x_3$	2	2	0	1	-1	0
x_2	3	-1	1	0	1	0
$\leftarrow x_5$	3.5	3.5	0	0	-3.5	1

Tableau 2

The only variable eligible to enter the basis is x_1, and any of x_3 and x_5 could be selected as leaving variable, because the two corresponding ratios $\mathbf{B}^{-1}b_i/\overline{\mathbf{A}}_{h_i}$, $i = (3, 5)$ are equal (see the two arrows next to the rows of the tableau). The choice between the two variables is arbitrary; the choice of x_3 implies the following tableau:

	b	x_1	x_2	x_3	x_4	x_5
z	9	0	0	1.5	0.5	0
x_1	1	1	0	0.5	−0.5	0
x_2	4	0	1	0.5	0.5	0
x_5	0	0	0	−1.75	−0.75	1
			Tableau 3			

The solution is optimal even if it is degenerate, because variable x_5 is a basic solution with null value. In fact, a basic solution with a null variable in basis is known as degenerate basis. From a geometric point of view, a degenerate corner is due to the presence of a redundant constraint. This is evident from Figure 4.12, showing the graphical solution for the LP problem (4.12): the redundancy of constraint C_3 stands out, in that it does not affect the feasible region.

The previous example shows that the algorithm still converges to the optimal solution. An example for cyclicality, proposed in Beale (1955), is:

$$\text{minimize} \quad z = \frac{3}{4}x_1 - 150x_2 + \frac{1}{50}x_3 - 6x_4$$

$$\text{subject to} \quad \frac{1}{4}x_1 - 60x_2 - \frac{1}{25}x_3 + 9x_4 \le 0$$

$$\frac{1}{2}x_1 - 90x_2 - \frac{1}{50}x_3 + 3x_4 \le 0$$

$$x_3 \le 1$$

$$x_i \ge 0, \quad \forall i = 1, \dots, 4.$$

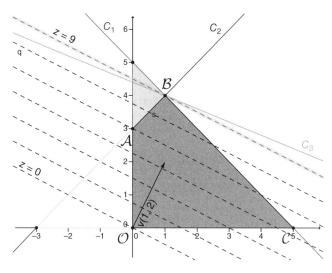

Figure 4.12 Graphical solution of the LP programming problem (4.12). The constraint C_3 is redundant because it does not affect the feasible region.

The application of the simplex algorithm on this problem triggers a cycle, as shown in the following tableaus (details for each iteration are omitted):

	b	x_1	x_2	x_3	x_4	x_5	x_6	x_7
z	0	−0.75	150	0.02	6	0	0	0
← x_5	0	0.25	−60	−0.04	9	1	0	0
x_6	0	0.5	−90	−0.02	3	0	1	0
x_7	1	0	0	1	0	0	0	1

Tableau 1

	b	x_1	x_2	x_3	x_4	x_5	x_6	x_7
z	0	0	−30	−0.14	33	3	0	0
x_1	0	1	−240	−0.16	36	4	0	0
← x_6	0	0	30	0.06	−15	−2	1	0
x_7	1	0	0	1	0	0	0	1

Tableau 2

	b	x_1	x_2	x_3	x_4	x_5	x_6	x_7
z	0	0	0	−0.08	18	1	1	0
← x_1	0	1	0	0.32	−84	−12	8	0
x_2	0	0	1	0.002	−0.5	−0.067	0.03	0
x_7	1	0	0	1	0	0	0	1

Tableau 3

	b	x_1	x_2	x_3	x_4	x_5	x_6	x_7
z	0	0.25	0	0	−3	−2	−3	0
x_3	0	3.12	0	1	−263	−37.5	25	0
← x_2	0	−0.006	1	0	0.25	0.0083	−0.016	0
x_7	1	3.12	0	0	263	37.5	−25	1

Tableau 4

	b	x_1	x_2	x_3	x_4	x_5	x_6	x_7
z	0	−0.5	120	0	0	−1	−1	0
← x_3	0	62.5	10500	1	0	50	50	0
x_4	0	0.25	40	0	1	0.33	−0.67	0
x_7	1	62.5	−10500	0	0	−50	150	1

Tableau 5

$\downarrow$

	b	x_1	x_2	x_3	x_4	x_5	x_6	x_7
z	0	-1.75	330	0.02	0	0	-2	0
x_5	0	-1.25	210	0.02	0	1	-3	0
$\leftarrow x_4$	0	0.167	-30	0.0067	1	0	0.33	0
x_7	1	0	0	1	0	0	0	1

Tableau 6

$\downarrow$

	b	x_1	x_2	x_3	x_4	x_5	x_6	x_7
z	0	-0.75	150	0.02	6	0	0	0
$\leftarrow x_5$	0	0.25	-60	-0.04	9	1	0	0
x_6	0	0.5	-90	-0.02	3	0	1	0
x_7	1	0	0	1	0	0	0	1

Tableau 7 $\equiv$ Tableau 1

This example shows how tableau 7 equates tableau 1, the process is cyclical, and the algorithm does not converge.

In general, if the LP problem is not degenerate, the simplex algorithm converges to the optimal solution in a finite number of iterations. Conversely, for degenerate problems, if the algorithm visits a degenerate solution and a feasible basic variable with null value leaves the basis, the algorithm could return to the same degenerate solution in a future step, triggering a cycle and hence not converging.

From a geometric point of view, the hyperplanes of $\mathbb{R}^n$ intersecting in a degenerate corner are more than n, that is, more than the minimum number of hyperplanes necessary to identify a point in the space. In other words, if $n + k$ hyperplanes intersect in a given point, with $k > 0$, then $\binom{n+k}{n}$ different combinations of such hyperplanes identify the same point. On the other hand, in case of a non–degenerate LP problem, each corner corresponds to a unique basic solution, and exactly n hyperplanes intersect. Hence, the m components of the feasible basic solution $\mathbf{x}_B$ are all nonzero, and switching from one solution to the next corresponds to moving along the vertices of a simplex. This is not the case for a degenerate solution, where one or more variables with null values are in basis, that is, one or more components of $\mathbf{x}_B$ are null. In this instance, more feasible basic solutions are associated with each corner: it is possible to move between basic solutions, though remaining in the same corner. A degenerate corner has p strictly positive components, with $p < m$, and $n - p$ null components. Since $n - p > n - m$, then $\binom{n-p}{n-m}$ is the number of possible choices of the $(n - m)$ not basic components of the corner with null value. Therefore, there are $\binom{n-p}{n-m}$ feasible basic solutions associated with different basis but defining the same degenerate corner. Although the basis changes over iterations, the algorithm could indefinitely cycle through a sequence of bases, and then through the corresponding eligible basic solutions, all associated with the same corner.

The degeneration of a basis is frequent in real LP problems, while the cyclicality occurs only occasionally in concrete problems. This is due to finite precision of the calculations, which is equivalent to introducing a small disturbance in the data. The use of specific counter-cyclical rules for the selection of the variables entering and leaving the basis is a very popular solution to face with cyclicality. The most common is the Bland's rule (Bland 1977), which chooses as entering/leaving variable the one with the lowest index among the entering/leaving candidates: the variable with the smaller index i is selected as entering variable if there are more negative reduced costs with identical minimum values; the same holds for the leaving variable when there are more equivalent minimum ratios. Using such a rule, the algorithm converges after $\binom{n}{m}$ iterations.

4.5   The revised simplex algorithm

In each iteration, the simplex algorithm presented in the previous sections makes use of only a part of the available information. The revised simplex is a variant aimed at limiting the amount of the stored information and the number of calculations to the minimum required. The standard simplex stores and updates a table of size $(m \times 1) \times (n \times 1)$ at each iteration. However, the essential steps of the simplex algorithm are the computation of the inverse matrix $\mathbf{B}^{-1}$ and of vector $\mathbf{c}_B^\top \mathbf{B}^{-1}$. The revised simplex focuses only on $\mathbf{B}^{-1}$ and $\mathbf{c}_B^\top \mathbf{B}^{-1}$, by means of a pivoting operation similar to the one presented for the tableau version of the algorithm.

To illustrate the revised simplex, consider the initial tableau extended through m auxiliary columns:

			x_1 ... x_n
0	0 ... 0		$\mathbf{c}^\top$
$\mathbf{b}$	$\mathbf{I}_m$		$\mathbf{A}$

Using the same notation for basic and nonbasic variables, the extended tableau can be written as follows:

			x_1 ... x_m	x_{m+1} ... x_n
0	0 ... 0		$\mathbf{c}_B^\top$	$\mathbf{c}_N^\top$
$\mathbf{b}$	$\mathbf{I}_m$		$\mathbf{B}$	$\mathbf{N}$

The rows of the tableau are expressed in canonical form with respect to the basis $\mathbf{B}$ through the premultiplication by $\mathbf{B}^{-1}$, while the reduced costs are obtained through the operation $\overline{\mathbf{c}}^{\mathsf{T}} = (\mathbf{c}^{\mathsf{T}} - \mathbf{c}_B^{\mathsf{T}}\mathbf{B}^{-1}\mathbf{A})$. Therefore, the resulting tableau in canonical form is[4]:

		x_1 $\cdots$ x_m	x_{m+1} $\cdots$ x_n
$-\mathbf{c}_B^{\mathsf{T}}\mathbf{B}^{-1}\mathbf{b}$	$\mathbf{0}^{\mathsf{T}} - \mathbf{c}_B^{\mathsf{T}}\mathbf{B}^{-1}\mathbf{I}_m = -\mathbf{c}_B^{\mathsf{T}}\mathbf{B}^{-1}$	$\mathbf{c}_B^{\mathsf{T}} - \mathbf{c}_B^{\mathsf{T}}\mathbf{B}^{-1}\mathbf{B} = \mathbf{0}^{\mathsf{T}}$	$\mathbf{c}_N^{\mathsf{T}} - \mathbf{c}_B^{\mathsf{T}}\mathbf{B}^{-1}\mathbf{N} = -\overline{\mathbf{c}}_N^{\mathsf{T}}$
$\mathbf{B}^{-1}\mathbf{b}$	$\mathbf{B}^{-1}$	$\mathbf{I}_m$	$\mathbf{B}^{-1}\mathbf{N}$

The auxiliary columns contain $\mathbf{B}^{-1}$ and $\overline{\mathbf{c}}_B^{\mathsf{T}}$. In particular row 0 of the auxiliary table contains $\mathbf{c}_B^{\mathsf{T}}\mathbf{B}^{-1}$, that is, the reduced costs corresponding to the nonbasic variables (reduced costs are null for the basic variables), and the other rows contain the inverse of matrix $\mathbf{B}$, that is, the matrix composed of the columns of matrix $\mathbf{A}$ that pertain to the basic variables. In the following, we pose $\mathbf{\Pi} = \mathbf{c}_B^{\mathsf{T}}\mathbf{B}^{-1}$.

The revised simplex stores only these parts relevant to computation, that is, column 0 and the m auxiliary columns. Therefore, the initial configuration of the table for the revised simplex is:

0	$\mathbf{0}^{\mathsf{T}}$
$\mathbf{b}$	$\mathbf{I}_m$

At each iteration:

① the algorithm computes the reduced costs $\overline{c}_i = c_i - \mathbf{c}_B^{\mathsf{T}}\mathbf{B}^{-1}\mathbf{A}_i = c_i - \mathbf{\Pi}\mathbf{A}_i$ for each $x_i \in \mathbf{x}_N$, that is, for the nonbasic variables. The reduced costs for the basic variables are indeed null. The entering variable x_h corresponds to the minimum reduced costs $\overline{c}_h$;

② the column $\begin{vmatrix} \overline{c}_h \\ \overline{\mathbf{A}}_h \end{vmatrix}$ is generated by stacking $\overline{c}_h$ (the reduced cost corresponding to the entering variable) with $\overline{\mathbf{A}}_h = \mathbf{B}^{-1}\mathbf{A}_h$. That is, the focus is limited only

---

[4] See Subsection 4.2.2 for the technical details.

to the column h of the matrix $\mathbf{A}$ corresponding to the entering variable. This column, associated with the entering variable x_h, is then placed side by side with the table composed of the column 0 and of the m auxiliary columns:

$$x_h$$

$-\mathbf{\Pi b}$	$-\mathbf{\Pi}$	$\bar{c}_h$
		$\bar{a}_{1,h}$
		$\vdots$
$\mathbf{B}^{-1}\mathbf{b}$	$\mathbf{B}^{-1}$	$\bar{a}_{k,h}$
		$\vdots$
		$\bar{a}_{m,h}$

③ the pivot element $\bar{a}_{k,h}$ (highlighted in the previous table) is selected using the criteria of the minimum ratio $b_i / \bar{a}_{i,h}$, and considering only the $\bar{a}_{i,h} > 0$;

④ the pivoting operation is carried out on the table detected at the previous step. The column corresponding to the variable x_h is deleted. The new table stores the information corresponding to the new basis, and the procedure is iterated.

As an example, consider again the 3–D LP problem (4.7), solved via the graphical method in subsection 4.1.3 and via the tableau algorithm in subsection 4.2.6. The corresponding tableau in canonical form with respect to the basis (x_4, x_5, x_6), the three slack variables, is here reported again for convenience:

	b	x_1	x_2	x_3	x_4	x_5	x_6
$\tilde{z}$	0	-300	-200	-350	0	0	0
x_4	40	1	1	1	1	0	0
x_5	120	4	2	3	0	1	0
x_6	20	0	0	1	0	0	1

Following the steps of the revised simplex to solve this LP problem (detailed comments are reported only for the first iteration).

initialization :

$$\mathbf{A} = \left[\mathbf{A}_1 \,\vert\, \mathbf{A}_2 \,\vert\, \mathbf{A}_3 \,\vert\, \mathbf{A}_4 \,\vert\, \mathbf{A}_5 \,\vert\, \mathbf{A}_6 \right] = \begin{bmatrix} 1 & 1 & 1 & 1 & 0 & 0 \\ 4 & 2 & 3 & 0 & 1 & 0 \\ 0 & 0 & 1 & 0 & 0 & 1 \end{bmatrix}$$

$$\mathbf{c}^\mathsf{T} = \left[\mathbf{c}_N^\mathsf{T} \,\vert\, \mathbf{c}_B^\mathsf{T} \right] = \begin{bmatrix} c_1 & c_2 & c_3 \,\vert\, c_4 & c_5 & c_6 \end{bmatrix}$$

$$= \begin{bmatrix} -300 & -200 & \boxed{-350} \,\vert\, 0 & 0 & 0 \end{bmatrix}$$

iteration 1 - step ①

The reduced costs for the three nonbasic variables (x_1, x_2, x_3) are computed:

$$\bar{c}_1 = c_1 - \mathbf{c}_B^{\mathsf{T}}\mathbf{B}^{-1}\mathbf{A}_1 = -300 - \begin{bmatrix} 0 & 0 & 0 \end{bmatrix} \begin{bmatrix} 1 & 0 & 0 \\ 0 & 1 & 0 \\ 0 & 0 & 1 \end{bmatrix} \begin{bmatrix} 1 \\ 4 \\ 0 \end{bmatrix} = -300$$

$$\bar{c}_2 = c_2 - \mathbf{c}_B^{\mathsf{T}}\mathbf{B}^{-1}\mathbf{A}_2 = -200 - \begin{bmatrix} 0 & 0 & 0 \end{bmatrix} \begin{bmatrix} 1 & 0 & 0 \\ 0 & 1 & 0 \\ 0 & 0 & 1 \end{bmatrix} \begin{bmatrix} 1 \\ 2 \\ 0 \end{bmatrix} = -200$$

$$\bar{c}_3 = c_3 - \mathbf{c}_B^{\mathsf{T}}\mathbf{B}^{-1}\mathbf{A}_3 = -350 - \begin{bmatrix} 0 & 0 & 0 \end{bmatrix} \begin{bmatrix} 1 & 0 & 0 \\ 0 & 1 & 0 \\ 0 & 0 & 1 \end{bmatrix} \begin{bmatrix} 1 \\ 3 \\ 1 \end{bmatrix} = \boxed{-350}$$

The variable x_3, whose reduced cost is minimum, is chosen as entering variable.

iteration 1 - step ②

The column $\overline{\mathbf{A}}_3$ corresponding to the entering variable x_3 is computed:

$$\overline{\mathbf{A}}_3 = \mathbf{B}^{-1}\mathbf{A}_3 = \begin{bmatrix} 1 & 0 & 0 \\ 0 & 1 & 0 \\ 0 & 0 & 1 \end{bmatrix} \begin{bmatrix} 1 \\ 3 \\ 1 \end{bmatrix} = \begin{bmatrix} 1 \\ 3 \\ 1 \end{bmatrix}$$

and placed side by side with the initial table composed using the known coefficients and the auxiliary columns. $\bar{c}_3$, computed in the previous step, is inserted in the row 0:

					x_3
z	0	0	0	0	-350
x_4	40	1	0	0	1
x_5	120	0	1	0	3
x_6	20	0	0	1	1

iteration 1 - step ③

The ratios between the known coefficients and the elements of the vector $\overline{\mathbf{A}}_3$ are computed for only the positive coefficients:

-350		-350
$40/1$	=	40
$120/3$		40
$20/1$		$\boxed{20}$

The variable x_6, whose ratio is minimum, is then selected as leaving variable.

iteration 1 - step ④

The pivoting operation is carried out so to obtain a value 1 at the pivot position and all 0s in the other position of the side vector x_3:

					x_3
$\boxed{\text{row}_0}$ $+350 \times \boxed{\text{pivot row}}$ $\rightarrow$ z	7000	0	0	350	0
$\boxed{\text{row}_1}$ $-1 \times \boxed{\text{pivot row}}$ $\rightarrow$ x_4	20	1	0	-1	0
$\boxed{\text{row}_2}$ $-3 \times \boxed{\text{pivot row}}$ $\rightarrow$ x_5	60	0	1	-3	0
$\boxed{\text{pivot row}}$ $= \boxed{\text{row}_3} / 1$ $\rightarrow$ x_6	20	0	0	1	1

The table now contains the information for the current basis (x_4, x_5, x_3). The litmus test is obtained replacing in matrix $\mathbf{B}$ the third column, corresponding to the leaving variable x_6, with the column $\mathbf{A}_3$, corresponding to the entering variable x_3 and computing its inverse:

$$\mathbf{B} = \begin{bmatrix} 1 & 0 & 1 \\ 0 & 1 & 3 \\ 0 & 0 & 1 \end{bmatrix} \implies \mathbf{B}^{-1} = \begin{bmatrix} 1 & 0 & -1 \\ 0 & 1 & -3 \\ 0 & 0 & 1 \end{bmatrix}$$

In row 0 the values of $\mathbf{\Pi} = \mathbf{c}_B^\top \mathbf{B}^{-1}$ are directly available for the current basis. The reduced costs for the new basis can be computed. The procedure is now iterated deleting the side column of x_3, now entered in basis.

iteration 2 - step ①

Computation of the reduced costs for the three nonbasic variables (x_1, x_2, x_6) and choice of the entering variable:

$$\bar{c}_1 = -300 - \begin{bmatrix} 0 & 0 & -350 \end{bmatrix} \begin{bmatrix} 1 \\ 4 \\ 0 \end{bmatrix} = \boxed{-300} \implies x_1 \text{ is the entering variable}$$

$$\bar{c}_2 = -200 - \begin{bmatrix} 0 & 0 & -350 \end{bmatrix} \begin{bmatrix} 1 \\ 2 \\ 0 \end{bmatrix} = -200$$

$$\bar{c}_6 = 0 - \begin{bmatrix} 0 & 0 & -350 \end{bmatrix} \begin{bmatrix} 0 \\ 0 \\ 1 \end{bmatrix} = 350$$

iteration 2 - step ②

Computation of the column $\overline{\mathbf{A}}_1$ corresponding to the entering variable x_1:

$$\overline{\mathbf{A}}_1 = \begin{bmatrix} 1 & 0 & -1 \\ 0 & 1 & -3 \\ 0 & 0 & 1 \end{bmatrix}\begin{bmatrix} 1 \\ 4 \\ 0 \end{bmatrix} = \begin{bmatrix} 1 \\ 4 \\ 0 \end{bmatrix}$$

The column $\overline{\mathbf{A}}_1$ is hence placed side by side with the table obtained in the last step of iteration 1, inserting $\overline{c}_1$ in row 0:

					x_1
z	7000	0	0	350	−300
x_4	20	1	0	−1	1
x_5	60	0	1	−3	4
x_3	20	0	0	1	0

iteration 2 - step ③

Choice of the leaving variable:

−300		−300
20/1	=	20
60/4		15
—		—

$\implies$ x_5 is the leaving variable

iteration 2 - step ④

The pivoting operation is carried out so to switch the role of x_1 and x_5:

							x_1
row$_0$ +300 × pivot row	$\rightarrow$	z	11500	0	75	125	0
row$_1$ −1 × pivot row	$\rightarrow$	x_4	5	1	−1/4	−1/4	0
pivot row = row$_2$ /4	$\rightarrow$	x_5	15	0	1/4	−3/4	1
row$_3$ −0 × pivot row	$\rightarrow$	x_3	20	0	0	1	0

The table contains now the information for the current basis (x_4, x_1, x_3).

iteration 3 - step ①

Computation of the reduced costs for the three nonbasic variables (x_2, x_5, x_6) and choice of the entering variable:

$$\bar{c}_2 = -200 - \begin{bmatrix} 0 & -75 & -125 \end{bmatrix} \begin{bmatrix} 1 \\ 2 \\ 0 \end{bmatrix} = \boxed{-50} \implies x_2 \text{ is the entering variable}$$

$$\bar{c}_5 = 0 - \begin{bmatrix} 0 & -75 & -125 \end{bmatrix} \begin{bmatrix} 0 \\ 1 \\ 0 \end{bmatrix} = 75$$

$$\bar{c}_6 = 0 - \begin{bmatrix} 0 & -75 & -125 \end{bmatrix} \begin{bmatrix} 0 \\ 0 \\ 1 \end{bmatrix} = 125$$

iteration 3 - step ②

Computation of the column $\overline{\mathbf{A}}_2$ corresponding to the entering variable x_2:

$$\overline{\mathbf{A}}_2 = \begin{bmatrix} 1 & -1/4 & -1/4 \\ 0 & 1/4 & -3/4 \\ 0 & 0 & 1 \end{bmatrix} \begin{bmatrix} 1 \\ 2 \\ 0 \end{bmatrix} = \begin{bmatrix} 1/2 \\ 1/2 \\ 0 \end{bmatrix}$$

The reduced cost $\bar{c}_2$ and the column $\overline{\mathbf{A}}_2$ are placed side by side to the table obtained in the last step of iteration 2:

					x_2
z	11500	0	75	125	-50
x_4	5	1	$-1/4$	$-1/4$	$1/2$
x_1	15	0	$1/4$	$-3/4$	$1/2$
x_3	20	0	0	1	0

iteration 3 - step ③

Choice of the leaving variable:

-50		-50
$\dfrac{5}{1/2}$	$=$	$\boxed{10}$
$\dfrac{15}{1/2}$		30
—		—

$\implies x_4$ is the leaving variable

iteration 3 - step ④

The pivoting operation is carried out so that x_2 enters and x_4 leaves the basis:

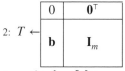

$$\boxed{\text{row}_0} +50\times\boxed{\text{pivot row}} \quad\rightarrow\quad z$$

$$\boxed{\boxed{\text{pivot row}}} = 2\times\text{row}_1 \quad\rightarrow\quad x_4$$

$$\boxed{\text{row}_2} -\tfrac{1}{2}\times\boxed{\text{pivot row}} \quad\rightarrow\quad x_1$$

$$\boxed{\text{row}_3} -0\times\boxed{\text{pivot row}} \quad\rightarrow\quad x_3$$

					x_2
z	12000	100	50	100	0
x_4	10	2	−1/2	−1/2	1
x_1	10	−1	1/2	−1/2	0
x_3	20	0	0	1	0

The table contains now the information for the current basis (x_2, x_1, x_3).

iteration 4 - step ①

Computation of the reduced costs for the three nonbasic variables (x_4, x_5, x_6):

$$\bar{c}_4 = 0 - \begin{bmatrix} -100 & -50 & -100 \end{bmatrix}\begin{bmatrix} 1 \\ 0 \\ 0 \end{bmatrix} = 100$$

$$\bar{c}_5 = 0 - \begin{bmatrix} -100 & -50 & -100 \end{bmatrix}\begin{bmatrix} 0 \\ 1 \\ 0 \end{bmatrix} = 50$$

$$\bar{c}_6 = 0 - \begin{bmatrix} -100 & -50 & -100 \end{bmatrix}\begin{bmatrix} 0 \\ 0 \\ 1 \end{bmatrix} = 100$$

The optimality condition is fulfilled because all the reduced costs are non–negative, and the algorithm stops. The optimal solution corresponds to the point $(x_1 = 10, x_2 = 10, x_3 = 20)$, which provides the objective function $z = 12000$, as previously shown with the other solution methods.

The revised simplex algorithm can be detailed as follows:

1: $\mathbf{B} \leftarrow$ initial basis   ▷ assume $\mathbf{B}$ is an initial feasible basic solution

2: $T \leftarrow$
0	$\mathbf{0}^{\mathsf{T}}$
$\mathbf{b}$	$\mathbf{I}_m$
        ▷ initial configuration of the table

3: *optimal* ← **false**   ▷ flag variable for checking optimality
4: *unbound* ← **false**   ▷ flag variable for checking unboundedness
5: **while** *optimal* = **false and** *unbound* = **false do**
6:    $\Pi = \mathbf{c}_B^{\mathsf{T}}\mathbf{B}^{-1}$
7:    $\bar{c}_i = c_i - \Pi\mathbf{A}_i, \forall x_i \in \mathbf{x}_N$ ▷ compute the reduced costs for the nonbasic variables
8:    **if** $\bar{c}_i \geq 0, \forall x_i \in \mathbf{x}_N$ **then**   ▷ optimality test
9:        *optimal* ← **true**
10:   **else**
11:       $h : \text{argmin}_{\bar{c}_i < 0}\, \bar{c}_i, \forall i$ related to non–basic variables   ▷ index of the entering variable
12:       $\bar{\mathbf{A}}_h = \mathbf{B}^{-1}\mathbf{A}_h$

13:     **if** $\bar{A}_{h_i} \leq 0, \forall i = 1, \ldots, m$ **then**

14:         *unbound* ← **true**

15:     **else**   ▷ change of the basis

16:         *sideColumn* ← $\begin{bmatrix} \bar{c}_h \\ \bar{A}_h \end{bmatrix}$ ▷ column used to perform the pivoting operation

17:         $k : \mathrm{argmin}_{\bar{A}_{h_i} > 0}\, b_i / \bar{A}_{h_i}, \forall i \in (1, \ldots, m)$ ▷ index of the leaving variable

18:         PIVOTING($k, T, sideVector$)   ▷ procedure for updating the table

19:     **end if**

20:   **end if**

21: **end while**

22: **procedure** PIVOTING($k, T, sideVector$) ▷ procedure for carrying out the pivoting operation

23:     m ← nRows(T)   ▷ compute the number of rows of the table

24:     n ← nCols(T)   ▷ compute the number of columns of the table

25:     pivot ← T[k]   ▷ pivot element

26:     **for** j = 1 to n **do**   ▷ compute the new pivot row dividing it for the pivot element

27:         $T[k,j] \leftarrow T[k,j]/pivot$

28:     **end for**

29:     **for** i = 0 to m **do**   ▷ for all the rows (row 0 is related to the objective function)

30:         **if** $i \neq k$ **and** $sideVector[i] \neq 0$ **then** ▷ except for the pivot row and when the entry on the sideVector is already null

31:             **for** j = 1 to n **do**   ▷ subtract from each row the new pivot row multiplied for the element on the pivot column

32:                 $T[i,j] \leftarrow T[i,j] - sideVector[i] * T[k,j]$

33:             **end for**

34:         **end if**

35:     **end for**

36: **end procedure**

The improved efficiency of the revised simplex with respect to the standard simplex depends on the fact that the pivoting operation is carried out only on the initial basis matrix **B** and on the vector of known coefficients. Essentially, the revised simplex method computes at each iteration only the coefficients needed to identify the pivot element rather than updating the entire tableau. The involved coefficients are the reduced costs for the nonbasic variables (used to detect the entering variable), the current known coefficients, and the updated coefficients of the entering variable (used to detect the leaving variable). The computational complexity is hence related to the number m of constraints and not to the number n of unknowns. Therefore, the larger the number of nonbasic variables with respect to the basic variables, that is, when $n \gg m$, the greater the efficiency. This is even more accentuated in the two-phase implementation: the introduction of the artificial variables along with the slack variables determines a higher convenience of the revised simplex with respect to the standard algorithm. This is the reason why the revised simplex is the basis of most computer codes in software for linear programming.

4.6   A summary of key points

- Linear programming is a special case of a mathematical programming problem, where both the objective function and the constraints are linear. Furthermore, the decisional variables are non–negative.

- All the linear programming problems can be easily formulated in standard form, namely in terms of a minimization problem whose structural constraints are expressed in equational form. This allows us to exploit basic properties of linear systems in order to search for the optimal solution.

- The simplex algorithm inspects the corners of the feasible region and chooses the one associated with the highest value of the objective function. It is based on the relationship between corners and optimal solutions, and on theoretical results ensuring the optimality of the solution in correspondence of one or more corners of the feasible region.

- The simplex algorithm restricts the possible solutions to the corners of the feasible region (feasible basic solutions). Although the number of such solutions is finite, it exponentially increases with the number of variables and constraints involved. An optimality condition allows us to avoid the full enumeration of all the feasible basic solutions.

- The revised simplex is a variant which reduces the amount of stored information and computations to the minimum. This is why it is the most popular method for solving linear programming problems, and quantile regression is one of them, as shown in the next chapter.

References

Beale, E. M. L. 1955 Cycling in the dual simplex algorithm. Naval Research Logistics, No. 2, pp. 269–275.

Bland, R. G. 1977 New finite pivoting rules for the simplex method. Mathematics of Operations Research, No. 2, pp. 103–107.

Hiller, F. K., and Lieberman, G. 2015 *Introduction to Operations Research*, Mc-Graw Hill, 10th Edition.

Matousek, J., and Gartner, B. 2005 *Understanding and Using Linear Programming*, Springer.

Strang, G. 2005 *Linear Algebra and Its Application*, Brooks/Cole INDIA, 4th Edition.

Strang, G. 2009 *Introduction to Linear Algebra*, Wellesley Cambridge Press, 4th Edition.

Vanderbei, R. J. 2014 *Linear Programming: Foundations and Extensions*, International Series in Operations Research & Management Science, Springer, 4th Edition.

5

Linear programming for quantile regression

Introduction

It is now time to see the simplex algorithm for solving the quantile regression problem. This chapter outlines the algorithm using a small data set, detailing the various steps of the procedure.

The machinery behind the use of linear programming for solving regression problems is first presented for the case of median regression and then extended to the more general quantile regression. A geometric interpretation of the minimization problem characterizing quantile regression is also outlined using the point/line duality introduced by Edgeworth in 1888 in his pioneer work on median methods for linear regression.

5.1 LP formulation of the L_1 simple regression problem

The L_1 simple linear regression problem, or median regression:

$$\min_{a_0, a_1} \sum_{i=1}^{n} |y_i - (a_0 + a_1 x_i)| \tag{5.1}$$

consists of fitting a straight line to a set of n points $(x_i, y_i), i = 1, \ldots, n$, minimizing the sum of the absolute deviations of the data from the line. The use of least absolute deviations as an alternative criterion to least squares for best fit is a recurring

Quantile Regression: Estimation and Simulation, Volume 2. Marilena Furno and Domenico Vistocco.
© 2018 John Wiley & Sons Ltd. Published 2018 by John Wiley & Sons Ltd.
Companion website: www.wiley.com/go/furno/quantileregression

subject in statistical literature. The least absolute criterion is even more ancient than the most popular least squares technique: median regression was early introduced by Boscovich in the 18–th century for estimating a bivariate linear model for the ellipticity of the earth. Laplace (1818) first and Edgeworth (1888, 1923) then, characterized Boscovich's work as a weighted median. Edgeworth, in particular, proposed a general formulation of median regression involving a multivariate vector of explanatory variables naming it plural median. This was the extension of the "methode de situation," the geometric approach of Laplace for solving Boscovich's problem. More recently, Koenker and Basset (1978) extended the problem to quantiles other than the median.

Stigler (1990) thoroughly presents the early proposals of median regression in his comprehensive book on the history of Statistics. The historical development of L_1 estimation procedures is also offered by Farebrother (1987), through a review of the contributions of the main scientists of the eighteenth century. The book of Farebrother (1999) provides an exhaustive description of the history of fitting linear relationships by the least squares (L_2), least absolute deviations (L_1), and minimax absolute deviations (L_∞) procedures. A quantile regression perspective on the topic is offered in a fascinating paper of Koenker (2000), where the historical issues are used to motivate recent developments.

Albeit different methods are available for solving the median regression, we focus on the simplex–based methods, since their use represent the turning point for the spread of the L_1 regression. The seminal work of Wagner (1959) presents the linear programming techniques for least absolute deviations in the mainstream statistical literature. Only few years later, Barrodale and Young (1966) first and Barrodale and Roberts (1970) after, treated the L_1 approximation to a real–valued function through linear programming in the literature on operation research and numerical mathematics.

To align with the LP treatment introduced in the previous chapter, there is a difference in notation between the LP problem and the regression problem. While we have denoted by x_i the decisional variables in the LP problem, here the x_i correspond to known coefficients and the y_i to known values. In fact, in the regression problem, the values x_i and y_i are available. The intercept a_0 and the slope a_1 of the regression line, that is, the parameters, are the variables in LP jargon. In line with the LP formulation, the cost function should be properly rewritten as:

$$\min_{a_0,a_1} \sum_{i=1}^{n} |y_i - (a_0 + x_i a_1)| \tag{5.2}$$

exchanging the position of a_1 and x_i in order to stress that x_i is known and a_1 is a decisional variable. The L_1 optimization problem is nonlinear and we need to reformulate it in a linear form to exploit LP. The transformation of the nonlinear cost function into a linear cost function can be obtained using the two different options presented in the following two subsections. The small data set reported in Table 5.1, where Y denotes a generic response variable and X the explicative variable, will be used for the example. This data, used in the paper of Barrodale and Roberts (1973), suffices to detail the application of the LP theory to the quantile regression setting, even if comprising

Table 5.1 The simple data set.

	X	Y
A	1	1
B	2	1
C	3	2
D	4	3
E	5	2

only 5 observations. A more realistic data set would become unmanageable, due to the large dimension of the corresponding tableau representation.

5.1.1 A first formulation of the L_1 regression problem

The first option for formulating the median regression problem (5.2) consists in treating the objective function as a sum of n nonnegative variables, through the absolute residuals $e_i^+ = |y_i - (a_0 + x_i a_1)|$. Hence, the corresponding LP formulation is:

$$\text{minimize} \quad \sum_{i=1}^{n} e_i^+$$
$$\text{subject to} \quad e_i^+ \geq y_i - a_0 - x_i a_1 \qquad i = 1, \dots, n$$
$$e_i^+ \geq -(y_i - a_0 - x_i a_1) \quad i = 1, \dots, n.$$

The LP formulation of the L_1 regression problem consists of $2 \times n$ constraints, since two constraints are added for each of the n available observations. The two sets of constraints guarantee that:

$$e_i^+ \geq \max\{y_i - a_0 - x_i a_1, -(y_i - a_0 - x_i a_1)\} = |y_i - a_0 - x_i a_1|$$

To keep the notation in line with the LP format, the known values y_i are hence isolated on the right side of the inequalities, playing the role of the b_i coefficients in the general LP problem :

$$\text{minimize} \quad \sum_{i=1}^{n} e_i^+$$
$$\text{subject to} \quad a_0 + x_i \, a_1 + e_i^+ \geq y_i \qquad i = 1, \dots, n$$
$$-a_0 - x_i \, a_1 + e_i^+ \geq -y_i \quad i = 1, \dots, n.$$

Finally, the second set of inequalities is multiplied by -1, to obtain:

$$\text{minimize} \quad \sum_{i=1}^{n} e_i^+$$
$$\text{subject to} \quad a_0 + x_i \, a_1 + e_i^+ \geq y_i \quad i = 1, \dots, n$$
$$a_0 + x_i \, a_1 - e_i^+ \leq y_i \quad i = 1, \dots, n.$$

This prevents negative known coefficients from being on the right side of the inequalities. Referring again to the LP notation, the vector $\mathbf{x}_i$ corresponds to the column of matrix $\mathbf{A}$ associated with the decisional variable a_1. All the constraint coefficients associated with the intercept a_0 are equal to 1. The coefficients associated with the e_i variables are instead equal to 1 for the first n constraints and to -1 for the other n constraints.

Even if the n unsigned $e_i^+, i = 1, \ldots, n$ are nonnegative by definition, both decisional variables a_0 and a_1 (regression coefficients) are unrestricted in sign, being allowed to assume both positive and negative values. A further modification is hence required. By recalling that a variable unrestricted in sign can always be expressed as difference of two nonnegative numbers (see subsection 4.1.1), the two regression coefficients are replaced with:

$$
\begin{aligned}
a_0 &= a_0' - a_0'' \\
a_1 &= a_1' - a_1''
\end{aligned}
$$

where $a_0', a_0'', a_1', a_1'' \geq 0$. Therefore, the LP formulation of the L_1 regression problem becomes:

$$
\begin{aligned}
\text{minimize} \quad & \sum_{i=1}^{n} e_i^+ \\
\text{subject to} \quad & (a_0' - a_0'') + x_i\,(a_1' - a_1'') + e_i^+ \geq y_i \quad i = 1, \ldots, n \\
& (a_0' - a_0'') + x_i\,(a_1' - a_1'') - e_i^+ \leq y_i \quad i = 1, \ldots, n.
\end{aligned}
\tag{5.3}
$$

All the variables involved are nonnegative and hence comply with the LP requirements. It is worth stressing that the e_i^+ have been defined as unsigned residuals. In order to determine the proper signs of the residuals at the end of the process, it is sufficient to compute the signed values $y_i - a_0 - x_i a_1$ for $i = 1, \ldots, n$, through the values obtained for the two decisional variables a_0 and a_1.

The equational form of the LP problem (5.3) involves the use of $2 \times n$ slack variables, n with coefficient $+1$ associated with the less–than–or–equal constraints, and n with coefficient -1 for the greater–than–or–equal constraint:

$$
\begin{aligned}
\text{minimize} \quad & \sum_{i=1}^{n} e_i^+ \\
\text{subject to} \quad & (a_0' - a_0'') + x_i\,(a_1' - a_1'') + e_i^+ - s_i = y_i \quad i = 1, \ldots, n \\
& (a_0' - a_0'') + x_i\,(a_1' - a_1'') - e_i^+ + s_i = y_i \quad i = 1, \ldots, n.
\end{aligned}
$$

Unlike the previous chapter, the slack variables are here denoted by s_i, $i = 1, \ldots, n$, since the x_i's in the regression framework are usually reserved to label the values of the explicative variables. This LP formulation is not in canonical form due to the negative coefficients associated with the four slack variables inserted to deal with the $\leq$ inequalities. To this end, n additional artificial variables are necessary in the first set of constraints. The artificial variables are here denoted with $t_i, i = 1, \ldots, n$, since the y_i have been adopted for the values of the response variable, as in the standard notation of regression:

$$
\begin{aligned}
\text{minimize} \quad & \sum_{i=1}^{n} e_i^+ \\
\text{subject to} \quad & (a_0' - a_0'') + x_i\,(a_1' - a_1'') + e_i^+ - s_i + t_i = y_i \quad i = 1, \ldots, n \\
& (a_0' - a_0'') + x_i\,(a_1' - a_1'') - e_i^+ + s_i = y_i \qquad\quad i = 1, \ldots, n.
\end{aligned}
\tag{5.4}
$$

The introduction of the additional variables t_i determines a system of equations in canonical form with respect to such n artificial variables along with the last n slack variables. The use of the artificial variables extends the feasible set to a new feasible set, characterized by a new artificial objective function. Such artificial problem shares the same constraint matrix with the initial problem. The objective function to minimize in the artificial problem is:

$$
w = \sum_{i=1}^{n} t_i
$$

The two–phase method in Section 4.3, detects an initial basis common to the artificial and the original problem and then switches back to the original problem, through a twofold application of the simplex algorithm. A common solution exists only when the optimal value of the artificial problem w^* equals 0. In such a case the corresponding solution can be used to trigger the simplex algorithm on the original problem.

Therefore, the LP formulation (5.3) in canonical form for the small data set is:

$$
\begin{aligned}
\text{minimize} \quad & z = e_1^+ + e_2^+ + e_3^+ + e_4^+ + e_5^+ \\
\text{subject to} \quad & a_0' - a_0'' + x_1\,a_1' - x_1\,a_1'' - e_1^+ + s_1 = y_1 && (C_1) \\
& a_0' - a_0'' + x_2\,a_1' - x_2\,a_1'' - e_2^+ + s_2 = y_2 && (C_2) \\
& a_0' - a_0'' + x_3\,a_1' - x_3\,a_1'' - e_3^+ + s_3 = y_3 && (C_3) \\
& a_0' - a_0'' + x_4\,a_1' - x_4\,a_1'' - e_4^+ + s_4 = y_4 && (C_4) \\
& a_0' - a_0'' + x_5\,a_1' - x_5\,a_1'' - e_5^+ + s_5 = y_4 && (C_5) \\
& a_0' - a_0'' + x_1\,a_1' - x_1\,a_1'' + e_1^+ - s_6 + t_1 = y_1 && (C_6) \\
& a_0' - a_0'' + x_2\,a_1' - x_2\,a_1'' + e_2^+ - s_7 + t_2 = y_2 && (C_7) \\
& a_0' - a_0'' + x_3\,a_1' - x_3\,a_1'' + e_3^+ - s_8 + t_3 = y_3 && (C_8) \\
& a_0' - a_0'' + x_4\,a_1' - x_4\,a_1'' + e_4^+ - s_9 + t_4 = y_4 && (C_9) \\
& a_0' - a_0'' + x_5\,a_1' - x_5\,a_1'' + e_5^+ - s_{10} + t_5 = y_5 && (C_{10}) \\
& e_1^+, e_2^+, e_3^+, e_4^+, a_0', a_0'', a_1', a_1'' \geq 0 \\
& s_1, s_2, s_3, s_4, s_5, s_6, s_7, s_8, s_9, s_{10} \geq 0 \\
& t_1, t_2, t_3, t_4, t_5 \geq 0
\end{aligned}
\tag{5.5}
$$

The canonical form involves the use of ten slack variables and five artificial variables. The introduction of the latter extends the feasible set to a new feasible set characterized by the following artificial objective function to minimize:

$$w = t_1 + t_2 + t_3 + t_4 + t_5$$

To express the artificial objective function w in terms of the variables of the original problem, the t_i can be derived from the constraints and the resulting expressions used to compute w is:

$$
\begin{aligned}
t_1 &= y_1 - a'_0 + a''_0 - x_1 a'_1 + x_1 a''_1 - e^+_1 + s_6 & \leftarrow (C_6)\\
t_2 &= y_2 - a'_0 + a''_0 - x_2 a'_1 + x_2 a''_1 - e^+_2 + s_7 & \leftarrow (C_7)\\
t_3 &= y_3 - a'_0 + a''_0 - x_3 a'_1 + x_3 a''_1 - e^+_3 + s_8 & \leftarrow (C_8)\\
t_4 &= y_4 - a'_0 + a''_0 - x_4 a'_1 + x_4 a''_1 - e^+_4 + s_9 & \leftarrow (C_9)\\
t_5 &= y_5 - a'_0 + a''_0 - x_5 a'_1 + x_5 a''_1 - e^+_5 + s_{10} & \leftarrow (C_{10})
\end{aligned}
$$

$$w = \sum_{i=1}^{5} y_i - 5\, a'_0 + 5\, a''_0 - \left(\sum_{i=1}^{5} x_i \right) a'_1 + \left(\sum_{i=1}^{5} x_i \right) a''_1 - \sum_{i=1}^{5} e^+_i + \sum_{i=6}^{10} s_i$$

Using the values in Table 5.1 for x_i and y_i, the expression of the artificial objective function becomes:

$$w = 9 - 5\, a'_0 + 5\, a''_0 - 15\, a'_1 + 15\, a''_1 - e^+_1 - e^+_2 - e^+_3 - e^+_4 - e^+_5$$
$$+ s_6 + s_7 + s_8 + s_9 + s_{10}$$

All the elements are now available for the initial tableau:

	y	a'_0	a''_0	a'_1	a''_1	e^+_1	e^+_2	e^+_3	e^+_4	e^+_5	s_1	s_2	s_3	s_4	s_5	s_6	s_7	s_8	s_9	s_{10}	t_1	t_2	t_3	t_4	t_5
w	-9	-5	5	-15	15	-1	-1	-1	-1	-1	0	0	0	0	0	1	1	1	1	1	0	0	0	0	0
z	0	0	0	0	0	1	1	1	1	1	0	0	0	0	0	0	0	0	0	0	1	1	1	1	1
s_1	1	1	-1	1	-1	-1	0	0	0	0	1	0	0	0	0	0	0	0	0	0	0	0	0	0	0
s_2	1	1	-1	2	-2	0	-1	0	0	0	0	1	0	0	0	0	0	0	0	0	0	0	0	0	0
s_3	2	1	-1	3	-3	0	0	-1	0	0	0	0	1	0	0	0	0	0	0	0	0	0	0	0	0
s_4	3	1	-1	4	-4	0	0	0	-1	0	0	0	0	1	0	0	0	0	0	0	0	0	0	0	0
s_5	2	1	-1	5	-5	0	0	0	0	-1	0	0	0	0	1	0	0	0	0	0	0	0	0	0	0
t_1	1	1	-1	1	-1	1	0	0	0	0	0	0	0	0	0	-1	0	0	0	0	1	0	0	0	0
t_2	1	1	-1	2	-2	0	1	0	0	0	0	0	0	0	0	0	-1	0	0	0	0	1	0	0	0
t_3	2	1	-1	3	-3	0	0	1	0	0	0	0	0	0	0	0	0	-1	0	0	0	0	1	0	0
t_4	3	1	-1	4	-4	0	0	0	1	0	0	0	0	0	0	0	0	0	-1	0	0	0	0	1	0
t_5	2	1	-1	5	-5	0	0	0	0	1	0	0	0	0	0	0	0	0	0	-1	0	0	0	0	1

The tableau contains two rows 0, one for the artificial problem and one for the original problem, labeled with w and z, respectively. If during the first phase, the pivoting operation is carried out on both rows 0, row w will be automatically ready for triggering the second phase. The initial tableau is in canonical form with respect to the slack variables s_1, s_2, s_3, s_4, s_5, and to the five artificial variables t_1, t_2, t_3, t_4, t_5, as highlighted through the boxed elements.

The first phase of the simplex algorithm starts with the selection of the entering variable: a'_1 is the first variable entering the basis since the corresponding coefficient is minimum. Using the same notation previously introduced, its column is marked using the symbol $\downarrow$ above the column label. The ratios between the known coefficients (first column of the tableau) and the column of the entering variable itself are computed for the positive entries. Both s_5 and t_5 could be selected as leaving variables:

$$\min\left\{\frac{1}{1},\frac{1}{2},\frac{2}{3},\frac{3}{4},\frac{2}{5},\frac{1}{1},\frac{1}{2},\frac{2}{3},\frac{3}{4},\frac{2}{5}\right\}$$
$$= \min\{1, 0.5, 0.67, 0.75, \boxed{0.4}, 1, 0.5, 0.67, 0.75, \boxed{0.4}\}$$

Since the aim of the first phase is to have all artificial variables as null, we choose t_5 for leaving the basis. Its row label is marked with the symbol $\leftarrow$. Therefore, the first iteration of the first phase is:

iteration	: 1 (first phase)
entering variable	: a'_1
leaving variable	: t_5
pivot	: 5
current solution	: $s_1 = 1,\quad s_2 = 1,\quad s_3 = 2,\quad s_4 = 3,\quad s_5 = 2$
	: $t_1 = 1,\quad t_2 = 1,\quad t_3 = 2,\quad t_4 = 3,\quad t_5 = 2$
objective function	: $w = -9$

	y	a'_0	a''_0	a'_1	a''_1	e_1^+	e_2^+	e_3^+	e_4^+	e_5^+	s_1	s_2	s_3	s_4	s_5	s_6	s_7	s_8	s_9	s_{10}	t_1	t_2	t_3	t_4	t_5
w	−9	−5	5	−15	15	−1	−1	−1	−1	−1	0	0	0	0	0	1	1	1	1	1	0	0	0	0	0
z	0	0	0	0	0	1	1	1	1	1	0	0	0	0	0	0	0	0	0	0	1	1	1	1	1
s_1	1	1	−1	1	−1	−1	0	0	0	0	1	0	0	0	0	0	0	0	0	0	0	0	0	0	0
s_2	1	1	−1	2	−2	0	−1	0	0	0	0	1	0	0	0	0	0	0	0	0	0	0	0	0	0
s_3	2	1	−1	3	−3	0	0	−1	0	0	0	0	1	0	0	0	0	0	0	0	0	0	0	0	0
s_4	3	1	−1	4	−4	0	0	0	−1	0	0	0	0	1	0	0	0	0	0	0	0	0	0	0	0
s_5	2	1	−1	5	−5	0	0	0	0	−1	0	0	0	0	1	0	0	0	0	0	0	0	0	0	0
t_1	1	1	−1	1	−1	1	0	0	0	0	0	0	0	0	0	−1	0	0	0	0	1	0	0	0	0
t_2	1	1	−1	2	−2	0	1	0	0	0	0	0	0	0	0	0	−1	0	0	0	0	1	0	0	0
t_3	2	1	−1	3	−3	0	0	1	0	0	0	0	0	0	0	0	0	−1	0	0	0	0	1	0	0
t_4	3	1	−1	4	−4	0	0	0	1	0	0	0	0	0	0	0	0	0	−1	0	0	0	0	1	0
$\leftarrow t_5$	2	1	−1	5	−5	0	0	0	0	1	0	0	0	0	0	0	0	0	0	−1	0	0	0	0	1

Tableau 1, Phase 1

The pivoting operation is carried out to switch the role of the entering variable a'_1 with the leaving variable t_5. The aim of the pivoting operation is to obtain a column of 0 for a'_1 except for the last entry, and it consists of the following steps:

- the pivot row (t_5) is divided by the pivot element (5) in order to have 1 on the pivot position,

- the resulting pivot row is multiplied by the value -15 (the value in the first row 0 in the pivot column), and hence subtracted from the row w,

- the same operation is repeated for all the other rows of the tableau (from the row z to the row t_4), in order to have 0 in all the other positions of the pivot column.

At the end of the pivoting operation, the column corresponding to the entering variable a_1' becomes a column of the identity matrix, and a_1' replaces the leaving variable t_5 in the canonical form. The resulting tableau is therefore used for the second iteration of the first phase:

iteration	: 2 (first phase)
entering variable	: a_0'
leaving variable	: t_2
pivot	: 0.6
current solution	: $s_1 = 0.6,\quad s_2 = 0.2,\quad s_3 = 0.8,\quad s_4 = 1.4,\quad s_5 = 0$
	: $t_1 = 0.6,\quad t_2 = 0.2,\quad t_3 = 0.8,\quad t_4 = 1.4,\quad a_1' = 0.4$
objective function	: $w = -3$

	y	a_0'	a_0''	a_1'	a_1''	e_1^+	e_2^+	e_3^+	e_4^+	e_5^+	s_1	s_2	s_3	s_4	s_5	s_6	s_7	s_8	s_9	s_{10}	t_1	t_2	t_3	t_4	t_5
w	-3	-2	2	0	0	-1	-1	-1	-1	2	0	0	0	0	0	1	1	1	1	-2	0	0	0	0	3
z	0	0	0	0	0	1	1	1	1	1	0	0	0	0	0	0	0	0	0	0	1	1	1	1	1
s_1	0.6	0.8	-0.8	0	0	-1	0	0	0	-0.2	1	0	0	0	0	0	0	0	0	0.2	0	0	0	0	-0.2
s_2	0.2	0.6	-0.6	0	0	0	-1	0	0	-0.4	0	1	0	0	0	0	0	0	0	0.4	0	0	0	0	-0.4
s_3	0.8	0.4	-0.4	0	0	0	0	-1	0	-0.6	0	0	1	0	0	0	0	0	0	0.6	0	0	0	0	-0.6
s_4	1.4	0.2	-0.2	0	0	0	0	0	-1	-0.8	0	0	0	1	0	0	0	0	0	0.8	0	0	0	0	-0.8
s_5	0	0	0	0	0	0	0	0	0	-2	0	0	0	0	1	0	0	0	0	1	0	0	0	0	-1
t_1	0.6	0.8	-0.8	0	0	1	0	0	0	-0.2	0	0	0	0	0	-1	0	0	0	0.2	1	0	0	0	-0.2
$\leftarrow t_2$	0.2	0.6	-0.6	0	0	0	1	0	0	-0.4	0	0	0	0	0	0	-1	0	0	0.4	0	1	0	0	-0.4
t_3	0.8	0.4	-0.4	0	0	0	0	1	0	-0.6	0	0	0	0	0	0	0	-1	0	0.6	0	0	1	0	-0.6
t_4	1.4	0.2	-0.2	0	0	0	0	0	1	-0.8	0	0	0	0	0	0	0	0	-1	0.8	0	0	0	1	-0.8
a_1'	0.4	0.2	-0.2	1	-1	0	0	0	0	0.2	0	0	0	0	0	0	0	0	0	-0.2	0	0	0	0	0.2

Tableau 2, Phase 1

The minimum coefficients in row 0 are associated with the variables a_0' and s_{10}. If we select the former, a_0', as entering variable, both variables s_2 and t_2 could be selected to leave the basis. Indeed, the ratios of the known coefficient to the positive entries in the pivot column are:

$$\min \left\{ \frac{0.6}{0.8}, \frac{0.2}{0.6}, \frac{0.8}{0.4}, \frac{1.4}{0.2}, \frac{0.6}{0.8}, \frac{0.2}{0.6}, \frac{0.8}{0.4}, \frac{1.4}{0.2}, \frac{0.4}{0.2} \right\}$$
$$= \min\{0.75, 0.33, 2, 7, 0.75, 0.33, 2, 7, 2\}$$

Again, we prefer the artificial variable t_2, hoping to close the first phase faster. The algorithm iterates through nine steps before the first phase stops, when the artificial objective function w equals 0. The entering and the leaving variables, along with the values of the objective function w in the next steps are:

iteration	leaving variable	entering variable	objective function w
3	s_7	s_2	-2.33
4	t_1	e_2^+	-2.33
5	s_5	s_{10}	-1.75
6	t_3	e_5^+	-1.75
7	s_3	s_8	-0.50
8	t_4	e_3^+	-0.50
9	—	—	0

The objective function changes only every two iterations. We are in the case of degenerate solutions (see Section 4.4), determined by null values of the minimum ratios associated with the leaving variables. Since the selected variables in steps 4, 6, and 8, enter the basis with a null value, the resulting solutions do not determine a movement toward another corner of the simplex, but only a change of the feasible solution in the basis. Hence the value of the objective function w does not change. At iteration 8, the last artificial variable t_4 exits the basis and the first phase ends, since the artificial objective function w equals 0. The columns associated with the five artificial variables t_1, t_2, t_3, t_4, t_5 are now deleted from the tableau, along with the row 0 associated with the artificial objective function w. The resulting tableau is the initial tableau of the second phase:

$\downarrow$

	y	a_0'	a_0''	a_1'	a_1''	e_1^+	e_2^+	e_3^+	e_4^+	e_5^+	s_1	s_2	s_3	s_4	s_5	s_6	s_7	s_8	s_9	s_{10}
z	-2.67	0	0	0	0	0.33	0	0	-1.33	0	0	1	1	0	1	0.67	0	0	2.33	0
s_1	0	0	0	0	0	-2	0	0	0	0	[1]	0	0	0	0	1	0	0	0	0
s_7	1.33	0	0	0	0	1.33	0	0	0.67	0	0	-1	0	0	0	-1.33	[1]	0	-0.67	0
← s_8	0.67	0	0	0	0	0.67	0	0	1.33	0	0	0	-1	0	0	-0.67	0	[1]	-1.33	0
s_4	0	0	0	0	0	0	0	0	-2	0	0	0	0	[1]	0	0	0	0	1	0
s_{10}	3.33	0	0	0	0	-0.67	0	0	2.67	0	0	0	0	0	-1	0.67	0	0	-2.67	[1]
e_2^+	0.67	0	0	0	0	0.67	[1]	0	0.33	0	0	-1	0	0	0	-0.67	0	0	-0.33	0
a_0'	0.33	[1]	-1	0	0	1.33	0	0	-0.33	0	0	0	0	0	0	-1.33	0	0	0.33	0
e_5^+	1.67	0	0	0	0	-0.33	0	0	1.33	[1]	0	0	0	0	-1	0.33	0	0	-1.33	0
e_3^+	0.33	0	0	0	0	0.33	0	[1]	0.67	0	0	0	-1	0	0	-0.33	0	0	-0.67	0
a_1'	0.67	0	0	[1]	-1	-0.33	0	0	0.33	0	0	0	0	0	0	0.33	0	0	-0.33	0

Tableau 1, Phase 2

Since it is already in canonical form with respect to the ten variables in basis, as highlighted through the boxed elements, the second phase of the algorithm starts:

iteration   : 1 (second phase)

entering variable : e_4^+

leaving variable   : s_8

pivot   : 1.33

current solution   : $s_1 = 0$, $s_7 = 1.33$, $s_8 = 0.67$, $s_4 = 0$, $s_{10} = 3.33$

         : $e_2^+ = 0.67$, $a_0' = 0.33$, $e_5^+ = 1.67$, $e_3^+ = 0.33$, $a_1' = 0.67$

objective function : $z = -2.67$

The variable e_4^+ is the only variable eligible to enter the basis, since the coefficients of all the other variables in row 0 are null or positive. For selecting the leaving variable, the ratios of the known coefficients to the positive entries of the pivot column are computed:

$$\min\left\{\frac{1.33}{0.67}, \frac{0.67}{1.33}, \frac{3.33}{2.67}, \frac{0.67}{0.33}, \frac{1.67}{1.33}, \frac{0.33}{0.67}, \frac{0.67}{0.33}\right\}$$
$$= \min\{2, \boxed{0.5}, 1.25, 2, 1.25, \boxed{0.5}, 2\}$$

Both s_8 and e_3^+ present a ratio equal to 0.5 and can be selected to exit the basis. By choosing s_8 as leaving variable, the algorithm stops with the last pivoting operation:

- the pivot row (s_8) is divided by the pivot element 1.33 in order to have a value 1 in the pivot position,

- the resulting row is therefore:

  - multiplied by -1.33 (the value of the pivot column) and subtracted by the row 0,

  - multiplied by 0.67 and subtracted by the row associated with s_7,

  - multiplied by -2 and subtracted by the row associated with s_4,

  - multiplied by 2.67 and subtracted by the row associated with s_{10},

  - multiplied by 0.33 and subtracted by the row associated with e_2^+,

  - multiplied by -0.33 and subtracted by the row associated with a_0',

  - multiplied by 1.33 and subtracted by the row associated with e_5^+,

  - multiplied by 0.67 and subtracted by the row associated with e_3^+,

  - multiplied by 0.33 and subtracted by the row associated with a_1'.

The resulting tableau fulfills the optimality condition since all the reduced costs are positive, and the algorithm stops:

iteration	: 2 (second phase)
entering variable	: —
leaving variable	: —
pivot	: —
current solution	: $s_1 = 0$, $s_7 = 1$, $e_4^+ = 0.5$, $s_4 = 1$, $s_{10} = 2$
	: $e_2^+ = 0.5$, $a_0' = 0.5$, $e_5^+ = 1$, $e_3^+ = 0$, $e_2^+ = 0.5$
objective function	: $z = -2$

	y	a'_0	a''_0	a'_1	a''_1	e^+_1	e^+_2	e^+_3	e^+_4	e^+_5	s_1	s_2	s_3	s_4	s_5	s_6	s_7	s_8	s_9	s_{10}
z	-2	0	0	0	0	1	0	0	0	0	0	1	0	0	1	0	0	1	1	0
s_1	0	0	0	0	0	-2	0	0	0	0	[1]	0	0	0	0	1	0	0	0	0
s_7	1	0	0	0	0	1	0	0	0	0	0	-1	0.5	0	0	-1	[1]	-0.5	0	0
e^+_4	0.5	0	0	0	0	0.5	0	0	[1]	0	0	0	-0.75	0	0	-0.5	0	0.75	-1	0
s_4	1	0	0	0	0	1	0	0	0	0	0	-1.5	0	[1]	0	-1	0	1.5	-1	0
s_{10}	2	0	0	0	0	-2	0	0	0	0	0	2	0	-1	2	0	-2	0	0	[1]
e^+_2	0.5	0	0	0	0	0.5	[1]	0	0	0	0	-1	0.25	0	0	-0.5	0	-0.25	0	0
a'_0	0.5	[1]	-1	0	0	1.5	0	0	0	0	0	-0.25	0	0	-1.5	0	0.25	0	0	0
e^+_5	1	0	0	0	0	-1	0	0	0	[1]	0	0	1	0	-1	1	0	-1	0	0
e^+_3	0	0	0	0	0	0	0	[1]	0	0	0	-0.5	0	0	0	0	0	-0.5	0	0
a'_1	0.5	0	0	[1]	-1	-0.5	0	0	0	0	0	0.25	0	0	0.5	0	-0.25	0	0	0

Tableau 2, Phase 2

By discarding the slack variables from the solution, it is possible to focus only on the decisional variables strictly related to the regression problem, namely the two regression coefficients a_0 and a_1, and the n absolute residuals $e^+_i, i = 1, \ldots, n$. Since $a'_0 = 0.5$ and a''_0 is out of the basis and hence equals 0, we have $a_0 = a'_0 - a''_0 = 0.5$. The same for the slope: $a_1 = a'_1 = 0.5$. With respect to the absolute residuals, from the tableau we have $e^+_2 = 0.5$, $e^+_3 = 0$, $e^+_4 = 0.5$, and $e^+_5 = 1$. The residual related to the first observation is out of the basis, and hence $e^+_1 = 0$. The sum of the absolute residuals is equal to 2, which is the value of the objective function z.

The e^+_i must be read as absolute residuals; for determining the proper signs of the residuals, it is sufficient to exploit the estimated regression coefficients, that is, to compute: $e_i = y_i - a'_0 - x_i a'_1$:

y_i	1	1	2	3	4	–
$a_0 + x_i a_1$	1	1.5	2	2.5	3	=
e_i	0	-0.5	0	0.5	-1	

The above solution is the same provided by any statistical software offering functions for quantile regression estimates. Here is the R code (R Core Team 2017) for computing the median regression for the small data set listed in Table 5.1. The code exploits the rq function contained in the quantreg package (Koenker 2017):

```
# loading the quantreg package
library(quantreg)
# storing the small data set
x <- c(1, 2, 3, 4, 5); y <- c(1, 1, 2, 3, 2)
# QR regression for the conditional median
median_reg <- rq(y ~ x, tau = 0.5)
```

```
# NOTE: the default value of the tau argument is set to 0.5
#      this command is hence equivalent to the previous one
median_reg <- rq(y ~ x)
# QR estimates for the conditional median
coefficient(median_reg)
        (Intercept)                x
            0.5                    0.5
Warning message:
In rq.fit.br(x, y, tau = tau, ...) : Solution may be nonunique
#QR residuals
residuals(median_reg)
        1         2         3         4         5
      0.0      -0.5       0.0       0.5      -1.0
```

The warning message in the previous code alerts the user about the non–uniqueness of the solution. As outlined in the paper of Barrodale and Roberts (1973), the solution is not unique for this example. The presence of a null variable in the basis is a clue even if not a sufficient condition for non–uniqueness. For this data set, we are in the very special case where the optimal solution does not correspond to a single vertex of the simplex but to an edge (see the geometrical interpretation of the non–uniqueness in subsection 4.1.3 and next Section 5.3 for the geometry of median regression).

In the previous example we assigned a higher priority to the artificial variables t_i for leaving the basis in the first phase. If on the contrary we would exploit the Bland's rule[1] for choosing among entering/leaving candidates, the entering and the leaving variables, and the values of the artificial objective function w in each step of the first phase of the simplex algorithm, become:

iteration	leaving variable	entering variable	objective function w
1	s_5	a'_1	-9
2	t_5	e_5^+	-3
3	s_2	a'_0	-3
4	t_2	e_2^+	-2.33
5	s_1	s_7	-2.33
6	t_1	e_1^+	-1.75
7	t_3	e_3^+	-1.75
8	t_4	e_4^+	-1.25
9	—	—	0

with the following tableau at the end of the first phase:

---

[1] In case of ambiguity we select the variable with the lowest index among the candidates (see Section 4.4).

	y	a'_0	a''_0	a'_1	a''_1	e^+_1	e^+_2	e^+_3	e^+_4	e^+_5	s_1	s_2	s_3	s_4	s_5	s_6	s_7	s_8	s_9	s_{10}	t_1	t_2	t_3	t_4	t_5
w	0	0	0	0	0	0	0	0	0	0	0	0	0	0	0	0	0	0	0	0	0	1	1	1	1
z	-2	0	0	0	0	0	0	0	0	0	0.5	1	0	0	1	0.5	0	1	1	0	0.5	1	0	0	1
s_7	0.5	0	0	0	0	0	0	0	0	0	0.75	-1	0	0	0.25	-0.75	[1]	0	0	-0.25	0.75	-1	0	0	0.25
a'_0	0.75	[1]	-1	0	0	0	0	0	0	0	0.625	0	0	0	-0.125	-0.625	0	0	0	0.125	0.625	0	0	0	-0.125
s_3	1	0	0	0	0	0	0	0	0	0	-0.5	0	[1]	0	-0.5	0.5	0	-1	0	0.5	-0.5	0	1	0	-0.5
s_4	2.5	0	0	0	0	0	0	0	0	0	-0.25	0	0	[1]	-0.75	0.25	0	0	-1	0.75	-0.25	0	0	1	-0.75
a'_1	0.25	0	0	1	-1	0	0	0	0	0	-0.125	0	0	0	0.125	0.125	0	0	0	-0.125	-0.125	0	0	0	0.125
e^+_1	0	0	0	0	0	[1]	0	0	0	0	-0.5	0	0	0	0	-0.5	0	0	0	0	0.5	0	0	0	0
e^+_2	0.25	0	0	0	0	0	[1]	0	0	0	0.375	-1	0	0	0.125	-0.375	0	0	0	-0.125	0.375	0	0	0	0.125
e^+_3	0.5	0	0	0	0	0	0	[1]	0	0	-0.25	0	0	0	-0.25	0.25	0	-1	0	0.25	-0.25	0	1	0	-0.25
e^+_4	1.25	0	0	0	0	0	0	0	[1]	0	-0.125	0	0	0	-0.375	0.125	0	0	-1	0.375	-0.125	0	0	1	-0.375
e^+_5	0	0	0	0	0	0	0	0	0	[1]	0	0	0	0	-0.5	0	0	0	0	-0.5	0	0	0	0	0.5

Tableau 9, Phase 1

Since the optimality condition is already satisfied for the row 0 associated with the initial objective function z, the second phase does not start, and the current solution is optimal also for the initial problem. Therefore the values $a'_0 = a_0 = 0.75$ and $a'_1 = a_1 = 0.25$ provide the same value 2 for the objective function z, obtained with the previous solution. The same for any couple of values $a_0 = 0.75 - k$ and $a_1 = 0.25 + k$, for $0 \le k \le 0.25$, as it is easy to verify through the following R code:

```
# set the values for the two coefficients
a0 <- 0.75
a1 <- 0.25
# compute the sum of absolute residuals for any linear combination
# of the points (a0 = 0.5, a1 = 0.5) and (a0 = 0.75, a1 = 0.25)
# i.e. for any points lying on the segment that join such two points
# in the plane (a0, a1)
for(k in seq(0, 0.25, 0.01))
    print(sum(abs(y - ((a0 - k) + (a1 + k) * x))))
you will see the value 2 printed 26 times
(where 26 are the values considered for k)
```

Section 5.3 will further detail this issue, focusing on its geometrical interpretation.

The same warning message concerning the non–uniqueness of the solution may result also for a different reason when computing quantile regression estimates in R. This is the case when there is an even number of observations, especially when the covariates are discrete. As in Koenker (2005, 2017):

All the possible solutions are correct. Just as any number between the two central order statistics is a median when the sample size is even and the order statistics are distinct. The main point here is that the differences between solutions are of order $1/n$, and the inherent uncertainty about the estimates is of order $1/\sqrt{(n)}$, so the former variability is essentially irrelevant.

And also:

> Should we worry about this? My answer would be no. Viewed from an asymptotic perspective, any choice of a unique value among the multiple solutions is a $1/n$ perturbation – with 2500 observations this is unlikely to be interesting. More to the point, inference about the coefficients of the model, which provides $O(1/\sqrt{(n)})$ intervals, is perfectly capable of assessing the meaningful uncertainty about these values.

5.1.2   A more convenient formulation of the L_1 regression problem

An alternative formulation of the L_1 problem in terms of LP directly provides the signed residuals. Such formulation has been proposed by Wagner (1959) in the statistical literature, and by Barrodale and Roberts (1970) in the LP literature. Barrodale and Roberts (1973) present an improved algorithm for the L_1 approximation that is implemented in the main software to compute median regression (see next subsection for details). Although this formulation could seem more verbose due to the handling of the sign of the residuals, it actually turns out to be more convenient in terms of the resulting tableau.

Two nonnegative variables, u_i and v_i, are introduced to treat the signed residuals $e_i = y_i - (a_0 + x_i a_1)$. The L_1 objective function (5.2) to minimize becomes:

$$\sum_{i=1}^{n} |y_i - (a_0 + x_i a_1)| = \sum_{i=1}^{n} |e_i| = \sum_{i=1}^{n} |u_i - v_i|$$

The u_i's are associated with points above the regression line while the v_i's are associated with points below the line. In fact, by posing:

$$u_i = \sum_{i=1}^{n} y_i - (a_0 + x_i a_1)$$

$$v_i = \sum_{i=1}^{n} (a_0 + x_i a_1) - y_i$$

it is evident that u_i and v_i are linearly dependent, being one the opposite of the other. Such linear dependence implies that in calculating a solution through the simplex method, at most one of either u_i and v_i, for any i, will be non–zero at every stage of the calculations. A theorem in Charnes et al (1953) ensures indeed that the column vectors selected at any simplex stage are linearly independent, and hence, the two following objective functions are equivalent:

$$\sum_{i=1}^{n} |u_i - v_i| = \sum_{i=1}^{n} (u_i + v_i)$$

Charnes et al (1955) exploits this trick in the context of a specific industrial application, albeit they stress its relevance to extend the use of LP to handle an important class of nonlinear problems. Through its use, the alternative LP formulation of the L_1 regression problem becomes:

$$\text{minimize} \quad \sum_{i=1}^{n}(u_i + v_i)$$
$$\text{subject to} \quad y_i - (a_0 + a_1 x_i) = u_i - v_i \quad i = 1, \dots, n$$

The n constraints can be rewritten to isolate the known value on the right side of the equation:

$$\text{minimize} \quad \sum_{i=1}^{n}(u_i + v_i)$$
$$\text{subject to} \quad (a_0 + a_1 x_i) + u_i - v_i = y_i \quad i = 1, \dots, n$$

Then, for dealing with the case of unrestricted in sign coefficient, the two coefficients are transformed as follows (see Subsection 5.1.1):

$$\text{minimize} \quad \sum_{i=1}^{n}(u_i + v_i)$$
$$\text{subject to} \quad (a_0' - a_0'') + x_i (a_1' - a_1'') + u_i - v_i = y_i \quad i = 1, \dots, n$$

and, finally we obtain:

$$\text{minimize} \quad \sum_{i=1}^{n}(u_i + v_i) \tag{5.6}$$
$$\text{subject to} \quad a_0' - a_0'' + x_i a_1' - x_i a_1'' + u_i - v_i = y_i \quad i = 1, \dots, n$$

The use of n additional variables for dealing with the signed residuals is more than balanced by the fact that the last formulation is already expressed in canonical form with respect to the u_i, thus avoiding the introduction of the slack and of the artificial variables, required in formulation (5.1.2). In fact, this formulation corresponds to the following condensed tableau consisting of $n + 1$ rows, $2n + 2k$ columns, where k is equal to the number of coefficients of the interpolating function, namely 2 for the simple regression problem:

	y	a_0'	a_0''	a_1'	a_1''	u_1	v_1	u_2	v_2	$\dots$	u_n	v_n
z	$\sum_{i=1}^{n} y_i$	n	$-n$	$\sum_{i=1}^{n} x_i$	$-\sum_{i=1}^{n} x_i$	0	-2	0	-2	$\dots$	-2	-2
u_1	y_1	1	-1	x_1	$-x_1$	$\boxed{1}$	-1	0	0	$\dots$	0	0
u_2	y_2	1	-1	x_2	$-x_2$	0	0	$\boxed{1}$	-1	$\dots$	0	0
$\vdots$	$\vdots$	$\vdots$	$\vdots$	$\vdots$	$\vdots$	$\vdots$	$\vdots$	$\vdots$	$\vdots$	$\vdots$	$\vdots$	$\vdots$
u_n	y_n	1	-1	x_n	$-x_n$	0	0	0	0	$\dots$	$\boxed{1}$	-1

The tableau is already in canonical form with respect to the $u_i, i = 1, \dots, n$, as highlighted through the boxed elements. The simplex algorithm can hence be directly carried out. The tableau representation can be further condensed by exploiting the relations:

$$a'_0 = -a''_0 \qquad a'_1 = -a''_1 \qquad u_i = -v_i, \ i = 1, \dots, n$$

and keeping in mind that the sum of the reduced costs of a'_0 and a''_0 is zero, the same for a'_1 and a''_1, while the sum of the costs of each u_i and v_i is always -2. This allows to reduce the tableau to the columns $a'_0, a'_1, u_1, \dots, u_n$, since the dropped columns are immediately available from the others. However, for an immediate understanding of the example, the steps detailed below for the small data set listed in Table 5.1 are shown using the full tableaus. In particular, the initial tableau for the median regression problem is:

	y	a'_0	a''_0	a'_1	a''_1	u_1	v_1	u_2	v_2	u_3	v_3	u_4	v_4	u_5	v_5
z	9	5	-5	15	-15	0	-2	0	-2	0	-2	0	-2	0	-2
u_1	1	1	-1	1	-1	$\boxed{1}$	-1	0	0	0	0	0	0	0	0
u_2	1	1	-1	2	-2	0	0	$\boxed{1}$	-1	0	0	0	0	0	0
u_3	2	1	-1	3	-3	0	0	0	0	$\boxed{1}$	-1	0	0	0	0
u_4	3	1	-1	4	-4	0	0	0	0	0	0	$\boxed{1}$	-1	0	0
u_5	2	1	-1	5	-5	0	0	0	0	0	0	0	0	$\boxed{1}$	-1

Tableau 1

Since it is already in canonical form with respect to the five variables u_i, the simplex can be carried out:

- a'_1 enters the basis, with the largest marginal cost 15;

- u_5 leaves the basis, as evident from the ratios of the known coefficients (first column of the tableau) to the column of the entering variable itself for the positive entries:

$$\min\left\{\frac{1}{1}, \frac{1}{2}, \frac{2}{3}, \frac{3}{4}, \frac{2}{5}\right\} = \min\{1, 0.5, 0.67, 0.75, \boxed{0.4}\}$$

Denoting again the entering variable with the symbol $\downarrow$ and the leaving variable with the symbol $\leftarrow$, the first iteration of the simplex algorithm is:

iteration	: 1
entering variable	: a'_1
leaving variable	: u_5
pivot	: 5
current solution	: $u_1 = 1, \quad u_2 = 1, \quad u_3 = 2, \quad u_4 = 3, \quad u_5 = 2$
objective function	: $z = 9$

	y	a'_0	a''_0	a'_1	a''_1	u_1	v_1	u_2	v_2	u_3	v_3	u_4	v_4	u_5	v_5
z	9	5	−5	15	−15	0	−2	0	−2	0	−2	0	−2	0	−2
u_1	1	1	−1	1	−1	1	−1	0	0	0	0	0	0	0	0
u_2	1	1	−1	2	−2	0	0	1	−1	0	0	0	0	0	0
u_3	2	1	−1	3	−3	0	0	0	0	1	−1	0	0	0	0
u_4	3	1	−1	4	−4	0	0	0	0	0	0	1	−1	0	0
← u_5	2	1	−1	5	−5	0	0	0	0	0	0	0	0	1	−1

Tableau 1

The pivoting operation to transform the a'_1 column in a column of the identity matrix is:

- the pivot row (u_5) is divided by the pivot element 5,
- the resulting pivot row is:
  - multiplied by 15 and subtracted from the row 0,
  - subtracted from the row associated with u_1,
  - multiplied by 2 and then subtracted from the u_2 row,
  - multiplied by 3 and then subtracted from the u_3 row,
  - multiplied by 4 and then subtracted from the u_4 row.

The resulting tableau is used for the second iteration of the algorithm:

iteration   : 2
entering variable   : a'_0
leaving variable   : u_2
pivot   : 0.6
current solution   : $u_1 = 0.6$, $u_2 = 0.2$, $u_3 = 0.8$, $u_4 = 1.4$, $a'_1 = 0.4$
objective function : $z = 3$

	y	a'_0	a''_0	a'_1	a''_1	u_1	v_1	u_2	v_2	u_3	v_3	u_4	v_4	u_5	v_5
z	3	2	−2	0	0	0	−2	0	−2	0	−2	0	−2	-3	1
u_1	0.6	0.8	−0.8	0	0	1	−1	0	0	0	0	0	0	−0.2	0.2
← u_2	0.2	0.6	−0.6	0	0	0	0	1	−1	0	0	0	0	−0.4	0.4
u_3	0.8	0.4	−0.4	0	0	0	0	0	0	1	−1	0	0	−0.6	0.6
u_4	1.4	0.2	−0.2	0	0	0	0	0	0	0	0	1	−1	−0.8	0.8
a'_1	0.4	0.2	−0.2	1	−1	0	0	0	0	0	0	0	0	0.2	−0.2

Tableau 2

The entering variable is now a'_0, with an associated reduced cost equal 2, providing the following values for determining the leaving variable:

$$\min\left\{\frac{0.6}{0.8}, \frac{0.2}{0.6}, \frac{0.8}{0.4}, \frac{1.4}{0.2}, \frac{0.4}{0.2}\right\} = \min\{0.75, 0.33, 2, 7, 2\}$$

The variable u_2 is hence selected as leaving variable and the following pivoting operation is carried out:

- the row u_2 is divided by the pivot element 0.6,

- the resulting pivot row is:

    - multiplied for -2, element of the pivot column in row 0, and then subtracted from the row 0 itself,

    - multiplied for 0.8, element of the pivot column in the row u_1, and then subtracted from the latter,

    - multiplied for 0.4, element of the pivot column in the row u_3, and then subtracted from the row u_3,

    - multiplied for 0.2, element of the pivot column in the row u_4, and then subtracted from the row u_4,

    - multiplied for 0.2, element of the pivot column in the row a'_1, and then subtracted from the last row.

The resulting tableau, where a'_0 and u_2 have switched their role in the basis, is the input of the next iteration:

iteration	: 3		
entering variable	: v_2		
leaving variable	: u_1		
pivot	: 1.33		
current solution	: $u_1 = 0.33$, $a'_0 = 0.33$, $u_3 = 0.67$, $u_4 = 1.33$, $a'_1 = 0.33$		
objective function	: $z = 2.33$		

$\downarrow$

	y	a'_0	a''_0	a'_1	a''_1	u_1	v_1	u_2	v_2	u_3	v_3	u_4	v_4	u_5	v_5
z	2.33	0	0	0	0	0	-2	-3.33	1.33	0	-2	0	-2	-1.67	-0.33
$\leftarrow u_1$	0.33	0	0	0	0	1	-1	-1.33	1.33	0	0	0	0	0.33	-0.33
a'_0	0.33	1	-1	0	0	0	0	1.67	-1.67	0	0	0	0	-0.67	0.67
u_3	0.67	0	0	0	0	0	0	-0.67	0.67	1	-1	0	0	-0.33	0.33
u_4	1.33	0	0	0	0	0	0	-0.33	0.33	0	0	1	-1	-0.67	0.67
a'_1	0.33	0	0	1	-1	0	0	-0.33	0.33	0	0	0	0	-1.67	-0.33

Tableau 3

The variable v_2 is the only one eligible to enter the basis, taking the place of u_1, selected to leave the basis once inspected the ratios:

$$\min\left\{\frac{0.33}{1.33}, \frac{0.33}{0.67}, \frac{1.33}{0.33}, \frac{0.4}{0.33}\right\} = \min\{0.25, 1, 4, 1\}$$

The pivoting operation is carried out on the pivot element 1.33:

- the row labeled u_1 is subtracted from the row 0,

- the row labeled u_1 is hence divided by the pivot element,

- the resulting pivot row is:

  - multiplied by -1.67 and subtracted from the row a'_0,

  - multiplied by 0.67 and subtracted from the row u_3,

  - multiplied by 0.33 and subtracted from the rows u_4 and a'_0.

The algorithm stops since the resulting tableau fulfills the optimality condition, being all the reduced costs negative:

iteration	: 4
entering variable	: —
leaving variable	: —
pivot	: —
current solution	: $v_2 = 0.25$, $a'_0 = 0.75$, $u_3 = 0.5$, $u_4 = 1.25$, $a'_1 = 0.25$
objective function	: $z = 2$

	y	a'_0	a''_0	a'_1	a''_1	u_1	v_1	u_2	v_2	u_3	v_3	u_4	v_4	u_5	v_5
z	2	0	0	0	0	-1	-1	-2	0	0	-2	0	-2	-2	0
v_2	0.25	0	0	0	0	0.75	-0.75	-1	[1]	0	0	0	0	0.25	-0.25
a'_0	0.75	[1]	-1	0	0	1.25	-1.25	0	0	0	0	0	0	-0.25	0.25
u_3	0.50	0	0	0	0	-0.50	0.50	0	0	[1]	-1	0	0	-0.5	0.5
u_4	1.25	0	0	0	0	-0.25	0.25	0	0	0	0	[1]	-1	0.25	-0.25
a'_1	0.25	0	0	[1]	-1	-0.25	0.25	0	0	0	0	0	0	-1.67	-0.33

Tableau 4

The solution with $a'_0 = a_0 = 0.75$ and $a'_1 = a_1 = 0.25$, providing an objective function equals to 2, interpolates the first and fifth data point, being u_1 and v_1, u_5 and v_5 out of the basis and hence equal 0. The median regression line lies above the second observations ($v_2 = 0.25$) and below the third and fourth point ($u_3 = 0.5$ and $u_4 = 1.25$). This solution coincides with the second optimal solution obtained in subsection 5.1.1, with the difference that the sign of the residuals is here directly provided. The nonnegative variables u_i and v_i can be interpreted as positive and negative deviations associated with the i–th observation. A geometrical interpretation for (one of) the optimal solution(s) $a_0 = 0.75$, and $a_1 = 0.25$ is provided in Figure 5.1: the median regression line interpolating the first and fifth observation is depicted using a ticker line, the vertical solid lines refer to the variables in basis while the vertical dashed lines to the variables out of the basis, and hence equal 0 in the simplex solution. If we refer, for example, to the 2nd observation, it lies below the median regression line: therefore only v_2 is not null and is equal to 0.25. Indeed:

$$e_2 = y_2 - a_0 - a_1 x_2 = -0.25 = 0 - 0.25 = u_2 - v_2 = -v_2.$$

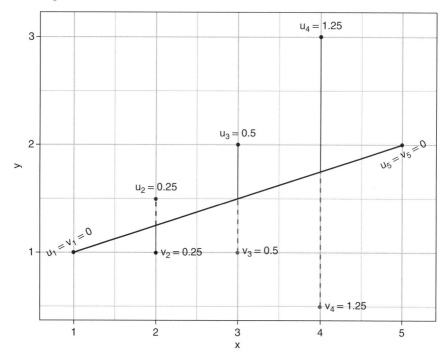

Figure 5.1 Geometrical representation of the two nonnegative variables u_i and v_i introduced in the LP problem (5.6) for the optimal solution $a_0 = 0.75$ and $a_1 = 0.25$.

On the contrary, the 3rd observation is above the line, v_3 is out of the basis, and is equal to 0, while $u_3 = 0.5$. In such a case:

$$e_3 = y_3 - a_0 - a_1 x_3 = 0.5 = 0.5 - 0 = u_3 - v_3 = u_3.$$

The same happens for the 4th observation, for which $e_4 = u_4 = 1.25$. Observations 1 and 5 are instead perfectly interpolated by the median regression line and hence $e_1 = e_5 = 0$.

5.1.3 The Barrodale–Roberts algorithm for L_1 regression

Exploiting the special structure of the LP formulation of the L_1 regression problem, Barrodale and Roberts (1973) proposed a slight modification of the simplex algorithm. This variant, implemented in most statistical software, passes through several neighboring simplex vertices in a single iteration. The original Fortran code is available in Barrodale and Roberts (1972).

Using the LP formulation (5.6) of the L_1 problem, an initial basis is provided by some n of the vectors u_i and v_i. In particular, when each y_i is nonnegative, an initial basis is provided by $u_1, u_2, \ldots, u_n$. Whenever a y_i is negative, the corresponding row in the tableau is multiplied by -1 and the u_i is replaced by v_i in the basis. In fact, $u_i = -v_i$. In the following we assume without loss of generality that each y_i is nonnegative. From the last relation, it follows that only n columns can be used to store the vectors in basis, since the current values of the corresponding negative vectors are readily available at each step starting from these columns. Recalling that $a'_j = -a''_j$, only m columns are needed to store the right–hand sides of the equality constraints of the LP problem (5.6). Thus, in considering any column vector of the tableau for the simplex algorithm, the corresponding negative vector must also be considered. Moreover, considering that the sum of the reduced costs associated with a'_j and a''_j is zero, as well as the sum of the reduced costs of u_j and v_j is -2, Barrodale and Roberts (1973) proposed a condensed form of the simplex tableau in which the basis is suppressed and only the $a'_1, a'_2, \ldots, a'_m$ columns are initially inserted. For the simple regression problem with $m = 2$, the condensed tableau is:

	y	a'_0	a'_1
z	$\sum_{i=1}^{n} y_i$	n	$\sum_{i=1}^{n} x_i$
u_1	y_1	1	x_1
u_2	y_2	1	x_2
$\vdots$	$\vdots$	$\vdots$	$\vdots$
u_n	y_n	1	x_n

The algorithm is implemented in two stages exploiting a recommendation provided in Barrodale and Roberts (1970), based on their computational experience. In particular, for the first m iterations (stage 1 of the algorithm) the choice of the pivotal column is restricted to the columns associated with the coefficients a'_j and a''_j. The choice of the variable entering the basis is always carried out using the standard rule of the simplex method, that is, looking for the largest nonnegative reduced cost at each iteration. The first stage consists in the selection of the proper signed regression coefficients to enter the basis. The basic a'_j and a''_j are not allowed to leave the basis during stage 2. Therefore, stage 2 involves interchanging nonbasic u_i or v_i with basic u_i or v_i. The main modification to the simplex method is the adoption of a special pivotal selection rule for choosing the vector u_i or v_i to leave the basis in both stage 1 and 2 of the algorithm. To ensure the maximum reduction in the objective function, the normal rule for determining the leaving variable is modified as follows. The normal rule to determine the pivotal row among the basic vectors u_i and v_i is firstly applied to locate the pivot. The pivot value is then subtracted twice from the reduced cost of the pivotal column: in case this operation yields a nonpositive result, this pivot is used to perform the simplex transformation, because the objective function cannot be further improved. Otherwise, in order to decrease further the value of the objective function, the vector u_i (or v_i) corresponding to the pivotal row is replaced in the basis by the vector v_i (or u_i). This is accomplished by subtracting twice the pivotal row from the reduced cost row, thus making the marginal cost of v_i (or u_i) zero, and then by multiplying the pivotal row by -1, in order to change the sign of the pivot. The first operation is based on the fact that the sum of each pair u_i and v_i is always -2: if u_i (or v_i) is a basic variable, its reduced cost is zero and hence the reduced cost of v_i (or u_i) is -2. The multiplication of the pivotal row by -1 exploits the relation $u_i = -v_i$ and interchanges the two variables in the basis. The normal rule for determining the pivotal row is applied again on the resulting configuration of the condensed tableau to locate a new pivot. Such procedure is iterated until the determination of a pivot that cannot be rejected. This pivot is then used to perform a simplex transformation. This special selection rule for determining the leaving variable allows to pass through several vertices in a single iteration. From a geometric point of view, this modified rule for determining the leaving variable usually entails a movement through several neighboring simplex vertices (Barrodale and Roberts 1973), as detailed in next Section 5.3.

It may happen that the chosen pivotal column contains no positive elements in correspondence of the basic u_i and v_i variables, and hence no suitable vector can be found to leave the basis. This occurs in stage 1 if the rank of the coefficient matrix is less than m. In such a case the current pivotal column a'_j (or a''_j) can be ignored in all future computations, and no simplex transformation is performed in this iteration. In stage 2, instead, a suitable pivotal row should be always available for any pivotal column, since a solution to the L_1 problem is guaranteed to exist. If this unexpected situation occurs, this is typically due to rounding errors or single-precision arithmetic, especially in presence of wide different magnitudes of the elements of the coefficient matrix. The implementation of the algorithm in multiple-precision arithmetic, as well as proper data transformation or the use of a small tolerance limit below which the magnitude of any quantity is considered to be zero should be applied when dealing with this case. The interested reader is referred to the Fortran implementation of the original algorithm (Barrodale and Roberts 1972).

The algorithm normally stops during stage 2 when no suitable vector can be designated as the next pivotal column, namely when there are no nonbasic variables u_i or v_i with a positive reduced cost. In case the final tableau contains basic vectors a'_j (or a''_j) with negative associated values, any such row must be multiplied by -1, and the basic vector a'_j (or a''_j) must be interchanged with the corresponding nonbasic vector a''_j (or a'_j) in order to yield a feasible, and hence optimal, solution. The output information is available on the first column of the final condensed tableau, which originally contained the y_i values. The number of iterations of the algorithm is related to the total number of simplex transformation performed: the counter is increased by one when a simplex transformation occurs (stage 1 or stage 2) or when the simplex transformation is bypassed (stage 2) because no suitable vector can leave the basis.

Figure 5.2 shows the flowchart of the Barrodale–Roberts algorithm. The proposed flowchart slightly differs from the original one, highlighting the main differences between stage 1 and stage 2, namely the set of variables among which the entering variable is selected and the different management in case no suitable vector is found as leaving variable. The algorithm is shown in action using the data of Table 5.1, the same of Barrodale and Roberts (1973).

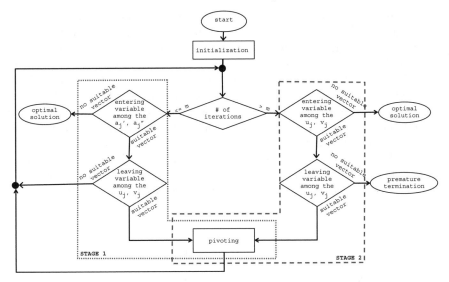

Figure 5.2   The general structure of the Barrodale–Roberts algorithm for L_1 regression.

Initialization

The initial condensed tableau for starting the algorithm follows:

	y	a'_0	a'_1
z	9	5	15
u_1	1	1	1
u_2	1	1	2
u_3	2	1	3
u_4	3	1	4
u_5	2	1	5

No additional transformation are needed since all the y_i are nonnegative.

Stage 1 - iteration 1

The vector a'_1 is selected to enter the basis, due to its largest reduced cost 15. Using the normal rule to determine the pivotal row, that is, by computing the ratios of the known coefficient to the positive values of the pivotal column for the nonnegative entries, the first candidate leaving variable is u_5:

	9		15	
u_1	1/1		1	
u_2	1/2	=	0.5	$\implies$ u_5 is the candidate leaving variable
u_3	2/3		0.67	
u_4	3/4		0.75	
u_5	2/5		0.4	

The use of the pivot value 5 (labeled with $*$ in the condensed tableau below), should involve a'_1 entering the basis with a value $\frac{2}{5} = 0.4$, and u_5 leaving the basis, and hence equals 0. Recalling that the sum of the reduced cost of u_i and v_i is always 0, such solution corresponds to an approximation which interpolates the fifth data point ($x_5 = 5, y_5 = 2$). However, by increasing a'_1 beyond the value $\frac{2}{5}$, the objective function could be further reduced. This makes u_5 negative and therefore the solution not feasible. In order to obtain an additional reduction in the objective function, u_5 is replaced in the basis by v_5. This is accomplished by multiplying the pivot row by -1. By subtracting twice the candidate pivot row from the row of the reduced costs (row 0), the marginal cost of a'_1 is now 5: a'_1 can be increased further so to improve the objective function. The first pivot ($5*$) is hence discarded, and the new condensed tableau to use for determining the pivotal row through the normal simplex rule is:

	y	a'_0	a'_1
z	5	3	5
u_1	1	1	1
u_2	1	1	2
u_3	2	1	3
u_4	3	1	4
v_5	2	-1	$-5*$

Condensed tableau 1
(pivot 5* discarded)

Using the normal simplex rule, u_2 is the new candidate leaving variable, as results from the following ratios:

	5		5
u_1	$1/1$		1
u_2	$1/2$	=	0.5
u_3	$2/3$		0.67
u_4	$3/4$		0.75
v_5	—		—

$\implies u_2$ is the candidate leaving variable

Also this second pivot is discarded, since the objective function can be improved exchanging u_2 with v_2 in the basis. This involves a reduction in the reduced cost of a'_1 to 1. The new condensed tableau, obtained by subtracting twice the candidate pivot row from row 0 and multiplying the candidate pivot row by -1 is:

	y	a'_0	a'_1
z	3	1	1
u_1	1	1	1
v_2	1	-1	$-2**$
u_3	2	1	3
u_4	3	1	4
v_5	2	-1	$-5*$

Condensed tableau 1
(pivot 2** discarded)

With v_2 in basis, the pivot 2** corresponds to a solution which interpolates the second data point. The third candidate pivot is u_3:

	3		3
u_1	1/1		1
v_2	—		—
u_3	2/3	=	0.67
u_4	3/4		0.75
v_5	—		—

$\implies u_3$ is the candidate leaving variable

It is actually used as leaving variable. Discarding it and interchanging u_3 and v_3, the marginal cost of a_1' would become indeed negative. The current iteration value ends with the simplex transformation performed using the pivot value 3***. By placing side by side the condensed tableau and the column corresponding to entering variable u_3:

	y	a_0'	a_1'		u_3
z	3	1	1		0
u_1	1	1	1		0
v_2	1	−1	−2 **		0
← u_3	2	1	3 ***		1
u_4	3	1	4		0
v_5	2	−1	−5 *		0
	Condensed tableau 1				
	(pivot 3 ** selected)				

the simplex transformation, which terminates the first iteration, is detailed below:

			y	a_0'	a_1'		u_3
$\boxed{\text{row}_0}$ −1 × $\boxed{\text{pivot row}}$	→	z	2.33	0.67	0		−0.33
$\boxed{\text{row}_1}$ −1 × $\boxed{\text{pivot row}}$	→	u_1	0.33	0.67	0		−0.33
$\boxed{\text{row}_2}$ +2 × $\boxed{\text{pivot row}}$	→	v_2	0.33	−0.33	0		0.67
$\boxed{\text{pivot row}}$ = $\boxed{\text{row}_3}$ /3	→	u_3	0.67	0.33	1		0.33
$\boxed{\text{row}_4}$ −4 × $\boxed{\text{pivot row}}$	→	u_4	0.33	−0.33	0		−1.33
$\boxed{\text{row}_5}$ +5 × $\boxed{\text{pivot row}}$	→	v_5	1.33	0.67	0		1.67

Stage 1 - iteration 2

The second condensed tableau exploits the column corresponding to the entering variable obtained after the simplex transformation in place of the column a_1' (leaving variable). The only eligible entering variable is a_0', with a positive reduced cost equals 0.67 (recall that in stage 1 of the algorithm the choice of the entering variable is restricted only to the columns a_0', a_0'', a_1', and a_1'', the latter already in basis). The normal rule for choosing the leaving variable signals u_1 as candidate leaving variable:

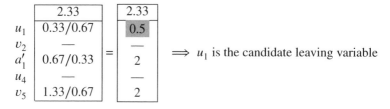

$$
\begin{array}{c|c}
 & 2.33 \\ \hline
u_1 & 0.33/0.67 \\
v_2 & - \\
a_1' & 0.67/0.33 \\
u_4 & - \\
v_5 & 1.33/0.67
\end{array}
=
\begin{array}{|c|}
\hline
2.33 \\ \hline
0.5 \\
- \\
2 \\
- \\
2 \\ \hline
\end{array}
\implies u_1 \text{ is the candidate leaving variable}
$$

No further improvement of the objective function can be obtained interchanging u_1 and v_1, and hence 0.67 is the actual pivot. The column corresponding to the entering variable u_1 is placed side by side with the condensed tableau:

	y	a_0'	u_3	u_1
z	2.33	0.67	−0.33	0
← u_1	0.33	0.67 *	−0.33	1
v_2	0.33	−0.33	0.67	0
u_3	0.67	0.33	0.33	0
u_4	0.33	−0.33	−1.33	0
v_5	1.67	0.67	1.67	0

Condensed tableau 2
(pivot 0.67* * selected)

Therefore the following simplex transformation ends the second iteration of the algorithm:

						y	a_0'	u_3	u_1
$\boxed{\text{row}_0}$ $- 1 \times \boxed{\text{pivot row}}$	→	z				2	−1	0	0
$\boxed{\text{pivot row}}$ $= \boxed{\text{row}_1}$ /0.67	→	u_1				0.5	1.5	1	−0.5
$\boxed{\text{row}_2}$ $+ 0.33 \times \boxed{\text{pivot row}}$	→	v_2				0.5	0.5	0	0.5
$\boxed{\text{row}_3}$ $- 0.33 \times \boxed{\text{pivot row}}$	→	u_3				0.5	−0.5	0	0.5
$\boxed{\text{row}_4}$ $+ 0.33 \times \boxed{\text{pivot row}}$	→	u_4				0.5	0.5	0	−1.5
$\boxed{\text{row}_5}$ $- 0.67 \times \boxed{\text{pivot row}}$	→	v_5				1	−1	0	2

The Stage 1 of the algorithm ends since $m = 2$ iterations have been performed. Here is the resulting condensed tableau, obtained inserting the column u_1 (leaving variable) in place of the column a_0' (entering variable):

	y	u_1	u_3
z	2	−1	0
a_0'	0.5	1.5	−0.5
v_2	0.5	0.5	0.5
a_1'	0.5	−0.5	0.5
u_4	0.5	0.5	−1.5
v_5	1	−1	2

Condensed tableau 3

It corresponds to an approximation that interpolates the first and third data points: u_1 and u_3 are in basis and hence equal zero, as well as their opposite v_1 and v_3. Since their reduced cost are -1 and 0, respectively, we have that also the reduced costs of v_1 and v_3 are negative (in particular equal to -1 and -2, recalling the condition on the sum of the costs of u_i and v_i). Therefore, the algorithm stops without the need of stage 2. The solution $a'_0 = 0.5$ and $a'_1 = 0.5$ coincides with the first solution of the standard simplex algorithm in the previous subsection. The presence of a null reduced cost in correspondence of u_3 is a clue, even if not a sufficient condition, for nonuniqueness of the solution: the algorithm indicates the possibility of nonuniqueness in presence of a zero reduced cost (Barrodale and Roberts 1972).

5.2   LP formulation of the quantile regression problem

The previous section introduced the median regression problem in case of univariate regression. The extension to the case of more regressors, multiple regression, is obtained considering a vector of explicative variables x and a vector of coefficients β in the minimization problem (5.1). More in detail, the least absolute residuals estimate $\hat{\beta}$ for the conditional median is obtained as the solution of the minimization problem:

$$\min_{\beta} \sum_{i=1}^{n} |y_i - \mathbf{x}_i^\top \beta|. \tag{5.7}$$

Let us denote again by $[x]_+$ the nonnegative part of x. By posing:

$$\mathbf{u} = [\mathbf{y} - \mathbf{X}\beta]_+$$
$$\mathbf{v} = [\mathbf{X}\beta - \mathbf{y}]_+$$

the original L_1 problem can be formulated as:

$$\min_{\beta}\{\mathbf{1}^\top \mathbf{u} + \mathbf{1}^\top \mathbf{v} | \mathbf{y} = \mathbf{X}\beta + \mathbf{u} - \mathbf{v}, (\mathbf{u}, \mathbf{v}) \in \mathbb{R}_+^{2n}\}.$$

Through the decomposition of the regression residual vector into its positive part ($\mathbf{u}$) and negative parts ($\mathbf{v}$), the original nonlinear problem is hence recast as minimization problem of a linear function of $2n$ vector ($\mathbf{u}, \mathbf{v}$) subject to n linear equality constraints along with the $2n$ linear inequality constraints (nonnegativity).

Koenker and Basset (1978) slightly modified the previous L_1 problem placing asymmetric weights on positive and negative residuals and introduced quantile regression:

$$\min_{\beta(\theta)} \sum_{i=1}^{n} \rho_\theta(y_i - \mathbf{x}_i^\top \beta(\theta))$$

where $\rho_\theta(.)$ denotes the following asymmetric absolute loss function:

$$\rho_\theta(y) = [\theta - I(y < 0)]y$$
$$= [(1 - \theta)I(y \leq 0) + \theta I(y > 0)]|y|.$$

Such loss function is a weighted sum of absolute deviations, whereas a $(1 - \theta)$ weight is assigned to the negative deviations and a θ weight is instead used for the positive deviations. This yields the modified linear program:

$$\min_{\beta}\{\theta \, \mathbf{1}^{\mathsf{T}}\mathbf{u} + (1 - \theta) \, \mathbf{1}^{\mathsf{T}}\mathbf{v} \mid \mathbf{y} = \mathbf{X}\beta + \mathbf{u} - \mathbf{v}, (\mathbf{u}, \mathbf{v}) \in \mathbb{R}^{2n}_{+}\}.$$

When $\theta = 1 - \theta = 0.5$ we are in the case of the median regression detailed above. Koenker and D'Orey (1987) adopted the Barrodale–Roberts algorithm presented in the previous section to compute regression quantiles. A detailed description of the involved technical aspects is provided in Koenker (2005) and in Davino et al (2014).

5.3 Geometric interpretation of the median and quantile regression problem: the dual plot

The dual plot (Edgeworth 1888) yields interesting insights for the geometric interpretation of median and quantile regression. Albeit limited to the case of bivariate regression, this "antique" graph is extremely valuable not only for its historical role in the literature on the computation of median regression but also for its aptitude in conveying the geometry underlying the problems. Since it represents the data and the solutions in the parameter space, it is also an essential tool for introducing additional but not less important issues. Among these, the elemental sets (Farebrother 1997; Mayo and Gray 1997) and the quantile regression process (Davino et al 2014; Koenker 2005). For an extensive discussion of the elemental sets in quantile regression see also Sections 3.1 and 3.2.

To the end of introducing the dual plot, let us start from the well–known scatterplot: Figure 5.3 depicts the simple data set used above to illustrate the estimation procedure of median regression (see Table 5.1). The labels (A, B, C, D, and E) easily allow to match the five points between the table and the graph. The two solutions of the median regression problem detected in the previous section, ($a_0 = 0.25, a_1 = 0.75$) and ($a_0 = 0.5, a_1 = 0.5$), are plotted using the dashed and the dotted line, respectively. Both solutions interpolate two data points, A and E for the first solution and A and C for the second one. In the terminology of linear programming, they are two basic solutions, while in the terminology of regression, they are two elements of the elemental sets. Since in the Barrodale–Roberts algorithm, k of the vectors u_i (or v_i) are removed from the basis in the first stage, and only the switch between an u_i and a v_i is allowed in the second stage, each simplex tableau corresponds to an approximation that interpolates k data points. In case of a degenerate tableau, the approximation interpolates more than k data points. Therefore, in terms of residuals, at least k residuals will be null. Recalling that k denotes the number of coefficients of the interpolating function, $k = 2$ in case of simple regression. See Bloomfield and Steiger (1983) and Koenker (2005) for technical details on such point. Furthermore, taking into account that $k = 2$ points lie on the median regression line and being there an odd number of points in the example, for both solutions, the number of points above the line is not perfectly equal to the number of points below the line.

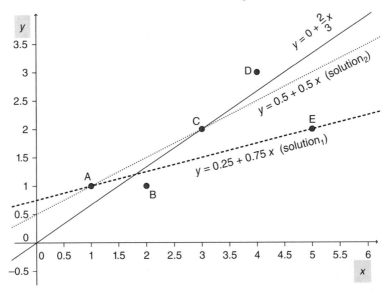

Figure 5.3 Data scatter for the simple data set in Table 5.1, along with two optimal solutions for the unconstrained median regression (solid and dotted line) and the solution when a_0 is forced to 0 (dashed line).

The solid line is instead the median solution when the intercept is constrained to be equal to 0. This constrained solution is the first historical proof (Boscovich 1757; Stigler 1990) of regression, and it directly involves least absolute deviations, and hence median regression. In particular Boscovich aimed at estimating the ellipticity of the earth and he suggested to exploit the sum of absolute errors subject to the constraint that the errors sum to zero, that is, the regression goes through the origin. Laplace (1818) showed later that this problem could be solved by computing a weighted median. In particular, the line with mean residuals that minimizes the sum of absolute residuals is the solution of the problem:

$$\text{minimize} \quad \sum_{i=1}^{n} |y_i - a_0 - a_1 x_i|$$

$$\text{subject to} \quad \bar{y} = a_0 + a_1 \bar{x}$$

The "Laplace Methode de Situation" consists in ordering the n candidate slopes and in finding the weighted median of these slopes, namely:

1: $b_i \leftarrow \frac{y_i - \bar{y}}{x_i - \bar{x}}$ ▷ compute the n candidate slopes

2: $w_i \leftarrow |x_i - \bar{x}|$ ▷ and their associated weights[2]

3: order the candidate slopes

4: compute the weighted median of the slopes

---

[2] Compare the weights of the elemental sets for OLS regression in Section 3.1 to see the different effect of outliers in OLS and in median regression.

Using an asymmetric weighted sum of absolute deviations, the algorithm can be easily generalized to the quantile regression estimates:

```
# QR generalization of the Laplace Methode de Situation
# (courtesy of Roger Koenker)
wquantile <- function(x, y, tau = 0.5) {
  o <- order(y/x)
  b <- (y/x)[o]
  w <- abs(x[o])
  k <- sum(cumsum(w) < ((tau - 0.5) * sum(x) + 0.5 * sum(w)))
  list(coef = b[k + 1], k = o[k+1])
}
```

The Laplace algorithm for the considered data set provides the solid line in Figure 5.3:

```
# the example data set
x <- c(1, 2, 3, 4, 5)
y <- c(1, 1, 2, 3, 2)
# solution for the median regression through the origin
wquantile(x, y)
$coef
[1] 0.6666667
$k
[1] 3
```

Edgeworth (1888) reverted to the same problem by dropping the zero-mean constraints on the residuals. He introduced the method of "double median" proposing to minimize the sum of absolute residuals in both intercepts and slope parameters. In particular, Edgeworth proposed a geometric approach on the dual plot that can be considered the fully fledged historical prelude to the simplex, as evident from the description of the method provided by Edgeworth:

The method may be illustrated thus: – Let $C - R$ (where C is a constant, [and R denotes the L_1 objective function]) represents the height of a surface, which will resemble the roof of an irregularly built slated house. Get on this roof somewhere near the top, and moving continually upwards along some one of the edges, or *arrêtes*, climb up to the top. The highest position will in general consist of a solitary pinnacle. But occasionally there will be, instead of a single point, a horizontal ridge or even a flat surface.

The geometric algorithm proposed by Edgeworth (1888) follows the path of steepest descent through points in the plane determined by the regression coefficients. This plane is the dual plot. Its axes are labeled a_0 and a_1, unlike the scatterplot (primal plot) where the axes (x, y) represent the regressor and the outcome variable.

Therefore, the dual plot transforms each line of the primal plot in a point, and each point of the primal plot in a line. That is, considering a generic line $\ell_i \hat{=} y = a_0 + a_1 x$, dualization transform it to the point $\mathcal{D}(\ell_i) = (a_0, a_1)$. Conversely, a point $\mathcal{P} = (x_i, y_i)$ is transformed through dualization in the line $\mathcal{D}(\mathcal{P})$, with equation $a_1 = y_i - x_i a_0$, that is, the set of all possible lines that pass through the point $\mathcal{P}$ in the primal plot. The elemental sets, that is, the lines joining all the possible pairs of points in the data scatter, play a remarkable role in the Edgeworth's proposal. If we consider the five points in Table 5.1 and in Figure 5.3 there are:

$$\frac{5(4-1)}{2} = 10$$

possible different pairs of points. The lines joining such 10 pairs of points are depicted in Figure 5.4 and labeled as $\ell_i, i = 1, \ldots, 10$. The corresponding coefficients are listed in Table 5.2: the first column reports the 10 possible pairs of points, and the second column reports the labels used for the corresponding lines. The points B, C, and D are collinear: the three lines ℓ_5, ℓ_6 and ℓ_8, respectively joining the pairs (B, C), (B, D), and (C, D), are hence coincident. Finally the two lines ℓ_1 and ℓ_9, respectively joining the two pairs of points (A, B) and (C, E), are parallel to the horizontal line, since the points share the same y–value. Following the previous Figure 5.3, the dashed line (ℓ_4) is again used to denote the first solution of the median regression problem and the dotted line (ℓ_2), the second solution. We have already discussed the nonuniqueness of the solution for this data set in Section 5.1. From a geometric point of view, the nonuniqueness is strictly related to the segment $\overline{CE}$. In particular, all the lines joining the point A with one of the points lying on such segment, that is, with any linear combination of the two points C and E, provide a sum of absolute residuals equal to 2.

Table 5.2   Intercepts and slopes (third and fourth column) for the lines depicted in Figure 5.4 and joining all the possible pairs of points in Table 5.1. The first column reports the pairs of points, and the second column shows the labels used for detecting the lines.

points	lines	a_0	a_1
(A, B)	ℓ_1	1	0
(A, C)	ℓ_2	0.5	0.5
(A, D)	ℓ_3	1/3	2/3
(A, E)	ℓ_4	3/4	1/4
(B, C)	ℓ_5	-1	1
(B, D)	ℓ_6	-1	1
(B, E)	ℓ_7	1/3	2/3
(C, D)	ℓ_8	-1	1
(C, E)	ℓ_9	2	0
(D, E)	ℓ_{10}	7	1

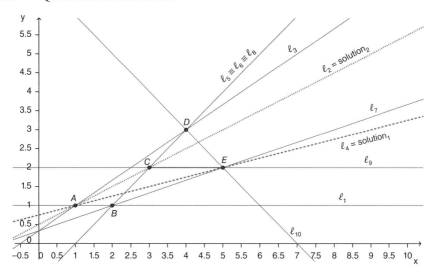

Figure 5.4   The $\binom{5}{2} = 10$ lines joining all the possible pairs of points in Table 5.1. The lines obtained by joining the point A with any of the points on the highlighted segment $\overline{CE}$ provide alternative optimal solutions to the median regression problem.

Figure 5.5 depicts the dual plot for the example, whereas the same label $\ell_i, i = 1, \ldots, 10$ is used to permit the immediate matching between the primal plot (lines) and the dual plot (points). The two coefficients of the 10 lines ℓ_i listed in the third and fourth column of Table 5.2 are computed by taking into account that the intercept and slope of a line passing through two points $P_i = (x_i, y_i)$ and $P_j = (x_j, y_j)$ are, respectively:

$$\frac{y_i \times x_j - x_i \times y_j}{x_j - x_i} \to \text{intercept}$$

$$\frac{y_j - y_i}{x_j - x_i} \to \text{slope}.$$

Moreover, if a line with a intercept and slope (a_0, a_1) passes through a point with coordinates (x_i, y_i), the following equation holds:

$$y_i = a_0 + a_1 x_i.$$

If we invert the equation in terms of a_1, we have that the point with coordinates (a_0, a_1) in the dual plane lies on the line:

$$a_1 = \frac{y_i}{x_i} - \frac{1}{x_i} a_0,$$

namely the line with intercept $\frac{y_i}{x_i}$ and slope $-\frac{1}{x_i}$. Starting from such relationship, it is easy to establish a matching between a point in the primal plane (x, y) and a line

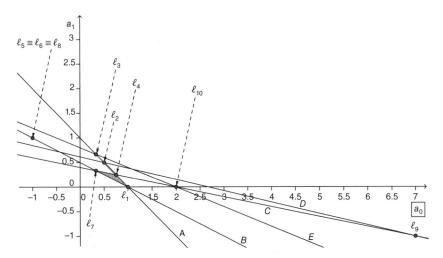

Figure 5.5   The dual plot of the data scatter in Figure 5.3. Each line of the latter corresponds to a point in the dual plot, and each line in the dual plot corresponds to a point in the scatterplot. Each point on the highlighted segment $\overline{\ell_2\ell_4}$ corresponds to a pair (a_0, a_1) that provides an optimal solution of the median regression problem.

in the dual plane (a_0, a_1). The five lines in the dual plot (Figure 5.5) correspond to the five data points in the scatterplot (Figure 5.3). The same labels, from A to E, are again used to allow the immediate matching between the two plots. Obviously, in case of observations with the same x value in the primal plot, we will find parallel lines in the dual plot: this is not the case in the considered example. The highlighted segment $\overline{\ell_2\ell_4}$ is the set of all the pairs of points (a_0, a_1) providing equivalent optimal solutions to the median regression problem, due to nonuniqueness. Namely, the points on the segment $\overline{\ell_2\ell_4}$ correspond to all the lines in the primal plot (Figure 5.3), which are obtained joining the point A with any of the points on the segment $\overline{CE}$. Finally, the five lines divide the dual plot into polygonal regions. The points (a_0, a_1) in any one of these regions correspond to a family of lines in the (x, y) plane that divide the points into two sets with the same cardinality. Geometrically speaking, the number of points above (below) one of such line is the same number of points above (below) any other line of the same family. Albeit the function $\sum_{i=1}^{n} |y_i - a_0 - a_1 x_i|$ is not linear from a global point of view, it is linear in each region. One of such polygonal regions is shaded in Figure 5.5. It is the projection on the (a_0, a_1) plane of a facet of the polyhedral surface corresponding to the objective function to minimize. Meanwhile, the vertices of such 3–D surface projects to points and the edges to segments in the dual plane. The algorithm for the minimization of the sum of absolute residuals starts at any one of the points of the dual plot and iteratively moves from point to point along the segment joining pairs of points, until attaining the minimum value for the objective function.

In terms of movement on the correspondent polyhedral surface, the algorithm proceeds from vertex to vertex moving along its edges. The descent direction involves the directional derivatives of the objective function: the point that leads to the minimum sum of absolute residuals is selected until arriving at the lowest vertex of the 3–D surface.

Figure 5.6 depicts the movements in the dual plot of the classic simplex algorithm to attain the optimal value. They correspond to the tableau reported in subsection 5.1.2, leading to the optimal solution ($a_0 = 0.75, a_1 = 0.25$). In particular, the algorithm starts from of the point labeled ⓪, being both the variables a_0 and a_1 out of the basis (tableau 1). In the first iteration, a_1 enters the basis with a value 0.4: the algorithm moves in point ① (see tableau 2). In the second iteration also a_0 enters the basis. The values of a_0 and a_1 in tableau 3 are both 0.33, leading the algorithm in point ②. Finally the optimal solution ③ is attained in the third iteration, with $a_0 = 0.75$ and $a_1 = 0.25$, as in tableau 4. The Barrodale–Roberts solution to the median regression problem, introduced in subsection 5.1.3, allows instead to pass through several neighboring points in a single iteration. This is depicted in Figure 5.7: unlike the classic simplex algorithm, which would attain the equivalent optimal solution $a_0 = 0.5$ and $a_1 = 0.5$ using several movements from point ⓪ to point ④, the Barrodale–Roberts algorithm passes through several neighboring points in a single iteration, reducing the total number of iterations. In particular, in the first iteration, the algorithm directly moves from point ⓪ to point ③, as already described numerically through the the condensed tableau of subsection 5.1.3.

The following box reports an implementation of the Barrodale–Roberts in R. Such implementation of the double median algorithm consists of the three following steps:

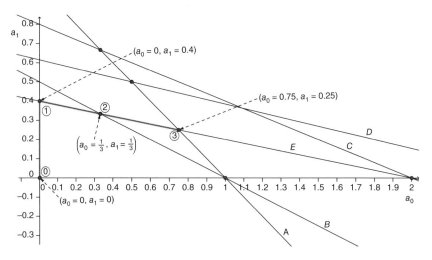

Figure 5.6 The movement of the classic simplex algorithm in the dual plot: the algorithm starts with both coefficients a_0 and a_1 equal to zero, and reaches the optimal solution $a_0 = 0.75$ and $a_1 = 0.25$ in three iterations. The movements correspond to the four tableaus reported in subsection 5.1.2.

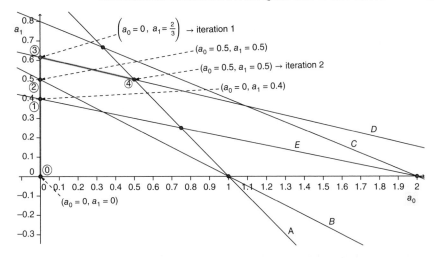

Figure 5.7   The movements of the Barrodale–Roberts algorithm in the dual plot: the algorithm starts with both coefficients a_0 and a_1 equal to zero, and reaches the optimal solution $a_0 = 0.5$ and $a_1 = 0.5$. Unlike the classic simplex algorithm, which attain the optimal solution using several movements, the algorithm passes through several neighboring points in a single iteration, moving from the initial solution $a_0 = 0$ and $a_1 = 0$ to the solution $a_0 = 0$ and $a_1 = 2/3$ in the first iteration. See the condensed tableaux in subsection 5.1.3.

- random selection of an initial basis,

- search for the direction of steepest descent along one of the edge,

- computation of the step length through Laplace's "methode de situation" (wquantile function introduced above).

The default value for the argument tau = 0.5 provides the median regression solution, a different value can also be used for the quantile regression estimates.

```
# Barrodale and Roberts -- lite
#(courtesy of Roger Koenker)
rqx <- function(x, y, tau = 0.5, max.it = 50) {
  p <- ncol(x); n <- nrow(x)
  h <- sample(1:n, size = p)
  # Phase I -- find a random (!) initial basis
  it <- 0
  repeat {
    it <- it + 1
    Xhinv <- solve(x[h, ])
    bh <- Xhinv %*% y[h]
    rh <- y - x %*% bh
    # find direction of steepest descent along one of the edges
    g <- -t(Xhinv)%*%t(x[-h, ])%*%c(tau-(rh[-h]<0))
```

```
    g <-c(g+(1-tau), -g+tau)
    ming <- min(g)
    if(ming >= 0 || it > max.it) break
    h.out <- seq(along = g)[g == ming]
    sigma <- ifelse(h.out <= p, 1, -1)
    if(sigma < 0) h.out <- h.out - p
    d <- sigma * Xhinv[, h.out]
    # find step length by one-dimensional wquantile minimization
    xh <- x %*% d
    step <- wquantile(xh, rh, tau)
    h.in <- step$k
    h <- c(h[ - h.out], h.in)
  }
  if(it > max.it) warning("non-optimal solution: max.it exceeded")
  return(bh)
}
```

The solution returned from this implementation on the example data coincides with the one obtained in subsection 5.1.3:

```
# the example data set
x <- cbind(1, 1:5)
y <- c(1, 1, 2, 3, 2)
# with solution
rqx(x, y)
        [,1]
[1,]    0.5
[2,]    0.5
```

Finally, the dual plot offers an immediate interpretation of the quantile regression process (Davino et al. 2014, Koenker 2005), which is the set of all the distinct quantile regression solutions for a given data set. The QR process is computed in the `quantreg` package (Koenker 2017) setting a value for the `tau` argument outside the unit interval (we use the value −1 below). For the previous example, it consists of the following four different solutions:

```
# QR regression process for the Barrodale-Roberts data
qr_process <- rq(y ~ x, tau = -1)
# the primal QR solution array
qr_process$sol
             [,1] [,2] [,3] [,4] [,5]
tau          0.00 0.30 0.5  0.70 1.00
Qbar         1.33 1.50 2.0  2.33 2.33
Obj.Fun      0.00 0.70 1.0  0.80 0.00
(Intercept)  0.33 0.75 0.5  0.33 0.33
x            0.33 0.25 0.5  0.67 0.67
```

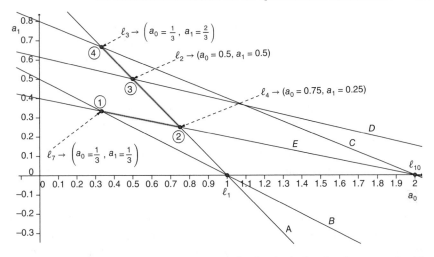

Figure 5.8 The quantile regression process in the dual plot for the example. The quantile regression process consists of all the possible distinct solutions which can be computed at various quantiles for a given data set.

The different solutions of the QR process are highlighted on the dual plot in Figure 5.8 through the four boxed numbers (from ① to ④). If we refer for example to the segment $\overline{\ell_7 \ell_4}$, any point lying on it can be obtained as a linear combination of $\ell_7 \hat{=} (a_0 = \frac{1}{3}, a_1 = \frac{1}{3})$ and $\ell_4 \hat{=} (a_0 = 0.75, 0.25)$ and provides the value 1.5 for the objective function (the row Qbar in the R code above). Still referring to the segment $\overline{\ell_7 \ell_4}$, in terms of quantile regression, the use of a conditional quantile between 0 and 0.3 returns the same estimates for the two coefficients ($a_0 = \frac{1}{3}, a_1 = \frac{1}{3}$). Exploring the whole path, that is, moving the value of the conditional quantile in the whole unit interval, it is possible to estimate through parametric linear programming all the distinct solutions computed in the code above and depicted in Figure 5.8. The limited number of different quantiles strictly depends on the small size of the data set. Obviously this number increases with the number of observations (Davino et al 2014; Koenker 2005).

The simplex approach here discussed is not the only possible approach for solving the quantile regression problem. In addition to the movement along the corner of the simplex (exterior–point method), it is indeed possible to exploit interior–point methods. Such methods allow to solve larger linear programming problems as quickly as least squares. Unlike exterior–point methods, which move along the edges of the feasible region, interior–point methods detect an initial point inside the feasible set and at each iteration move from the current solution to a better feasible solution. They are also referred to as barrier methods, since the constraints determining the boundary of the feasible region acts as a barrier to limit the search inside the feasible region. Albeit interior–point methods have their roots in the seminal paper of Karmarkar (1984), in this case as for the simplex, there exists an historical prelude in an earlier paper (Frisch 1956):

> My method is altogether different than simplex. In this method we work sys-
> tematically from the interior of the admissible region and employ a logarithmic
> potential as a guide – a sort of radar – in order to avoid crossing the boundary.

For a detailed treatment of interior–point methods, see Hiller and Lieberman (2011); Wright (1992). For their application to the QR framework, see instead Koenker (2000) and the literature mentioned therein.

Finally, several alternative QR estimators have been recently proposed. Among these, we point out the optimal quantization approach (Charlier et al 2015a,b) and the Bayesian approach (Yu and Moyeed 2001). The former, implemented in the R package *QuantifQuantile* (Charlier et al 2015c), replaces the typical continuous covariate X with its discretized version obtained by projecting it on a grid of N points. The resulting conditional quantile estimators compete very well with the classical QR estimators. The Bayesian approach to QR, implemented in the R package *bayesQR* (Benoit and Van den Poel 2017), exploits the asymmetric Laplace distribution (Yu and Zhang 2005) as likelihood function. The authors show that the use of improper uniform priors yield in any case a proper joint posterior. This approach has the merit to embed QR in the likelihood framework. It is exploited by Geraci and Bottai (2014) for the treatment of linear quantile mixed models and by Bianchi et al (2018) for M-quantile regression. The approach based on the asymmetric Laplace distribution offers also a unified framework for quantile, M–quantile (see Section 2.3) and expectile regression (see Section 2.1). It is indeed possible to show that the three loss functions can be obtained by properly setting the coefficients of the asymmetric Laplace distribution.

5.4   A summary of key points

- L_1 regression dates back to 1755, earlier than the widespread least squares regression. Boscovich, Laplace, and Edgeworth anticipated the simplex method through a geometric approach to the median regression. Koenker and Basset in 1978 introduced the use of an asymmetric loss function, leading to the definition of quantile regression.

- L_1 regression can be formulated in a LP problem by expressing the loss function as a sum of n nonnegative variables through the absolute residuals. The decisional variables associated with the regression coefficients are instead treated as variables unrestricted in sign. At the end of the process, the proper signs of the residuals are available through the regression equation.

- An alternative formulation of the L_1 problem works with variables unrestricted in sign both for the regression coefficients and the residuals. Even if it is more verbose in terms of the variables involved, this formulation is more convenient in terms of the tableau.

- Barrodale and Roberts proposed a slight modification of the simplex algorithm. This variant is implemented in most statistical software for computing the median regression. It is a two-stage algorithm: in the first stage, the choice of the pivotal column is restricted to the signed regression coefficients, while in the second stage, the proper signed residuals are selected to enter the basis. This procedure entails a movement through several neighboring simplex vertices in one step of the algorithm. Koenker and D'Orey extended the Barrodale–Roberts algorithm to compute regression quantiles.

- The dual plot proposed by Edgeworth represents the data and the solution in the parameter space. Each point (line) in the original scatterplot corresponds to a line (point) in the dual plot. The Laplace "Methode de Situation" is a geometric approach on the dual plot which can be considered the fully fledged historical prelude to the simplex. The dual plot offers also an immediate interpretation of the QR process, namely the set of all the distinct quantile regression solution for a given data set.

References

Bloomfield, P., and Steiger, W. L. 1983 *Least Absolute Deviations: Theory, Applications, and Algorithms*. Boston, Birkhauser.

Barrodale, I., and Roberts, F. D. K. 1970 Application of mathematical programming to l_p approximation. *Nonlinear Programming*, J. B. Rosen, O. L. Mangasarian, and K. Ritter (Eds.), Academic Press, New York, 205–215.

Barrodale, I., and Roberts, F. D. K. 1972 Solution of an over–determined system of equations in the l_1 norm. *Mathematics Dept. Report 69*, University of Victoria, Victoria, British Columbia.

Barrodale, I., and Roberts, F. D. K. 1973 An improved algorithm for discrete l_1 linear approximation. *SIAM Journal on Numerical Analysis*, Vol. 10, 15, 839–848.

Barrodale, I., and Young, A. 1966 Algorithms for best L_1 and L_∞ linear approximation on a discrete set. *Numerische Mathematik*, Vol. 8, 295–306.

Benoit, D. F., and Van den Poel, D. 2017 bayesQR: A Bayesian approach to quantile regression. *Journal of Statistical Software*, 76(7), 1–32.

Bianchi, A., Fabrizi, E., Salvati, N., and Tzavidis, N. 2018 M-quantile regression: diagnostics and parametric representation of the model, *International Statistical Review*, in press, DOI 10.1111/insr.12267.

Boscovich, R. J. 1757 De litteraria expeditione per pontificiam ditionem et synopsis amplioris operis ..., *Bononiensi Scientiarum et Artum Instituto atque Academia Commentarii*, 4, 353–396. Reprinted by the Institute for Higher Geodesy, Zagreb.

Charlier, I., Paindaveine, D., and Saracco, J. 2015a Conditional quantile estimation based on optimal quantization: from theory to practice. *Computational Statistics & Data Analysis*, 91, 20–39.

Charlier, I., Paindaveine, D., and Saracco, J. 2015b Conditional quantile estimation through optimal quantization. *Journal of Statistical Planning and Inference*, 156, 14–30.

Charlier I, Paindaveine, D., and Saracco, J. 2015c QuantifQuantile: an R package for performing quantile regression through optimal quantization *The R Journal*, 7/2, 65–80.

Charnes, A., Cooper, W. W., and Henderson, A. 1953 *An Introduction to Linear Programming*, John Wiley and Sons, New York.

Charnes, A., Cooper, W. W., and Ferguson, R. O. 1955 Optimal estimation of executive compensation by linear programming. *Management Science*, Vol. 1, 2, 138–151.

Davino, C., Furno, M., and Vistocco, D. 2014 *Quantile Regression: Theory and Applications*, Wiley Series in Probability and Statistics, Wiley.

Edgeworth, F. Y. 1888 On a new method of reducing observations to several quantiles. *Philosophical Magazine*, Vol. 25, 184–191.

Edgeworth, F. Y. 1923 On the use of medians for reducing observations relating to several quantities. *Philosophical Magazine*, Vol. 46, 1074–1088.

Farebrother, R. W. 1987 The historical development of L_1 and L_∞ estimation procedures. *Statistical Data Analysis Based on the L_1–Norm and Related Methods*, Dodge Y (Ed.), North–Holland, Amsterdam, 37–64.

Farebrother, R. W. 1997 Notes on the early history of elemental set methods. *Lecture Notes–Monograph Series*, L_1 Statistical Procedures and Related Topics, 161–170.

Farebrother, R. W. 1999 *Fitting Linear Relationships: A History of the Calculus of Observations 1750-1900*, Springer series in Statistics, Springer.

Frisch, R. 1956 *La résolution des problèmes de programme linéaire par la méthode du potential logarithmique, Cahiers du Séminaire d'Econometrie* 4, 7–20.

Geraci, M., and Bottai, M. 2014 Linear quantile mixed models. *Statistics and Computing*, 24(3), 461–479.

Hiller, F. K., and Lieberman, G. 2015 *Introduction to Operations Research*, Mc-Graw Hill, 10th Edition.

Karmarkar, N. 1984 A new polynomial time algorithm for linear programming. *Combinatorica*, 4, 373–395.

Koenker, R. 2000 Galton, Edgeworth, Frisch, and prospects for quantile regression in econometrics. *Journal of Econometrics*, 95, 347–374.

Koenker, R. 2005 *Quantile Regression*. Cambridge University Press.

Koenker, R. 2017 *quantreg: Quantile Regression*, R package version 5.34, url = https://CRAN.R-project.org/package=quantreg.

Koenker, R., and Basset, G. 1978 Regression Quantiles. *Econometrica*, Vol. 46, No. 1, 33–50.

Koenker, R. W., and D'Orey, V. 1987 Algorithm AS 229: Computing Regression Quantiles. *Journal of the Royal Statistical Society. Series C (Applied Statistics)*, Vol. 36, No. 3, pp. 383–393.

Laplace, P. S. 1818 *Théorie analytique des probabilités*, Supp. 2me, Oeuvres Compl<U+00E8>s 7, Gauthier–Villars, Paris.

Mayo, M. S., and Gray, J. B. 1997 Elemental subsets: the building blocks of regression *The American Statistician*, Vol. 51, No. 2, 122–129.

R Core Team 2017 *R: A Language and Environment for Statistical Computing*, R Foundation for Statistical Computing, Vienna, Austria, url = https://www.R-project.org/.

Stigler, S. M. 1990 *The History of Statistics: The Measurement of Uncertainty Before 1900*, Harvard University Press.

Yu, K., and Moyeed, R. 2001 Bayesian quantile regression. *Statistics and Probability Letters*, 54, 437–447.

Yu, K., and Zhang, J. 2005 A three–parameter asymmetric Laplace distribution and its extension. *Communications in Statistics – Theory and Methods*, 34, 1867–1879.

Wagner, H. M. 1959 Linear programming techniques for regression analysis. *Journal of the American Statistical Association*, Vol. 54, 285, 206–212.

Wright, M. H. 1992 Interior methods for constrained optimization. *Acta Numerica*, 341–407.

6

Correlation

Introduction

The chapter considers estimation and inference in case of stationary and non-stationary autoregressive processes as estimated by quantile regressions. The logarithm of the annual change in the consumer price index, the three-month inflation rate, and a few simulated series provide the empirical examples. Tests of stationarity are implemented together with other closely related tests, although the latter are not specifically defined for the quantile regression model. The case of spurious regression and of cointegrated variables are discussed in simulated data sets and for the consumption function. The test for cointegration brings to the analysis of changing coefficient models and to the test functions defined to detect them. An example considers the student performance on an international proficiency test, the OECD-PISA test, together with the analysis of simulated data. The quantile regression conditionally heteroskedastic model concludes the chapter by further analyzing the inflation rate series.

6.1 Autoregressive models

In time series the general autoregressive model is defined as

$$y_t = a_0 + a_1 y_{t-1} + a_2 y_{t-2} + a_3 y_{t-3} + \ldots + a_q y_{t-q} + e_t$$

which is a q-order autoregressive process, AR(q), where the values of the past q elements of y influence its actual value, plus an i.i.d. error term e_t. The simplest version of this model is the AR(1) process, where only one lagged term, y_{t-1}, influences the actual value of y_t:

$$y_t = a_0 + a_1 y_{t-1} + e_t$$

Quantile Regression: Estimation and Simulation, Volume 2. Marilena Furno and Domenico Vistocco.
© 2018 John Wiley & Sons Ltd. Published 2018 by John Wiley & Sons Ltd.
Companion website: www.wiley.com/go/furno/quantileregression

In the quantile regression framework, Koenker and Xiao (2006) discuss the quantile regression estimator for autoregressive processes. The quantile regression objective function estimating the unknown parameters a_0 and a_1 of the quantile autoregressive process QAR(1) is

$$\sum_{y_t > a_0 + a_1 y_{t-1}} \theta |y_t - a_0 - a_1 y_{t-1}| + \sum_{y_t < a_0 + a_1 y_{t-1}} (1 - \theta)|y_t - a_0 - a_1 y_{t-1}|$$

$$= \sum_{t=1,\dots,n} \rho(y_t - a_0 - a_1 y_{t-1}) = \sum_{t=1,\dots,n} \rho(e_t) = min$$

where the check function ρ is defined as $\rho(e_t) = e_t[\theta - 1(e_t < 0)]$ and the vector of gradient is

$$\sum_{t=1,..n} \psi(e_t) = 0$$

$$\sum_{t=1,..n} \psi(e_t)y_{t-1} = 0$$

with $\psi(e_t) = [\theta - 1(e_t < 0)]$, in a sample of size n. The QAR(q) model allows to analyze the process not only on average, as in the AR(q) approach, but also in the tails, at various quantiles. Koenker and Xiao (2006) show that, for any fixed θ the quantile regression estimators of $a^T(\theta) = \begin{bmatrix} a_0(\theta) & a_1(\theta) \end{bmatrix}$ is distributed as $f(F^{-1}(\theta))\Omega^{1/2}\sqrt{n}(\hat{a}(\theta) - a(\theta)) \implies B(\theta)$, where $B(\theta)$ is a 2-dimensional Brownian bridge[1], the matrix Ω is defined as $\Omega = E(x_t^T x_t)$ and the vector x_t is given by $x_t = \begin{bmatrix} 1 & y_{t-1} \end{bmatrix}$. By definition, for any fixed quantile θ, $B(\theta)$ is a normal distribution with parameters N(0, $\theta(1 - \theta)I_{q+1})$, where I_{q+1} is a $(q + 1)$-dimensional identity matrix. In the QAR(1) case it is I_2.

The following example considers the series of the annual change in the consumer price index ($\Delta cpi = cpi_t - cpi_{t-1}$) in Italy from 1955 to 2011, comprising $n = 57$ observations.[2] Figure 6.1 presents the pattern of the original series Δcpi, which approximates the inflation rate, and of its logarithm, $log\Delta cpi$. The summary statistics for $log\Delta cpi = \pi$ are in Table 6.1. The series is estimated as a QAR(1) process

$$\pi_t = a_0 + a_1 \pi_{t-1} + e_t$$

where the actual value of π depends on its previous value. Table 6.2 presents the quantile regression estimates at the quartiles $\theta = .25, .50, .75$. In the last column, the table includes the OLS estimates of the AR(1) process as term of comparison. All the estimated coefficients are statistically different from zero. The slope coefficient in QAR(1) decreases with the quantile: while at the first quartile the proportionality between π_t and π_{t-1} is 0.89, it becomes 0.73 at the upper quartile. At large values of π_t the past has a lower impact.

Next, the approach to select the order q of the autocorrelation process is analyzed. In the OLS framework, to define q there are two possible methods which can be both

---

[1] In the general case of a QAR(q) process, B(θ) has dimension $q + 1$.
[2] Source ISTAT at http://seriestoriche.istat.it/fileadmin/allegati/Prezzi/tavole/Tavola _21.8.xls

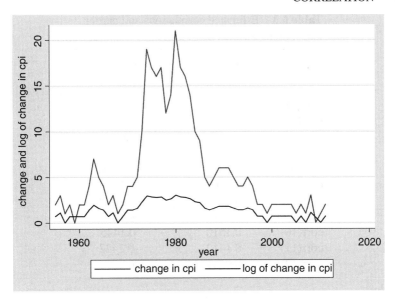

Figure 6.1 Italian data on changes in the consumer price index, $\Delta cpi = cpi_t - cpi_{t-1}$, and of its logarithm, $log\Delta cpi = \pi$, annual data from 1955 to 2011, sample size $n = 57$. The top graph depicts the variable Δcpi, while the bottom graph presents its logarithm, π, characterized by a smoother pattern.

Table 6.1   Descriptive statistics for $\pi = log\Delta cpi$.

	mean	std.dev.	25th	50th	75th	skewness	kurtosis
π	1.388	0.882	0.693	1.386	1.945	0.250	2.111

Table 6.2   QAR(1) and AR(1) estimates, $\pi = log\Delta cpi$.

	QAR(1)			AR(1)
	$\theta = .25$	$\theta = .50$	$\theta = .75$	
π_{t-1}	0.890	0.831	0.734	0.838
	(0.12)	(0.11)	(0.07)	(0.08)
constant	−0.046	0.302	0.693	1.26
	(0.24)	(0.18)	(0.11)	(0.39)

Note: Standard errors in parenthesis, sample size $n = 57$

implemented in the median regression as well, one based on correlations and the other on overfitted models. The first computes the correlation coefficients at many lags: the correlation at lag one is given by the covariance between π_t and π_{t-1} divided by the variance of π_t, corr(1)$=\frac{cov(\pi_t, \pi_{t-1})}{var(\pi_t)}$; the correlation at lag two considers the link between the actual value and the value assumed by the series at two previous

Table 6.3   Estimated correlations and partial correlations for the AR(q) process $\pi = log\Delta cpi$.

	correlation	partial correlation
corr(1)	0.8068	0.8264
corr(2)	0.7080	0.2346
corr(3)	0.6441	−0.2521
corr(4)	0.5372	0.0068
corr(5)	0.4533	0.0141
corr(6)	0.3791	0.1439
corr(7)	0.3061	−0.0406
corr(8)	0.2670	−0.1089
corr(9)	0.1916	−0.2508
corr(10)	0.1510	0.1445
corr(11)	0.1412	−0.2422
corr(12)	0.0231	−0.2512

periods, and it is defined as corr(2)=$\frac{cov(\pi_t,\pi_{t-2})}{var(\pi_t)}$, while the correlation at lag three is given by corr(3) = $\frac{cov(\pi_t,\pi_{t-3})}{var(\pi_t)}$, and so forth. The first column of Table 6.3 reports the first twelve correlations and the graph in Figure 6.2 presents the first 25 estimated correlations for π. The first correlations in the table are large, but afterward, the estimated correlations get smaller and slowly decline toward zero.

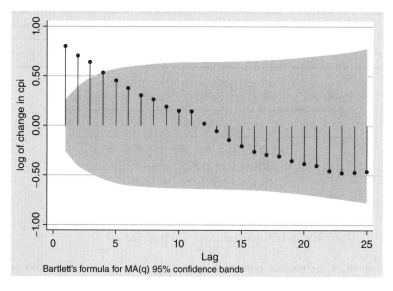

Figure 6.2   Estimates of correlations at various lags, corr(π_t, π_{t-h}) = $\frac{cov(\pi_t,\pi_{t-h})}{var(\pi_t)}$ for $h = 1, 2, 3, ..., 25$, $\pi = log\Delta cpi$, in a sample of size $n = 57$. The shaded area provides a confidence interval around zero. The values outside this area are statistically different from zero. A slowly declining pattern of the correlations is typical of an autoregressive process.

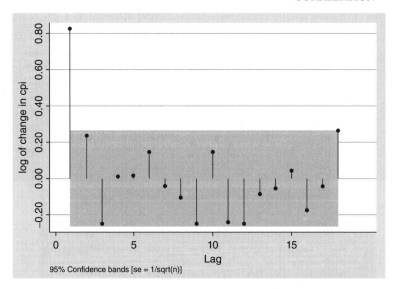

Figure 6.3   Estimates of partial correlations at various lags, which measure the link between π_t and π_{t-h} excluding the influence of all the intermediate lags from $t-1$ to $t-h+1$, for the variable $\pi = log\Delta cpi$, in a sample of size $n = 57$. The shaded area provides the confidence interval around zero. The values outside this area are statistically different from zero, and only the first partial correlation is outside the shaded area. This signals that the autoregressive process is of the first order.

The graph of Figure 6.3 presents the first 25 estimated values of the partial autocorrelation. The partial correlations measure the link between π_t and the lagged value π_{t-h}, excluding the impact of all the intermediate lags π_{t-1}, π_{t-2},.. π_{t-h+1}. The first partial autocorrelation coefficient is given by the slope coefficient in the regression of π_t on π_{t-1}; the second partial correlation coefficient is obtained by regressing π_t on the variables π_{t-1}, π_{t-2} and it coincides with the estimated coefficient of π_{t-2}; the third partial correlation is the estimated coefficient of π_{t-3} in the regression of π_t on the variables π_{t-1}, π_{t-2}, π_{t-3}; and so forth. The second column of Table 6.3 reports the estimated partial correlations computed at lags 1 to 12 for $\pi = log\ \Delta cpi$. These estimates are reported in the graph of Figure 6.3, where the preponderance of the first value is quite clear. Indeed this is the sole element outside the shaded area that represents the confidence interval around zero and thus is the only one to be statistically different from zero. Thus, the autoregressive model is of the first order, $q = 1$.

The final step consists in a check on the estimated model and looks at the residuals. The correlation at various lags of the residuals allows to verify if there is any systematic component of the estimated autoregressive model that has been left unexplained. Figure 6.4 reports the autocorrelations of the residuals of the estimated AR(1) model, computed at the conditional mean. In this graph the estimated error correlations are inside the confidence interval and do not statistically differ from zero. This implies that the AR(1) estimated model is a valid interpretation of the π process and that it does not forgo any systematic component. This same check can easily be implemented in the QAR(1) model as estimated, for instance, at the median. Figure 6.5

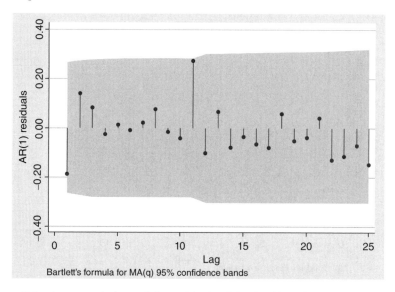

Figure 6.4   Autocorrelations of the residuals of the AR(1) estimated model. All the estimated correlations are within the shaded area and do not differ from zero, thus the AR(1) estimated model does not forgo any systematic component of the π series.

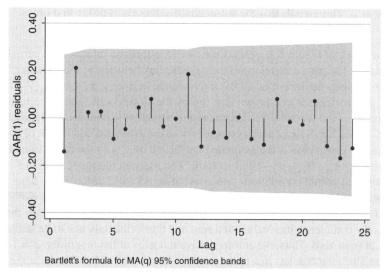

Figure 6.5   Autocorrelations of the residuals of the QAR(1) estimated model. All the estimated correlations are inside the shaded area and are not statistically relevant. Thus the QAR(1) estimated model does not forgo any systematic component of the π series.

depicts the estimated correlations in the residuals of the $QAR_{\theta=.50}(1)$, and these values are all inside the confidence interval around zero: the $QAR_{\theta=.50}(1)$ yields good estimates.

Alternatively, the order of an autoregressive process can be determined via a general-to-specific sequential rule. This is the second approach to define the order of autoregression. It basically consists in overfitting the model and then discarding the statistically irrelevant lags. For instance for the π series a QAR(6) can be estimated as a starting model

$$\pi_t = a_0 + a_1\pi_{t-1} + a_2\pi_{t-2} + a_3\pi_{t-3} + a_4\pi_{t-4} + a_5\pi_{t-5} + a_6\pi_{t-6}$$

which states that six previous periods realizations have an impact on the actual value of π. The standard errors will help defining the number of lags statistically relevant. Table 6.4 reports the estimated coefficients at the conditional median and at the conditional mean of the autoregressive model for $\pi = log\Delta cpi$. In the table only the π_{t-1} coefficient in $QAR_{\theta=.50}(6)$, and the π_{t-1} and π_{t-2} coefficients in the AR(6) case, are significantly different from zero. The remaining variables can be safely dropped.

For a final check a second order autoregressive model is computed. At the median, for $\theta = .50$, the estimates of QAR(2) are $\pi_t = 0.32 + 0.537\pi_{t-1} + 0.292\pi_{t-2}$, with $\hat{se}(\hat{a}_1(.50)) = 0.18$ and $\hat{se}(\hat{a}_2(.50)) = 0.19$. To compare results, the AR(2) model is estimated as well yielding $\pi_t = 1.18 + 0.617\pi_{t-1} + 0.262\pi_{t-2}$, with $\hat{se}(\hat{a}_1) = 0.14$ and $\hat{se}(\hat{a}_2) = 0.15$. In both QAR(2) and AR(2) the π_{t-2} coefficient does not significantly differ from zero and the π_{t-2} variable can be discarded.

Next, a controlled experiment is considered, where a simulated first-order autoregressive series is generated as $\zeta_t = 0.6\ \zeta_{t-1} + e_t$, with $e_t \sim N(0, 1)$. Then the QAR model is estimated for the artificial data ζ_t and ζ_{t-1} so that the results can be

Table 6.4   Estimates of $QAR_{\theta=.50}(6)$ and $AR(6)$ for $\pi_t = log\Delta cpi$.

	$QAR_{\theta=.50}(6)$	AR(6)
π_{t-1}	0.552	0.625
	(0.26)	(0.17)
π_{t-2}	0.622	0.423
	(0.35)	(0.19)
π_{t-3}	−0.364	−0.107
	(0.35)	(0.19)
π_{t-4}	−0.043	−0.145
	(0.30)	(0.26)
π_{t-5}	0.065	0.017
	(0.34)	(0.19)
π_{t-6}	0.044	0.057
	(0.28)	(0.12)
constant	0.130	1.21
	(0.28)	(0.47)

Note: Standard errors in parenthesis

compared with the true model. The original series is depicted in Figure 6.6, while Figures 6.7 and 6.8 present the correlations. From these figures, it is quite evident that the order of correlation of the series is $q = 1$, and a QAR(1) process can be estimated. At the median it yields

$$\zeta_t \quad = \quad 0.305 \quad + \quad 0.615 \quad \zeta_{t-1}$$
$$(se = 0.16) \qquad (se = 0.10)$$

where the slope is significant and very close to the true value while the intercept is not statistically relevant at the 5% level, and indeed there is no intercept in the true data-generating process. The dashed line in Figure 6.6 reports the QAR(1) estimated process. Then, to check the validity of the estimates, the correlations of the residuals are considered. Figure 6.9 presents the graph of these correlations, and they are within the confidence interval around zero. One can conclude that the estimated process does not leave unexplained any systematic component of the process.

So far the QAR(q) process has been estimated at the median. To analyze an autoregressive process away from the median, at $\theta \neq .50$, Li et al. (2015) extend the idea of autocorrelation and partial autocorrelation to define the quantile correlation function, QACF, and the quantile partial correlation function, QPACF. In the sample these statistics are computed as

$$\widehat{QACF} = \frac{1}{n(\theta - \theta^2)\sigma_{e,\theta}^2} \sum_{t=k+1}^{n} \psi_\theta(\hat{e}_t)[\hat{e}_{t-k} - E(\hat{e}_t)]$$

$$\widehat{QPACF} = \frac{1}{n(\theta - \theta^2)\sigma_y^2} \sum_{t=k+1}^{n} \psi_\theta(\hat{e}_t)y_{t-k}$$

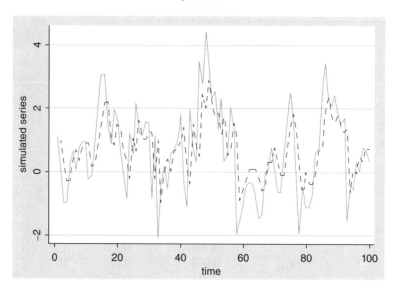

Figure 6.6   Simulated autoregressive series of the first order, $\zeta_t = 0.6\zeta_{t-1} + e_t$, sample size $n = 100$. This series fluctuates around zero and is stationary. The dashed line shows the QAR(1) estimated model and is very close to the true process.

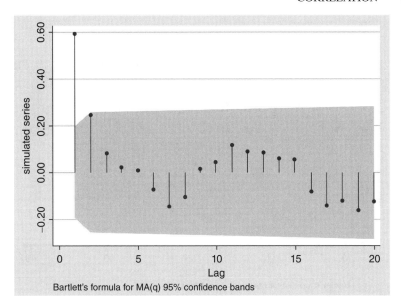

Figure 6.7 Autocorrelations for the simulated data, $\zeta_t = 0.6\zeta_{t-1} + e_t$, sample size $n = 100$. The first correlation is large while the others are smaller and fluctuate around zero, thus signaling a first-order autoregressive process.

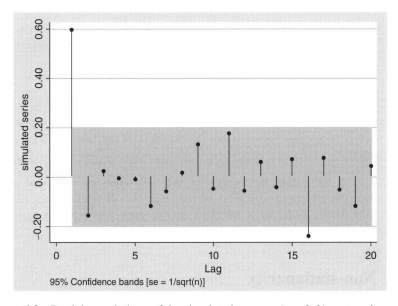

Figure 6.8 Partial correlations of the simulated process $\zeta_t = 0.6\zeta_{t-1} + e_t$, in a sample of size $n = 100$. The first partial correlation is well outside the shaded area, signaling a first- order autoregressive process with $q = 1$.

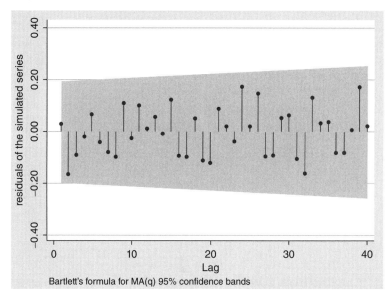

Bartlett's formula for MA(q) 95% confidence bands

Figure 6.9   Correlations of the residuals of the simulated process ζ as estimated at the median by QAR(1), in a sample of size $n = 100$. In this graph all the correlations are inside the shaded area, signaling that there is no unexplained component in the estimated model.

where $\psi_\theta(\widehat{e}_t) = \theta - 1(\widehat{e}_t < 0)$. For instance, the simulated process ζ_t is analyzed at the 75th quantile and provides the following estimated $\text{QAR}_{\theta=.75}(1)$ model

$$\zeta_t = 0.864 + 0.635\ \zeta_{t-1}$$
$$(se = 0.20)\quad (se = 0.13)$$

The QACF at lag one for the residuals of the above estimated equation is $\widehat{\text{QACF}}(e_t, e_{t-1}) = \dfrac{-2.379}{99*1.18*(0.75-0.5625)} = -0.108$, while at lag two $\widehat{\text{QACF}}(e_t, e_{t-2}) = \dfrac{1.159}{99*1.18*(0.75-0.5625)} = 0.053$. Both estimates for the $\widehat{\text{QACF}}(e_t, e_{t-1})$ and the $\widehat{\text{QACF}}(e_t, e_{t-2})$ are within the bounds of the 95% confidence interval around zero. When the estimated QACF at various lags are all within the confidence interval, the model estimated at the third quartile does not forgo any systematic component of the true data-generating process.

6.2   Non-stationarity

One crucial assumption in the AR(1) model is that the coefficient a_1 is in absolute value smaller than one. This is the stationarity condition and represents a relevant issue: it curbs the influence of past values on the present, thus ensuring reliable estimates. Consider for instance the first- order process $y_t = a_0 + a_1 y_{t-1} + e_t$.

The same equation is valid for y_{t-1} as well, which can be expressed as $y_{t-1} = a_0 + a_1 y_{t-2} + e_{t-1}$. Substituting this in the initial equation yields

$$y_t = a_0 + a_1(a_0 + a_1 y_{t-2} + e_{t-1}) + e_t = a_0(1 + a_1) + a_1^2 y_{t-2} + (a_1 e_{t-1} + e_t)$$

$$= \alpha_0 + a_1^2 y_{t-2} + \varepsilon_t$$

where the constant term is $\alpha_0 = a_0(1 + a_1)$, and the errors are $\varepsilon_t = (a_1 e_{t-1} + e_t)$. If $a_1 = 1$ the impact of y_{t-2} on y_t is large and, above all, it does not decrease over time. More in general, the lagged value y_{t-2} represents the impact of the past on y_t, which persists instead of decaying with time if $a_1 = 1$. Indeed, by substituting y_{t-2} with the right term of the equation $y_{t-2} = a_0 + a_1 y_{t-3} + e_{t-2}$, one has

$$y_t = a_0(1 + a_1) + a_1^2(a_0 + a_1 y_{t-3} + e_{t-2}) + a_1 e_{t-1} + e_t$$

$$= [a_0(1 + a_1 + a_1^2)] + a_1^3 y_{t-3} + (a_1^2 e_{t-2} + a_1 e_{t-1} + e_t)$$

$$= \alpha_0 + a_1^3 y_{t-3} + \varepsilon_t$$

where now the constant term is $\alpha_0 = a_0(1 + a_1 + a_1^2)$ and the errors are $\varepsilon_t = (a_1^2 e_{t-2} + a_1 e_{t-1} + e_t)$. Then, as a matter of course, y_{t-3} can be in turn replaced by its definition, and so forth: the back substitution can go far back to the starting value of the series, y_0, and the exponent of a_1 grows accordingly.

When $|a_1| < 1$ the model is stationary since the impact of past values, identified by the term $a_1^h y_{t-h}$, expires after some time. When $a_1 = 1$ the influence of the past observations persists, causing the variance of the process to increase over time since $\text{var}(\varepsilon_t) = \text{var}(a_1^2 e_{t-2} + a_1 e_{t-1} + e_t)$. As time goes to infinity the variance goes to infinity as well.

The presence of an estimated coefficient equal to one suggests the need of a deeper analysis. Indeed, when the estimated coefficient of autocorrelation is equal to one, the basic assumption on the stationarity of the model is violated, and the asymptotic distribution of both OLS and quantile regression estimators is nonstandard. Without loss of generality, the intercept can be dropped. In the simple QAR(1) model without intercept, $y_t = a_1 y_{t-1} + e_t$, Koenker and Xiao (2004) show that when $a_1 = 1$, the term $n(\hat{a}_1(\theta) - a_1(\theta)) = n(\hat{a}_1(\theta) - 1)$ converges to $\frac{1}{f(F^{-1}(\theta))}\left[\int_0^1 B_e^2\right]^{-1} \int_0^1 B_e \, dB_\psi$, where B_e is a demeaned Brownian motion and, for fixed θ and e, $B_\psi(e)$ is normally distributed.

6.2.1   Examples of non-stationary series

The three-month US inflation rate, $\pi_{3t} = \frac{(cpi_t - cpi_{t-3})}{cpi_{t-3}}$, is here analyzed in a sample of $n = 462$ observations going from February 1950 to July 1988.[3] Figure 6.10 presents the series while the summary statistics can be found in Table 6.5. The graphs in Figures 6.11 and 6.12 report the correlations and the partial correlations of this series, while Table 6.6 presents their estimates. In this table and in the graph of Figure 6.12 the value of the first partial correlation is sizably greater than the others

---

[3] Source: Bureau of Labor Statistics at ftp://ftp.bls.gov/pub/special.requests/cpi/cpiai.txt

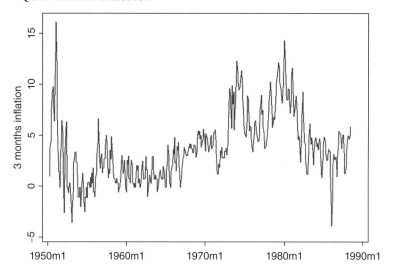

Figure 6.10   Three-month US inflation rate π_{3t}, in a sample of size $n = 462$. The series presents many peaks and does not fluctuate around zero. It is a non-stationary process where the impact of past shocks does not decay and the variability grows over time.

Table 6.5   Descriptive statistics of the three-month US inflation π_{3t}.

	mean	std.dev.	.25	.50	.75	skewness	kurtosis
π_{3t}	3.956	3.315	1.439	3.591	5.434	0.707	3.371

and is very close to one. All the other partial correlations are smaller and fluctuate around zero. Following the overfitting approach, a QAR(6) model is estimated at the median yielding

$$\begin{aligned}
\pi_{3t} &= 0.302 &&+1.184\pi_{3t-1} &&-0.120\pi_{3t-2} &&-0.614\pi_{3t-3}\\
&(se = 0.09) &&(se = 0.05) &&(se = 0.08) &&(se = 0.07)\\
&+0.559\pi_{3t-4} &&-0.058\pi_{3t-5} &&-0.030\pi_{3t-6}\\
&(se = 0.07) &&(se = 0.08) &&(se = 0.05)
\end{aligned}$$

In the above equation only the first four lags are statistically relevant, and the π_{3t-1} estimated coefficient is greater than one, which implies that the series is not stationary. A QAR(4) process can be computed, keeping in mind that the correlation at lag one is very large. The quantile regression estimates of a QAR(4) model for this series at the first, second, and third quartile, together with the AR(4) results, are presented in Table 6.7. The estimated slope of π_{3t-1} increases across quartiles and is slightly above one at all the quartiles here considered, confirming the suggestions gathered from the graphs of Figures 6.11 and 6.12: the series is characterized by persistence, that is by a regression coefficient $a_1 = 1$, and the past influence does not decrease over time, causing an increasing variability in the series. Figure 6.13 reports the original and the

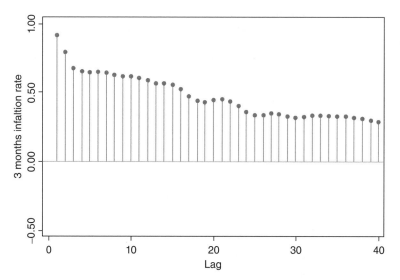

Figure 6.11 Correlations of the three-month US inflation rate π_{3t}, in a sample of size $n = 462$. The first correlation is very close to one while the other correlations are smaller and slowly decrease toward zero.

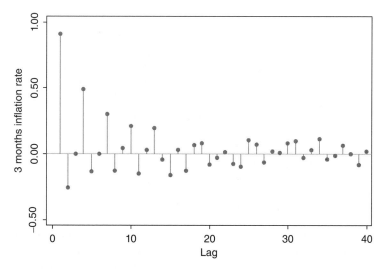

Figure 6.12 Partial correlations of the three-month US inflation rate π_{3t}, sample size $n = 462$. The first partial correlation is very close to one while all the others are much smaller.

Table 6.6   Estimated correlations and
partial correlations for the three- month US
inflation rate π_{3t}.

	correlation	partial correlation
corr(1)	0.9129	0.9137
corr(2)	0.7925	−0.2586
corr(3)	0.6757	−0.0033
corr(4)	0.6484	0.4843
corr(5)	0.6433	−0.1318
corr(6)	0.6447	0.0037
corr(7)	0.6389	0.3024
corr(8)	0.6266	−0.1309
corr(9)	0.6177	0.0417
corr(10)	0.6118	0.2121
corr(11)	0.6038	−0.1519
corr(12)	0.5871	0.0256

Table 6.7   Three-month US inflation rate π_{3t},
QAR(4) and AR(4) estimates.

	.25	.50	.75	AR(4)
π_{3t-1}	1.117	1.142	1.206	1.147
se	(0.05)	(0.05)	(0.06)	(0.03)
π_{3t-2}	−0.094	−0.125	−0.177	−0.124
se	(0.08)	(0.08)	(0.08)	(0.05)
π_{3t-3}	−0.527	−0.540	−0.574	−0.571
se	(0.08)	(0.07)	(0.07)	(0.05)
π_{3t-4}	0.447	0.451	0.495	0.491
se	(0.05)	(0.05)	(0.05)	(0.03)
constant	−0.464	0.240	0.796	4.00
se	(0.10)	(0.11)	(0.11)	(0.91)

Note: Standard errors in parenthesis, $n = 462$.

estimated series at the median, as a QAR(4) process. The two series are very close
to one another, and the QAR(4) estimates approximate well the π_{3t} series. At the
lower section of the graph are the residuals. Although the estimated series provides
a good approximation of the original one, the process is not stationary, inference is
not standard, and the graph of the residuals present some sudden peaks. Indeed, by
computing the correlations of the residuals of the QAR(4) estimates, there are some
correlations outside the shaded area that are statistically relevant, as can be seen in
Figure 6.14. This implies that the residuals may embed some systematic component
that has not been modeled by the estimated QAR(4).

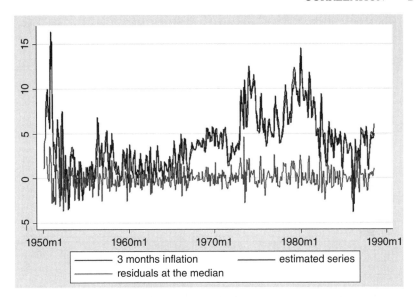

Figure 6.13 Three-month US inflation π_{3t}. The original and the estimated series, as computed at the median by a QAR(4) model, coincide almost everywhere. The bottom line depicts the residuals, which present some sudden higher peaks.

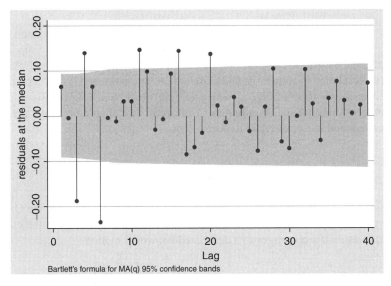

Figure 6.14 Correlations at many lags of the residuals of the QAR(4) estimates for the π_{3t} process. There are a number of values outside the confidence intervals. There is some systematic component that is not captured by the estimated model reported in Table 6.7.

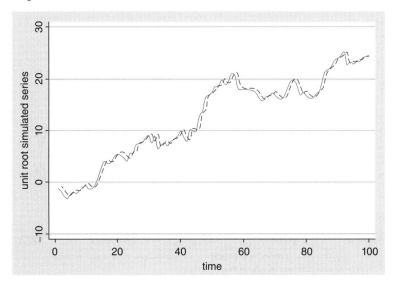

Figure 6.15   The solid line reports the simulated unit root series, $\xi_t = \xi_{t-1} + e_t$, in a sample of size $n = 100$. Compared to the simulated process in Figure 6.6, where the ζ_t series fluctuates around zero, this graph shows an increasing pattern which is due to the non-stationarity of the process. The dashed line depicts the QAR(1) estimated series.

As term of comparison, a controlled experiment considers the simulated series $\xi_t = \xi_{t-1} + e_t$, where $e_t \sim N(0, 1)$, in a sample of size $n = 100$. Figure 6.15 reports the original and the QAR(1) estimated series while Figure 6.16 presents the correlations. The first correlation is quite large and close to one. The estimated QAR(1) process at the median is $\xi_t = 0.46 + 0.99\xi_{t-1}$ with $\widehat{se}(\widehat{a}_0(.50)) = 0.26$ and $\widehat{se}(\widehat{a}_1(.50)) = 0.018$. The intercept does not statistically differ from zero, as it should be, while the slope is significant and is almost equal to one.

6.3   Inference in the unit root model

The case $a_1 = 1$ yields the so called unit root model. As mentioned, for the QAR(1) model $y_t = a_1 y_{t-1} + e_t$ with $a_1 = 1$, Koenker and Xiao (2004) show that the quantile regression estimator converges to a demeaned Brownian motion

$$n(\widehat{a}_1(\theta) - a_1(\theta)) = n(\widehat{a}_1(\theta) - 1) \rightarrow \frac{1}{f(F^{-1}(\theta))}\left[\int_0^1 B_e^2\right]^{-1}\int_0^1 B_e dB_\psi$$

The presence of unit root modifies inference, and the usual t statistic has a nonstandard distribution, as discussed by Dickey and Fuller (1979) in the OLS case.

The unit root model, $y_t = y_{t-1} + e_t$, can be written as $y_t - y_{t-1} = e_t$, which implies that, although y_t is not stationary, the difference $y_t - y_{t-1}$ is stationary. Indeed

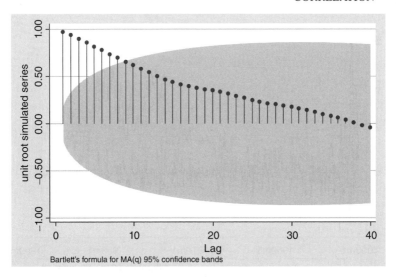

Figure 6.16   Correlations of the unit root simulated process $\xi_t = \xi_{t-1} + e_t$, in a sample of size $n = 100$. The first correlation is very close to one, and the other correlations slowly decline. This plot characterizes a first-order autoregressive process with unit root.

$\Delta y_t = y_t - y_{t-1}$ is function of the sole error term e_t, which by assumption is i.i.d. and thus independent from its own past values. The series Δy_t is termed first differenced since only one lag is considered in the difference operator Δ. An alternative way to indicate persistence in the series is to say that the process is integrated of order one, I(1). While the original series is non-stationary, taking first differences, that is, by integrating the series, the differenced series is stationary. Figure 6.17 reports the plots of the original and of the first differenced series of the US three-month inflation rate π_{3t}, already analyzed in levels in the previous section. The differenced series fluctuates around zero and is stationary, while the original series presents many peaks and is non-stationary. Table 6.8 provides a quick check on the stationarity of the differenced series. It reports the estimated correlations up to five lags of $\Delta\pi_{3t}$, and these values are all smaller than one, thus confirming that the $\Delta\pi_{3t}$ series is indeed stationary.

   In order to test for unit root, the lagged dependent variable can be subtracted, and the model is modified as follows

$$y_t = a_1 y_{t-1} + e_t$$

$$\Delta y_t = a_1 y_{t-1} - y_{t-1} + e_t$$

$$\Delta y_t = (a_1 - 1)y_{t-1} + e_t$$

The usual estimated value of the t statistics can be considered to verify the null H_0: $a_1 - 1 = 0$, although the statistic is no longer distributed as a Student-t. The unit root null, H_0: $a_1(\theta) = 1$, against the alternative of stationarity, H_1: $a_1(\theta) < 1$, is still tested

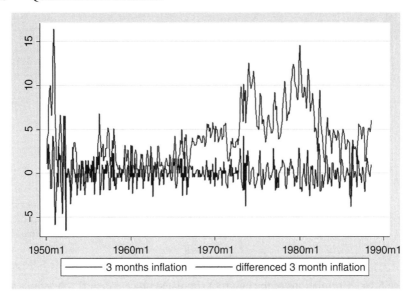

Figure 6.17 The top graph depicts the original series π_{3t} of the three-month US inflation rate, while the bottom graph reports the first differenced series $\Delta \pi_{3t} = \pi_{3t} - \pi_{3t-1}$. While π_{3t} is non-stationary and presents various peaks, $\Delta \pi_{3t}$ fluctuates around zero and is stationary.

Table 6.8 Estimated correlations and partial correlations for $\Delta \pi_{3t}$

	correlation	partial correlation
corr(1)	0.2053	0.2055
corr(2)	−0.0188	−0.0637
corr(3)	−0.5109	−0.5161
corr(4)	−0.1236	0.1079
corr(5)	−0.0358	0.0304

by the usual test function $t(\theta) = \frac{a_1(\theta)-1}{se(a_1(\theta))}$, but this ratio has a different, nonstandard distribution, which is approximated by simulations. The critical values are tabulated in Hansen (1995) and reported in Koenker and Xiao (2004).

The previous model can be generalized as

$$y_t = a_1 y_{t-1} + \varphi_1 \Delta y_{t-1} + e_t$$

where the additional term $\Delta y_{t-1} = (y_{t-1} - y_{t-2})$ allows to verify if there is further correlation in the series besides the unit root coefficient. Analogously, by defining

$\Delta y_{t-2} = y_{t-2} - y_{t-3}$, $\Delta y_{t-3} = y_{t-3} - y_{t-4}$, ..., $\Delta y_{t-h} = y_{t-h} - y_{t-h+1}$, the model can comprise more than one lag in the differenced variable

$$y_t = a_1 y_{t-1} + \varphi_1 \Delta y_{t-1} + \varphi_2 \Delta y_{t-2} + \varphi_3 \Delta y_{t-3} + \ldots + e_t$$

$$y_t = a_1 y_{t-1} + \sum_{i=1..h} \varphi_i \Delta y_{t-i} + e_t$$

When estimated by quantile regression, this model provides the quantile regression analogue of the Augmented Dickey-Fuller test (Dickey and Fuller, 1979). The number of lags in Δy to be included in the equation is selected by looking at the partial correlations or by implementing a general- to-specific sequential rule, the overfitting approach. The test function $t(\theta) = \frac{a_1(\theta)-1}{se(a_1(\theta))}$ under the null of unit root converges to $\frac{1}{\sqrt{\theta(1-\theta)}} \left[\int_0^1 B_w^2 \right]^{-1/2} \int_0^1 B_w \, dB_{\psi}$, with $w_t = y_t - y_{t-1} = \Delta y_t$, and B_w a demeaned Brownian motion. The critical values depend upon the correlation between B_{ψ} and B_w, comprised in the parameter $\delta^2 = \frac{cov^2(B_{\psi}, B_w)}{var(B_w)\theta(1-\theta)}$. The latter can be approximated by the covariance between $w_t = y_t - y_{t-1} = \Delta y_t$ and $\psi_{\theta}(e_t) = [\theta - 1(e_t < 0)]$ divided by the standard deviation of Δy_t, and can be estimated by a kernel function.[4]

We can implement inference now in the three-month US inflation rate model. Table 6.9 presents the estimates of the model $\pi_{3t} = a_1 \pi_{3t-1} + e_t$ to test for unit root. The $t(\theta)$ values have to be compared with the critical values in Koenker and Xiao (2004). Rejection of the null H_0: $a_1(\theta) = 1$ would imply that the model is stationary and that the alternative H_1: $|a_1(\theta)| < 1$ is true. At the 5% level the critical values are in the range of -2.81 and -2.12 (Koenker and Xiao (2004), Appendix, Table 1), depending on the estimated δ and the null is rejected for smaller values. In Table 6.9 it is $t(.25) = \frac{\hat{a}_1(.25)-1}{\hat{se}(\hat{a}_1(.25))} = \frac{0.887-1}{0.022} = -5.13$ at the first quartile, $t(.50) = \frac{\hat{a}_1(.50)-1}{\hat{se}(\hat{a}_1(.50))} = \frac{0.903-1}{0.023} = -4.21$ at the median and $t(.75) = \frac{\hat{a}_1(.75)-1}{\hat{se}(\hat{a}_1(.75))} = \frac{0.959-1}{0.020} = -2.05$ at the third quartile. These results lead to reject the unit root

Table 6.9   Three-month US inflation rate, π_{3t}, test for unit root

	.25	.50	.75	OLS
π_{3t-1}	0.887	0.903	0.959	0.913
se	(0.022)	(0.023)	(0.020)	(0.019)
constant	−0.308	0.327	0.932	0.351
se	(0.115)	(0.119)	(0.100)	(0.097)

Note: Standard errors in parenthesis, sample size $n = 462$

---

[4] The computation of δ can be avoided when the estimated t value is too small (large) to be inside (outside) the acceptance region.

Table 6.10   Three months US inflation rate π_{3t}, augmented Dickey Fuller test.

	.25	.50	.75
π_{3t-1}	0.926	0.927	0.955
se	(0.02)	(0.02)	(0.02)
$\Delta\pi_{3t-1}$	0.255	0.273	0.309
se	(0.07)	(0.05)	(0.06)
$\Delta\pi_{3t-2}$	0.108	0.107	0.074
se	(0.05)	(0.04)	(0.05)
$\Delta\pi_{3t-3}$	−0.472	−0.466	−0.526
se	(0.06)	(0.04)	(0.05)
$\Delta\pi_{3t-4}$	0.141	0.119	0.095
se	(0.06)	(0.05)	(0.06)

Note: Standard errors in parenthesis

hypothesis at the first two quartiles but not at the upper one, regardless of the estimated value of δ.

To estimate the augmented version of the unit root test, the lagged values of $\Delta\pi_{3t}$ should be introduced. Table 6.10 presents the results of the augmented Dickey Fuller test as estimated at the median and the first and the third quartiles. Following the overfitting approach, the model includes terms up to the fourth lagged difference, $\pi_t = a_0 + a_1\pi_{3t-1} + \varphi_1\Delta\pi_{3t-1} + \varphi_2\Delta\pi_{3t-2} + \varphi_3\Delta\pi_{3t-3} + \varphi_4\Delta\pi_{3t-4} + e_t$. The t-value of the π_{3t-1} coefficient under the null H_0: $a_1(\theta) = 1$ is $t(.25) = \frac{0.926-1}{0.022} = -3.36$ at the first quartile, $t(.50) = \frac{0.927-1}{0.020} = -3.65$ at the median, and $t(.75) = \frac{0.955-1}{0.022} = -2.04$. The latter t-value, when compared to the Koenker and Xiao (2004) critical values, does not reject the null of unit root at the 5% level for any value of δ, while the t values at the first and second quartiles reject the null at the 5% level for any value of δ, just as in the previously implemented non-augmented unit root test.

For a controlled experiment, the unit root tests can be implemented in the simulated unit root process ξ_t considered in the previous section. The simple unit root test in the regression $\xi_t = 0.46 + 0.99\xi_{t-1}$ yields $t(.50) = \frac{\hat{a}_1(\theta)-1}{\hat{se}(\hat{a}_1(\theta))} = \frac{0.99-1}{0.018} = -0.555$ at the median, and the null of unit root is not rejected. At the first quartile the estimated equation is $\xi_t = -0.39 + 1.003\xi_{t-1}$ and yields $t(.25) = \frac{1.0035-1}{0.014} = 0.25$, while at the upper quartile $\xi_t = 0.99 + 0.99\xi_{t-1}$ yields $t(.75) = \frac{0.99-1}{0.017} = -0.588$. The null of unit root is not rejected at all the estimated quantiles.

6.3.1   Related tests for unit root

Consider the simple AR(1) model $y_t = a_1 y_{t-1} + e_t$ without intercept. There are additional tests proposed in the unit root literature for the OLS estimator, which, although not specifically designed for quantile regressions, are closely related to it.

Campbell and Dufour (1995) test for unit root analyzes the product between dependent and explanatory variable. The focus is on the OLS estimator of a_1, and the goal is to improve its robustness. Their test function considers the product between the first differenced series, $\Delta y_t = y_t - y_{t-1}$, and the lagged value y_{t-1}. The authors look at the number of nonnegative terms of this product. Under some regularity conditions[5] define the function

$$C = \sum_{t=2,\dots,n} I(\Delta y_t \, y_{t-1})$$

where $I(g)$ is an indicator function assuming value $I(g) = 1$ for $g \geq 0$ and $I(g) = 0$ otherwise. C follows a binomial distribution regardless of the form of the y_t distribution, its moments, the presence of heteroskedasticity, and even in case of an infinite variance. The term $\sum_{t=2,\dots,n} I(\Delta y_t \, y_{t-1})$ is a pivotal quantity and is distributed as a binomial $B(n, 1/2)$. The null of unit root is rejected for low values of the C statistic, in the left tail of the binomial distribution. The link with the median regression is quite immediate, since the sign of the products $\Delta y_t \, y_{t-1}$ defines the normal equation for the median regression estimator of a_1 in the QAR(1) model $y_t = a_1 y_{t-1} + e_t$. In the median regression, both positive and negative signs of the term $\Delta y_t \, y_{t-1}$ are considered, while the C test focuses only on the number of nonnegative signs of the Δy_t y_{t-1} products.

In the example of the three-month US inflation, the number of non-negative signs of the product $\Delta \pi_{3t} \, \pi_{3t-1}$ is $C = 219$, which in the binomial distribution $B(n = 461; p=1/2)$ is close to the mean of 230.5, and the null of unit root cannot be rejected at the median.

So and Shin (2001) test for unit root is based on the sign transform as well. Under the null H_0: $a_1 = 1$ they define the following test function

$$S = \sum_{t=2,\dots,n} sgn(y_t - y_{t-1}) sgn(y_{t-1} - med_{t-1})$$

where $sgn(g) = 1$ if $g > 0$, $sgn(g) = 0$ if $g = 0$, $sgn(g) = -1$ for $g < 0$, and med_{t-1} is the running median, that is, the median of the variable computed up to time $t - 1$. With respect to C, the lagged variable is considered in terms of its distance from the running median, in order to ensure transformation invariance. Under the null of unit root, assuming some regularity conditions,[6] S follows a binomial distribution, $(S + n)/2 \sim B(n, 1/2)$. The null is rejected for $S \leq 2B(\alpha)-n$, α being the lower α-quantile of the binomial distribution. Here the link with the median regression is even stronger, since S does not exclude the negative sign elements. However, the y_{t-1} term in the gradient of the median regression is replaced by $(y_{t-1} - med_{t-1})$ in S, that is, by its distance from the running median.

---

[5] The assumptions are that y_t is symmetric, has no probability mass at zero, and that y_t and Δy_t have zero median.

[6] The regularity conditions state that y_t has no probability mass at zero; that $sgn(\Delta y_t)$ is a martingale difference sequence; that $P(\Delta y_t = 0 \mid I_{t-1}) = 0$, with $I_{t-1} = (y_{t-1}, \dots, y_0)$ being the information set at time $t - 1$.

In the example of the US three-month inflation $S = \sum_{t=2,..,n} sgn(\pi_{3t} - \pi_{3t-1})$ $sgn(\pi_{3t-1} - med_{t-1}) = -49$ and $(S+n)/2 = 206$, which is close to the mean of the binomial B(n=461; p=1/2), equal to 230.5. Once again the null of unit root is not rejected at the median.

The sign transform is also used in a stationarity test defined by de Jong et al. (2007), which is a ratio-type test comparing the sum of squares of cumulative terms and the long term variance σ^2:

$$ J = n^{-2}\sigma^{-2} \sum_{t=1,..,n} \left[\sum_{s=1,..,t} sgn(y_s - med) \right]^2 $$

where med is the median in the entire sample. Under the null of stationarity, J can be compared with the critical values of the tables in Kwiatkowski et al. (1992). In J the test function drastically differs from the gradient of the median regression and, although using the sign function, the link with quantile regressions is not as close as in the S and C tests.

In the US three-month inflation example, this test function assumes the value $J = 1.176$, to be compared with the critical values of 0.347 at the 10% level, 0.463 at the 5% level, and 0.739 at 1%. The null of stationarity is rejected.

6.4   Spurious regression

The presence of unit root, besides causing a nonstandard distribution of the t test, has an additional relevant implication in a regression model. Granger and Newbold (1974), in a Monte Carlo experiment, show that in the linear regression model $y_t = \beta_0 + \beta_1 x_t + \epsilon_t$ the slope coefficient β_1 can be significantly different from zero even if x_t and y_t are independent from one another. This occurs when the two independent series are both characterized by unit root. When this is the case, the slope coefficient does not mirror the proportionality between X and Y but it simply reflects the unit root shared by both processes. This is a case of spurious regression, where the regression coefficients are statistically significant solely because each variable is characterized by unit root, but there is no true correlation between dependent and independent variable. An example of spurious regression is provided by the following controlled experiment. In a sample of size $n = 100$, two AR(1) processes with roots very close to 1 are independently generated as follows

$$ y_t = 0.99y_{t-1} + \epsilon_t $$

$$ x_t = 0.97x_{t-1} + e_t $$

The terms ϵ_t and e_t are the realizations of two independent standard normals. Figure 6.18 depicts the realizations of both processes, and it is quite evident that the two variables have a similar pattern.

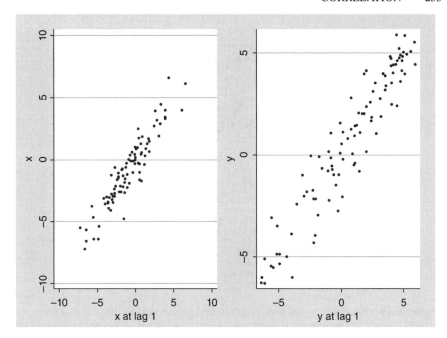

Figure 6.18 The graphs report two AR(1) processes defined as $x_t = 0.97x_{t-1} + e_t$ and $y_t = 0.99y_{t-1} + \varepsilon_t$, generated independently from one another, each one with correlation coefficient very close to 1, in a sample of size $n = 100$. The two processes have a similar pattern.

Table 6.11 Estimates of a regression model relating two independent unit root variables.

	OLS	.25	.50	.75
slope	−0.648	−0.605	−0.519	−0.677
se	(0.11)	(0.23)	(0.15)	(0.13)
intercept	0.777	−1.445	0.634	2.361
se	(0.30)	(0.66)	(0.43)	(0.38)

Note: Standard errors in parenthesis

Next, a regression model where y_t is a linear function of x_t is computed. The model $y_t = \beta_0 + \beta_1 x_t + u_t$ is estimated by OLS and by quantile regressions and the results are presented in Table 6.11.

The estimated coefficients are statistically relevant in both OLS and quantile regressions, although the variables have been independently generated. The common

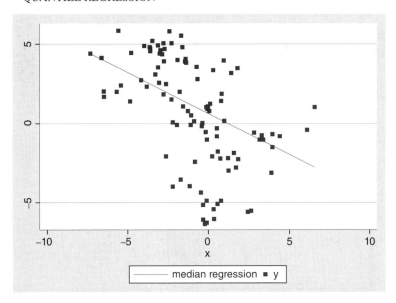

Figure 6.19 Spurious regression as estimated at the median for the equation $y_t = \beta_0 + \beta_1 x_t + u_t$. The presence of a unit root in both dependent and explanatory variable yields an estimated slope significantly different from zero, although the two variables have been independently generated. The histogram of these residuals is reported in the left panel of Figure 6.23. Sample size $n = 100$.

unit root in each series, as depicted in Figure 6.18, is the sole cause of a statistically significant slope coefficient in this equation. The failure to reject the null H_0: $\beta_1 = 0$, as it should be since the two variables are independent from one another, is caused by the common unit root that characterizes each variable of the model. Figure 6.19 depicts the estimated median regression.

To eliminate the unit root characterizing both variables of the model, y_t and x_t can be considered in first differences: the process $y_t = a_1 y_{t-1} + \varepsilon_t$ by taking first differences becomes $\Delta y_t = y_t - y_{t-1} = \varepsilon_t$, and the series is stationary since ε_t is i.i.d. standard normal. Analogously for x_t the differenced variable, $\Delta x_t = x_t - x_{t-1} = e_t$, is stationary since e_t is an i.i.d. standard normal.

Once the variables of the model are turned into stationary processes, the regression model is re-estimated. Figure 6.20 presents the estimated median regression and Table 6.12 reports the results for the model $\Delta y_t = \beta_0 + \beta_1 \Delta x_t + \eta_t$, defined in terms of the stationary variables $\Delta y_t = y_t - y_{t-1}$ and $\Delta x_t = x_t - x_{t-1}$, as estimated by OLS and quantile regressions. In Figure 6.20 the estimated line is less steep than in Figure 6.19, and the results in Table 6.12 show that the slope coefficient is not statistically different from zero in both the OLS and the quantile estimated regressions. Figures 6.21 and 6.22 present respectively the histograms of the original and of the first-differenced variables. The comparison of these graphs shows that first differencing has greatly

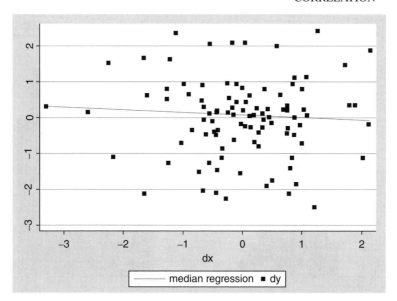

Figure 6.20 Median regression where the unit root variables have been differenced to gain stationarity, $\Delta x = x_t - x_{t-1} = dx$ in the graph, $\Delta y = y_t - y_{t-1} = dy$ in the graph. The estimated slope of the model $\Delta y_t = \beta_0 + \beta_1 \Delta x_t + \eta_t$ is almost flat and is not significantly different from zero. The histogram of these residuals is reported in the right panel of Figure 6.23.

Table 6.12 Estimates of the regression model relating first-differenced independent variables, $\Delta y_t = \beta_0 + \beta_1 \Delta x_t + \eta_t$.

	OLS	.25	.50	.75
slope	−0.019	0.127	−0.071	−0.106
se	(0.11)	(0.22)	(0.11)	(0.18)
intercept	−0.135	−0.599	0.078	0.551
se	(0.11)	(0.21)	(0.10)	(0.18)

Note: Standard errors in parenthesis

improved their approximation to normality. Finally, Figure 6.23 compares the histograms of the residuals of the median regression estimated with the original variables, in the left graph, and of the residuals of the median regression computed with first-differenced variables, on the right-hand side graph. The latter distribution is less skewed and provides a better approximation to a normal than the former.

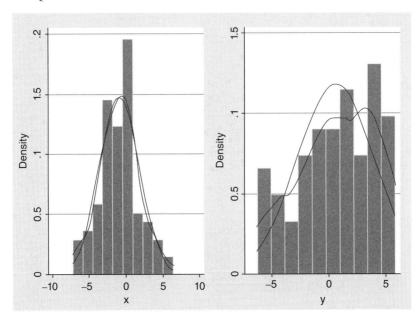

Figure 6.21   Histograms of the original unit root variables, y_t in the right plot and x_t in the left one, together with the normal density and the Epanechnikov kernel density. The kernel density in the left graph presents two peaks, while in the right plot, the histogram shows a right tail with probability greater than normal.

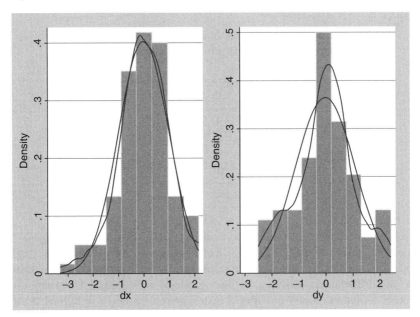

Figure 6.22   Histograms of the first differenced series $\Delta y_t = \varepsilon_t$ and $\Delta x_t = e_t$ together with the normal density and the Epanechnikov kernel density. Compared with the graphs in Figure 6.21, it can be seen that first-differencing greatly improves the approximation of the variables to the normal density. The kernel density of Δy is not bimodal.

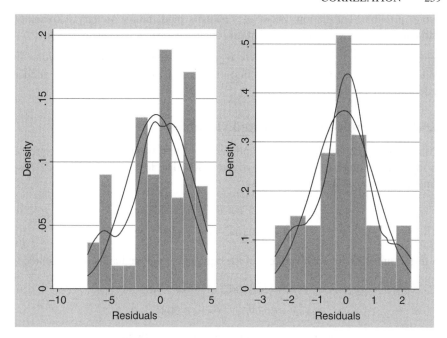

Figure 6.23   In the left panel is the histogram of the errors u_t as estimated at the median in the regression relating the original unit root variables y_t and x_t. These residuals have a large peak in the left tail. In the right panel is the histogram of the errors η_t as estimated at the median regression with first-differenced stationary variables, Δy_t and Δx_t. This histogram is less skewed than the former and provides a better approximation to the normal. The curves in the graphs depict the normal density and the Epanechnikov kernel density.

6.5   Cointegration

The opposite case of a spurious regression is a model where the variables, even in the presence of unit root, present co-movements. This implies that, although non-stationary variables tend to diverge and do not display a long- run equilibrium, their linear combination does conform to a long-run equilibrium, diverging from it only in the short run. This is the case of cointegrated variables, and when this occurs the errors of the cointegrating equation are stationary. Xiao (2009) analyzes the quantile regression cointegrating model. In the presence of a unit root in x_t and of a stationary Δx_t variable, Xiao proposes to compute the linear model $y_t = \beta_0 + \beta_1 x_t + e_t$ by the following cointegrating regression at the selected quantile

$$y_t = \beta_0 + \beta_1 x_t + \sum_{j=-h}^{h} \varphi_j \Delta x_{t-j} + e_t$$

where, as discussed in Saikkonen (1991) for the OLS model, h lags and leads of the differenced explanatory variable Δx_t are introduced in order to avoid possible correlation between residuals and non-stationary explanatory variables. The asymptotic distribution of the slope β_1 is $n(\hat{\beta}_1(\theta) - \beta_1(\theta)) \rightarrow \frac{1}{f(F^{-1}(\theta))} \left[\int_0^1 B_{\Delta x}^2 \right]^{-1} \int_0^1 B_{\Delta x} dB_\psi$, where $B_{\Delta x}$ is a demeaned Brownian motion. The addition of lags and leads in the first-differenced explanatory variable induces independence between $B_{\Delta x}$ and B_ψ.

6.5.1   Example of cointegrated variables

Consider the series on Italian consumption, annual data from 1970 to 2009 expressed in log of real values in a sample of size $n = 40$.[7] The log of consumption, c_t, as a function of gnp_t, at the median yields

$$c_t \quad = \quad 8215 \quad + \quad 0.7851 \quad gnp_t$$
$$(se = 11949) \quad (se = 0.012)$$

A test of unit root for each variable of the equation at the median is implemented

$$c_t \quad = \quad 23387 \quad + \quad 0.9872 \quad c_{t-1}$$
$$(se = 6574) \quad (se = 0.0084)$$

$$gnp_t \quad = \quad 45160 \quad + \quad 0.9747 \quad gnp_{t-1}$$
$$(se = 11601) \quad (se = 0.012)$$

and both variables present estimated slopes very close to one. The unit root test for the QAR(1) consumption process is $\frac{0.9872-1}{0.0084} = -1.52$ and for the QAR(1) gnp process it is $\frac{0.9747-1}{0.012} = -2.10$, to be compared with the Koenker and Xiao table. These values do not allow to reject the null at the 5% level regardless of the value of δ, since at the 5% level the tabulated critical values vary between -2.81 and -2.12. To verify whether the estimated consumption function is a case of spurious regression, the cointegrating equation can be computed by adding lags of the first differenced explanatory variable. At the median the estimated cointegrating equation is

$$c_t \quad = \quad 16290 \quad + \quad 0.7829 \quad gnp_t \quad - \quad 0.3904 \quad \Delta gnp_t \quad - \quad 0.2984 \quad \Delta gnp_{t-1}$$
$$(se = 17424) \quad (se = 0.013) \quad (se = 0.109) \quad (se = 0.163)$$
$$+ \quad 0.0954 \quad \Delta gnp_{t-2} \quad + \quad 0.0501 \quad \Delta gnp_{t-3}$$
$$(se = 0.164) \quad (se = 0.182)$$

where all the lagged coefficients Δgnp_{t-i} for $i = 1, 2, 3$ are not statistically relevant. These variables can be dropped, and the more parsimonious version is

$$c_t \quad = \quad 14301 \quad + \quad 0.7860 \quad gnp_t \quad - \quad 0.4908 \quad \Delta gnp_t$$
$$(se = 8961) \quad (se = 0.008) \quad (se = 0.075)$$

---

[7] Source: ISTAT at http://timeseries.istat.it/

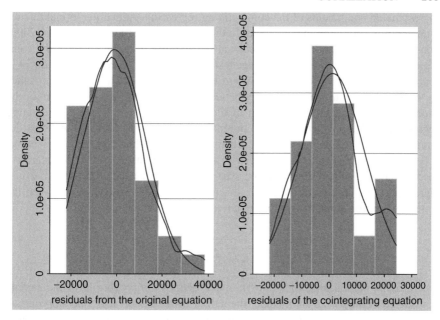

Figure 6.24   The left graph plots the residuals from the simple consumption function, $c_t = \alpha + \beta\,gnp_t + e_t$, while the right graph plots the residuals of the cointegrating equation, $c_t = \alpha + \beta gnp_t + \gamma\Delta gnp_t + e_t$. The former presents a skewed histogram. The curves in the graphs depict the normal density and the Epanechnikov kernel density.

Contrarily to the example discussed for the spurious regression model in y_t and x_t, in this case the estimated coefficients are statistically relevant, and the equation shows the existence of co-movements between c_t, gnp_t and Δgnp_t. Figure 6.24 compares the residuals of the initial consumption equation and of the cointegrating equation. In the latter, the skewness is greatly reduced.

A final quick empirical check can be implemented by repeating the experiment performed in Section 6.4 on the simulated unit root variables, where y and x were replaced by the stationary Δy and Δx. Here c and gnp can be replaced by Δc and Δgnp. The regression between the first-differenced variables measures the impact of changes in income on changes in consumption. This is not of great interest per se, but it checks the existence of the link between the stationary variables: if the regression between differenced variables is statistically relevant, then the correlation is not spurious. At the median, the regression $\Delta c_t = \alpha + \beta\Delta gnp_t$ yields a slope of $\widehat{\beta}(.50) = 0.34$ with $t(\widehat{\beta}(.50)) = 4.63$. The null $H_0 : \beta(\theta) = 0$ is rejected, and it is possible to conclude that the correlation between consumption and income is not spurious.

6.5.2   Cointegration tests

In the above example on the consumption function, the cointegrating model has been estimated, and cointegration has been verified by checking the statistical relevance of the estimated coefficients of the cointegrating regression. Actually a formal test for cointegration should be implemented. In the quantile regression model Xiao (2009) considers a cointegration test that looks at the stability of the gradient of the cointegrating equation in a fluctuation-type test. The test function relies on the quantile regression gradient $E\psi(e_t(\theta)) = 0$, where $\psi(e_t(\theta)) = [\theta - 1(e_t(\theta) < 0)]$. The validity of the cointegrating equation may be tested by looking at the partial sum of the process $\psi(e_t(\theta))$, the sign transform of the residuals of the quantile cointegrating regression, and at its fluctuation as additional observations are included in the partial sum. In case of cointegration, the residuals should be stable, and their fluctuations reflect only equilibrium errors. Otherwise, when the variables are not cointegrated, the fluctuations in the residuals can be expected to be large and the partial sum would significantly diverge from the total sum, the gradient. Thus, cointegration can be tested by looking at the partial sum of $\psi(e_t(\theta))$, that is, at the term

$$T = \frac{1}{\omega\sqrt{n}} \sum_{t=1}^{\lambda n} \psi(e_t(\theta))$$

where ω^2 is the variance of $\psi(e_t(\theta))$, and λ defines the proportion of the sample considered in the partial sum. Under the null of cointegration and under appropriate regularity conditions, the T test function converges to a standard Brownian motion, while under the alternative of non-cointegration, T diverges. In the empirical analysis, the terms ω and $e_t(\theta)$ are replaced by their estimates in the sample. However, Xiao (2009) does not provide further details, stating that "In principle, any metric that measures the fluctuation in $y_n(\theta)$ is a natural candidate. The classical Kolmogoroff - Smirnoff type or Cramer - von Mises type measures are of particular interest."

6.6 Tests of changing coefficients

The Xiao (2009) fluctuation test for cointegration recalls a test for changing coefficients defined for stationary variables by Qu (2008) in the quantile regression model and further generalized by Oka and Qu (2011) to the case of multiple changes. In the simple linear regression $y_t = \beta_0 + \beta_1 x_t + e_t$ with e_t being i.i.d. errors, in a sample of size n, it is possible to relax the assumption that the regression coefficients remain constant in the entire sample. This amounts to allow the structure of the equation to change in response, for instance, to a change in economic policy or to a legislative change, to a generational gap, or to other characteristics of the sample not explicitly modeled in the equation, such as different funding or type of contracts. Although they seem two distinct issues, stationarity and stability of the regression coefficients are related. The presence of changing coefficients affects the behavior of the unit root tests and failure to discriminate one from the other leads to incorrect inference. For instance a shift in the stationary coefficients can be erroneously signaled as unit root. To model changes in the regression coefficients, the equation becomes

$$I) \; y_t = \beta_0 + \beta_1 x_t + \beta_2 z_t + \varepsilon_t$$

where $z_t = d_t x_t$ and d_t is a dummy variable having unit value after the change to model the impact, incremental or decremental, of the new policy or of the new legislation on the regression model. The term β_2 measures the impact of a policy introducing temporary contracts or of a generational gap on the dependent variable.

As an alternative, instead of introducing the dummy variable in z_t, the sample can be split into two subgroups and two regressions are separately estimated, one before and another after the policy change:

$$II)\ y_{ta} = \beta_{0a} + \beta_{1a} x_{ta} + \varepsilon_{ta} \qquad t = 1, .., \lambda n, \qquad 0 < \lambda < 1$$

$$y_{tb} = \beta_{0b} + \beta_{1b} x_{tb} + \varepsilon_{tb} \qquad t = \lambda n + 1, .., n$$

where λ defines the proportion of the sample before the change and the subscripts "a" and "b" refer to the subsets into which the sample is divided, up to λn in subset "a" and after λn in subset "b." The two subsets model the behavior of the equation before and after the legislative change, in the case of temporary versus permanent contracts, for young versus old generations, and so forth. The linear model $y_t = \beta_0 + \beta_1 x_t + e_t$ estimated in the full sample provides the constrained equation, where the regression coefficients are not allowed to change. The test of changing coefficients implies the null hypothesis $H_0 : \beta_2 = 0$ in model (I), or $H_0 : \beta_{1a} = \beta_{1b}$ when model (II) is considered. Under the alternative, the unconstrained equation allows for changes in the parameters of the model and the hypothesis is given by $H_1 : \beta_2 \neq 0$ or $H_1 : \beta_{1a} \neq \beta_{1b}$, depending on which one of the two models, (I) or (II), is analyzed.

In the simple linear regression $y_t = \beta_0 + \beta_1 x_t + e_t$, that is, in the constrained model, the gradient of the quantile regression coefficients is the building block of the Qu (2008) test. In detail, for the slope coefficient it is:

$$S_n(\theta) = n^{-1/2} \sum_{t=1}^{n} x_t \psi(e_t) = n^{-1/2} \sum_{t=1}^{n} x_t \psi(y_t - \beta_0 - \beta_1 x_t)$$

where as usual $\psi(e_t) = \theta - 1(e_t < 0)$. Then the gradient is evaluated in a subset of size λn, for $0 < \lambda \leq 1$, and the functions $S_n(\lambda, \theta)$, $H_{\lambda,n}(\theta)$ are defined as

$$S_n(\lambda, \theta) = n^{-1/2} \sum_{t=1}^{\lambda n} x_t \psi(e_t)$$

$$H_{\lambda,n}(\theta) = (\mathbf{X}^T \mathbf{X})^{-1/2} \sum_{t=1}^{\lambda n} x_t \psi(e_t)$$

where $\mathbf{X}$ is the (n, p) matrix of explanatory variables, and in this equation $p = 2$. Qu (2008) compares $H_{\lambda,n}(\theta)$ and $H_{1,n}(\theta)$ in the test function

$$Q_\theta = sup_\lambda \| [\theta(1 - \theta)]^{-1/2} [H_{\lambda,n}(\theta) - \lambda H_{1,n}(\theta)] \|$$

where the sup norm selects the maximum difference between $H_{\lambda,n}(\theta)$ and $H_{1,n}(\theta)$ within the entire vector of coefficients. If the regression coefficients do not change, $H_{\lambda,n}(\theta)$ is a good proxy for $H_{1,n}(\theta)$, while in the opposite case of changing parameters this is no longer true, $H_{\lambda,n}(\theta)$ sizably differ from $H_{1,n}(\theta)$, and the model (I) or (II) should be estimated. The test function Q_θ converges to a p vector of independent Brownian bridge processes on [0, 1]. The critical values for the Q_θ test function are computed via simulations and depend on the number of regressors of the model (Table 1, Qu, 2008). Once the null is rejected, Q_θ provides an additional valuable information, the location of the change point, λn, which is generally unknown.

Actually Qu (2008) considers the above statistics to test the presence of a change across quantiles as well. This implies to search for the sup norm not only across coefficients, but also across quantiles

$$QQ_\theta = sup_\theta sup_\lambda \|[\theta(1 - \theta)]^{-1/2}[H_{\lambda,n}(\theta) - \lambda H_{1,n}(\theta)]\|$$

A recent article (Oka and Qu, 2011) extends the Qu test to verify the presence of multiple changes. It allows to determine the number of changes by repeatedly implementing the Q_θ and QQ_θ test functions. The sample is partitioned according to an assumed number of changes, and then the Q_θ and QQ_θ tests are implemented within each subset of the sample to verify the presence of additional shifts. The main difference relies in the definition of the $H_{\lambda,n}(\theta)$ function, which is now computed within each subset. For instance, by implementing Q_θ or QQ_θ in the full sample one change is found at point m. Then Q_θ or QQ_θ are implemented again in each of the two subsets, where the first subset goes from 1 to m and the second one from $m+1$ to n. The purpose is to verify if within each subset there is an additional change point. If in one of the two subsets, for example the second one, an additional parameters shift is signaled at point h, the second subset is split in two parts, the first comprising data from $m+1$ to h and the other comprising the observations from $h + 1$ to n. Within each of them Q_θ or QQ_θ are once again implemented until the null of no change in each subset is not rejected.

Besides the Qu (2008) test for structural shifts, a robust version of the Chow (1960) test can be considered (Furno, 2007). This test is based on the comparison of the quantile regression objective functions estimated under the null of constant coefficients and under the alternative of changing coefficients. The null assumes stability of the equation as provided by the constrained model where the coefficients are not allowed to change. The alternative allows the regression coefficients to change as in model (II), which represents the unconstrained model where the coefficients are free to change from one subset to the other. The test verifies whether the coefficients estimated in the two subsets differ from the parameters computed in the entire sample. This involves the comparison of the value of the objective function under the null of stable coefficients , $V(\theta) = \sum_{t=1,..n} \rho(y_t - \beta_0 - \beta_1 x_t)$, with the sum of the two objective functions computed under the alternative (II), $V_a(\theta) = \sum_{t=1,}^{\lambda n} \rho(y_t - \beta_{0a} - \beta_{1a} x_t)$ and $V_b(\theta) = \sum_{t=\lambda n+1}^{n} \rho(y_t - \beta_{0b} - \beta_{1b} x_t)$. The test function is given by:

$$C_\theta = \frac{[V(\theta) - (V_a(\theta) + V_b(\theta))]/d_1}{(V_a(\theta) + V_b(\theta))/d_2}$$

In case of changes in coefficients, the worsening of the fit due to the unnecessary constraints causes the increase of the objective function $V(\theta)$, making it significantly different from the objective functions of the unconstrained model, $V_a(\theta) + V_b(\theta)$. In the numerator, the degrees of freedom are equal to the number of constraints, that is, the number of parameters remaining constant, $d_1 = p$. At the denominator, d_2 is given by the sample size minus the number of estimated coefficients under the alternative, $d_2 = n - 2p$.

C_θ is asymptotically distributed as F_{d_1, d_2} since both numerator and denominator are independently distributed as χ^2 (Koenker and Machado, 1999; Koenker and Bassett, 1982).

The C_θ test reveals two different characteristics of the model: 1) a comparison of C_θ across quantiles signals at what level of the dependent variable the change is more effective; 2) the equations estimated to compute C_θ allow to check which coefficient or group of coefficients are more affected by the shift, thus signaling the occurrence of a partial or a global change in the coefficients, that is, whether the change affects some or all the regression coefficients (Furno and Vistocco, 2013).

6.6.1   Examples of changing coefficients

The first example considers the score in an international test on reading proficiency, the OECD-PISA test, of Italian fifteen-year-old students in a sample of size $n = 59558$. The analyzed sample comprises year 2000, 2003, 2006, and 2009. The score of the test is related to explanatory variables describing school characteristics such as school size, funding, and student-teacher ratio. The purpose is to relate student performance to school efficiency. There is a long-lasting debate on the role of school variables on student proficiency. Growth analysis emphasizes school attainment and shows its high correlation to differences in economic growth across countries. Hanushek (2006) relates one standard deviation difference of student performance to a one percent difference in the annual rate of growth of per-capita gross domestic product. However, higher spending in school resources does not necessarily involve higher test scores (Hanushek and Woessmann, 2011).

To implement the Q_θ test, the partial sum of the gradient is compared with the gradient in the full sample. In case of changing coefficients, that is, under the alternative H_1: $\beta_{1a} \neq \beta_{1b}$, the $H_{\lambda,n}(\theta)$ function will substantially differ from $\lambda H_{1,n}(\theta)$. Under the null of constant coefficients H_0: $\beta_{1a} = \beta_{1b}$, the difference $H_{\lambda,n}(\theta) - \lambda H_{1,n}(\theta)$ will be statistically irrelevant.

Consider the following equation explaining students reading proficiency as estimated at the median regression

$$reading = -75.7927 \ vocational - 38.6104 \ boy +$$
$$(se = 1.56) \qquad\qquad (se = 1.36)$$
$$+ 3.8993 \ private + 3.2536 \ stratio +$$
$$(se = 3.59) \qquad\qquad (se = 0.229)$$
$$+ 0.015746 \ schlsize + 0.05287 \ computer + 477.4652$$
$$(se = 0.002) \qquad\qquad (se = 0.0096) \qquad\qquad (se = 2.619)$$

where the reading score, *reading*, is explained by the variables: *vocational*, a dummy variable assuming unit value for students enrolled in vocational schools; *boy*, which assumes unit value for boys and zero otherwise; *private*, having unit value for students enrolled in privately funded schools and zero elsewhere; *stratio*, which is equal to the student-teacher ratio in the school; *schlsize*, describing the number of students enrolled in the school; *computer*, which is equal to the number of computers in the school to mirror school facilities. All the estimated coefficients are statistically relevant but the one for the privately funded schools variable. Focusing on the changing coefficients issue, the Qu test is here implemented by looking at the results for each coefficient, that is, without taking the supremum across coefficients so that the behavior of partial and total sum of the gradient of each coefficient can be analyzed. Figure 6.25 shows the plots of the differences between partial and total sum of the gradient for each regression coefficient. The peak values for each coefficient at the median are $Q_{.5,student-teacher\ ratio} = 12.13$; $Q_{.5,school\ size} = 10.90$; $Q_{.5,boy} = 6.59$; $Q_{.5,vocational} = 5.91$; $Q_{.5,intercept} = 5.71$; $Q_{.5,computer} = 4.81$; $Q_{.5,private} = 1.69$. All of them are greater than the critical value at the 5% level, equal to 1.650 (Qu, 2008, Table 1), and reject the null of stable coefficients. This is a case of a global change, affecting all the regression coefficients.

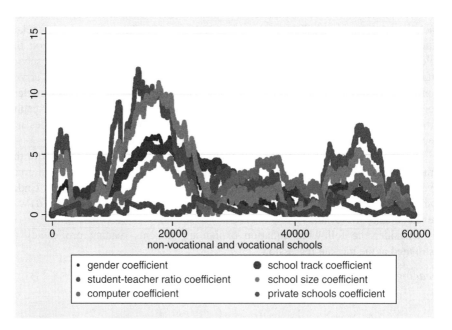

Figure 6.25   Difference between partial and total sum of the gradient for each regression coefficient. The student–teacher ratio and the school size coefficient present the highest peak, and this peak is located at about one third of the sample. OECD–PISA data on the reading scores of fifteen years old Italian students, sample size $n = 59558$.

Next consider the C_θ test for changing coefficients. The model estimated in the entire sample can be compared with the same equation independently estimated in two subsets. Figure 6.25 shows that the Qu test functions present a peak at about one third of the sample, and the two subsets are selected accordingly, comprising respectively $n_a = 19858$ and $n_b = 39706$ observations. Subset "a" comprises data of year 2000, 2003, and part of 2006, while subset "b" includes the majority of year 2006 and year 2009 data. Basically the test verifies whether there is a change over time in the impact of the explanatory variables on reading proficiency.

The estimates at the median are reported in Table 6.13, where in the first column is the constrained model, while in the second and third columns are the estimates in the two subsets defining the unconstrained model. Comparing the estimated coefficients, the variables *stratio, computer, schlsize, boy* present the widest variations in going from one subset to the other and from the unconstrained to the constrained model, by and large confirming the findings of the Qu test in Figure 6.25. The C_θ test assumes the value

$$C_{.50} = \frac{[V(\theta) - (V_a(\theta) + V_b(\theta))]/d_1}{(V_a(\theta) + V_b(\theta))/d_2} = \frac{[18815768 - (10762586 + 7943995)]/7}{(10762586 + 7943995)/(59558 - 14)}$$

$$= 49.65$$

and the null of stable coefficients is rejected. Thus the impact of the covariates on students proficiency has changed over time

As a counterexample, consider a controlled experiment: simulated data comprising $n = 100$ observations of an independent variable, x_i, are drawn from a

Table 6.13 Estimated coefficients at the median to verify changing coefficients, OECD-PISA data set.

	$n = 59558$	$n_a = 19858$	$n_b = 39706$
vocational	−75.79	−73.13	−74.54
se	(1.86)	(2.57)	(2.38)
boy	−38.61	−32.99	−43.29
se	(1.62)	(2.46)	(1.89)
private	3.90	5.22	−6.33
se	(3.58)	(6.92)	(6.86)
stratio	3.25	2.04	7.955
se	(0.33)	(0.12)	(0.43)
schlsize	0.016	0.019	−0.0086
se	(0.002)	(0.003)	(0.003)
computer	0.053	0.033	0.136
se	(0.011)	(0.018)	(0.015)
intercept	477.5	487.3	441.1
se	(3.46)	(3.18)	(4.40)
$V(\theta)$	18815768	10762586	7943995

χ_4^2 distribution. The dependent variable is computed as $y_i = 3 + 0.7x_i + e_i$, and the error term e_i follows a standard normal distribution. This is a case of a fixed coefficients model, and at the median, the estimated equation is $y_i = 3.19 + 0.67x_i$, with $se(intercept) = 0.25$ and $se(slope) = 0.05$. The largest value of the Qu test for the intercept is $Q_{.5,intercept} = 0.4$ and for the slope is $Q_{.5,slope} = 1.15$, to be compared with the critical value of 1.43 at the 5% level. The null of constant coefficients is not rejected by the Qu test. The C test assumes the value $C_{.50} = \frac{[32.82-(15.02+17.09)]/2}{(15.02+17.09)/96} = 1.05$ to be compared with the critical value $F_{2,100} = 3.09$ at the 5% level. Thus in the $C_{.50}$ test, the null of stability is not rejected as well.

Next, the experiment is modified to model changing coefficients. The sample is split in two halves, and the first 50 observations are generated as $y_i = 3 + 0.7x_i + e_i$, while the second half of the sample is generated as $y_i = 3 + 0.1x_i + e_i$. In this case, which models changing coefficients, the largest value for the Qu test is $Q_{.5,intercept} = 3.80$ at the intercept and $Q_{.5,slope} = 6.90$ at the slope, both larger than the 1% critical value of 1.69, while the C_θ test yields the value $C_{.50} = \frac{[66.60-(15.02+17.09)]/2}{(15.02+17.09)/96} = 51.54$, which is well above the critical value of 3.09. The null of stable coefficients is rejected by both tests. It is worth noting the worsening of the estimated objective function of the constrained model in going from the fixed to the changing coefficients example: $\hat{V}(\theta) = 32.82$ in the fixed coefficients model becomes $\hat{V}(\theta) = 66.60$ due to the unrealistic constraints imposed to the changing coefficients model. Figure 6.26 plots the difference between partial and total sum of the gradient when the coefficients are kept constant while Figure 6.27 plots the same statistics when the simulated model

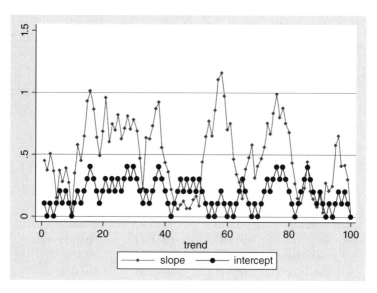

Figure 6.26 The Qu test computed for each coefficient at the median regression in a simulated data set with constant coefficients, $y_i = 3 + 0.7x_i + e_i$. The null of stable coefficients is not rejected. Sample size $n = 100$.

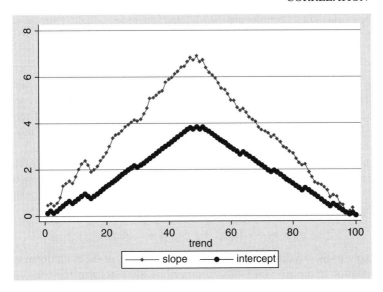

Figure 6.27   The Qu test computed for each coefficient at the median regression in a simulated data set with changing coefficients. The changing coefficients model is defined as $y_i = 3 + 0.7x_i + e_i$ in $n = 1, ..., 50$ and $y_i = 3 + 0.1x_i + e_i$ in $n = 51, ..., 100$. The null of stability is rejected. The graph shows a large peak which correctly signals the change point occurring at the middle of the sample. The scale of the vertical axis is bigger than in the previous graph of Figure 6.26. Sample size $n = 100$.

is characterized by changing coefficients. The latter graph shows quite clearly the position of the change point, which occurs at the middle of the sample.

6.7   Conditionally heteroskedastic models

While in the unit root example an early peak in a time series would grow and would not disappear, time series can also be characterized by sudden and temporary peaks caused by an increase in variability. The main feature of these series is a changing dispersion, which can be captured by an autoregressive conditional heteroskedasticity model, ARCH, first discussed by Engle (1982). This model describes the error variance as function of its own past values. Consider the autoregressive process $y_t \sim \mathrm{AR}(q)$ with error term u_t, the ARCH(r) conditional variance of u_t is defined as

$$E(u_t^2 \mid I_t) = \sigma_t^2 = \gamma_0 + \gamma_1 u_{t-1}^2 + \gamma_2 u_{t-2}^2 + ... + \gamma_r u_{t-r}^2$$

where I_t is the information set at time t, and the r past squared errors concur to determine its actual value. The model can be estimated by a two-step approach. In the first stage the $y_t \sim \mathrm{AR}(q)$ model is estimated and the residuals are saved. In the second stage, the squared residuals become the dependent variable of an auxiliary regression, where the explanatory variables are the r lagged squared residuals. Thus the auxiliary equation is estimated as

$$\widehat{u}_t^2 = \gamma_0 + \gamma_1 \widehat{u}_{t-1}^2 + \gamma_2 \widehat{u}_{t-2}^2 + ... + \gamma_r \widehat{u}_{t-r}^2$$

where $\widehat{u}_t$ are the residuals of the AR(q) estimated process and $\widehat{u}_{t-1}, ..., \widehat{u}_{t-r}$ its lagged values. The Student-t test of the estimated γ's verifies the validity of each coefficient, under the null H_0: $\gamma_h = 0$ for $h = 1, .. r$, while the F test on the entire model verifies the null that all the γ coefficients but the intercept γ_0 are equal to zero. The F test verifies the presence of ARCH when the null H_0:$\gamma_1 = \gamma_2 = .. = \gamma_r = 0$ is rejected.[8]

Koenker and Zhao (1996) extend a modified version of the ARCH(r) model to quantile regressions where, rather than modeling the variance, they focus on the standard deviation. The quantile regression ARCH(r) model, QARCH(r), considers the following definition

$$\sigma_t = \gamma_0 + \gamma_1 |u_{t-1}| + \gamma_2 |u_{t-2}| + ... + \gamma_r |u_{t-r}|$$

which explains the conditional standard deviation instead of the conditional variance. The estimation approach is a two-step procedure analogous to the one previously described: the QAR(q) model is estimated and the absolute value of the residuals become the dependent variable of the auxiliary equation

$$|\tilde{u}_t| = \gamma_0 + \gamma_1 |\tilde{u}_{t-1}| + \gamma_2 |\tilde{u}_{t-2}| + ... + \gamma_r |\tilde{u}_{t-r}|$$

where the vector comprising the absolute value of the quantile regression errors at the various lags, $U = (1 \ |u_{t-1}| \ ... \ |u_{t-r}|)^T$, is approximated by the quantile regression residuals $|\tilde{u}_{t-h}|$ for $h = 1, .. r$, and the dependent variable is given by the absolute value of the quantile regression residual at time t, $|\tilde{u}_t|$. The objective function of the auxiliary quantile regression is

$$\sum \rho(|\tilde{u}_t| - \gamma_0 - \gamma_1 |\tilde{u}_{t-1}| - \gamma_2 |\tilde{u}_{t-2}| - .... - \gamma_r |\tilde{u}_{t-r}|)$$

Under the assumption of bounded and continuous error density f having $f(F^{-1}(\theta)) > 0$, the QARCH(r) estimator of the coefficient vector estimated at the quantile θ, $\gamma(\theta) = (\gamma_0(\theta) \ \gamma_1(\theta) \ ... \ \gamma_r(\theta))^T$, is asymptotically normal:

$$\sqrt{n}(\tilde{\gamma}(\theta) - \gamma(\theta)) \rightarrow N\left(0; \frac{\theta(1 - \theta)}{f^2(F^{-1}(\theta))} D_1^{-1} D_0 D_1^{-1}\right)$$

where $D_1 = E[\sigma_1^{-1} U U^T]$ and $D_0 = E[U U^T]$.

---

[8] Instead of a two-step approach, Francq and Zakoian (2004) define a quasi-maximum likelihood approach to compute a generalized ARCH model, GARCH. Huang et al. (2008) consider the LAD estimator, that is, the median quantile regression, as a version of the quasi-maximum likelihood for the GARCH model.

To verify the presence of conditional heteroskedasticity in quantile regressions, Koenker and Zhao (1996) apply the heteroskedasticity test for quantile regressions to the auxiliary equation. Thus the proposed test looks at the coefficients of the auxiliary equation estimated at k different quantiles. The null of absence of QARCH(r) implies that the vector $\gamma(\theta) = (\gamma_1(\theta) \,.\,.\, \gamma_r(\theta))^T$ does not change across quantiles, H_0: $\gamma(\theta_i) = \gamma(\theta_j)$. However, in the single quantile regression, the usual Student-t statistic is a valid test function to verify the null H_0: $\gamma_h(\theta) = 0$ for $h = 1, .. r$, and $h \neq 0$.

A generalization of the ARCH model is given by the GARCH(r, q) process, discussed by Bollerslev (1986), which is defined as

$$\sigma_t^2 = \gamma_0 + \gamma_1 u_{t-1}^2 + \gamma_2 u_{t-2}^2 + ... \gamma_r u_{t-r}^2 + \delta_1 \sigma_{t-1}^2 + \delta_2 \sigma_{t-2}^2 + ... + \delta_q \sigma_{t-q}^2$$

In the above equation the lagged values of the variance are added to the lagged values of the squared errors to define σ_t^2. In order to solve this model, the lagged variances on the right-hand side of the equation have to be replaced by a valid proxy. For instance in a GARCH (1,1) model with $\sigma_t^2 = \gamma_0 + \gamma_1 u_{t-1}^2 + \delta_1 \sigma_{t-1}^2$, the lagged variance σ_{t-1}^2 can be replaced by $\sigma_{t-1}^2 = \gamma_0 + \gamma_1 u_{t-2}^2 + \delta_1 \sigma_{t-2}^2$, and

$$\sigma_t^2 = \gamma_0 + \gamma_1 u_{t-1}^2 + \delta_1 (\gamma_0 + \gamma_1 u_{t-2}^2 + \delta_1 \sigma_{t-2}^2)$$
$$\sigma_t^2 = \gamma_0 (1 + \delta_1) + \gamma_1 u_{t-1}^2 + \gamma_1 \delta_1 u_{t-2}^2 + \delta_1^2 \sigma_{t-2}^2$$

which implies that after repeated back substitution, the final definition of the conditional variance does not depend on any lagged variance term, and the definition of the GARCH (1,1) process becomes ARCH(∞)

$$\sigma_t^2 = \frac{\gamma_0}{1 - \delta_1} + \gamma_1 (u_{t-1}^2 + \delta_1 u_{t-2}^2 + \delta_1^2 u_{t-3}^2 + \delta_1^3 u_{t-4}^2 ....)$$

In the above equation, the two-step approach described for the ARCH(r) model can be once again implemented. Along these lines, Xiao and Koenker (2009) generalize the QARCH(r) model to comprise the lagged dependent variable terms, yielding the QGARCH(r, q) process. Again, their model does not focus on the variance but on the standard deviation, which is defined as

$$\sigma_t = \gamma_0 + \gamma_1 |u_{t-1}| + \gamma_2 |u_{t-2}| + ... + \gamma_r |u_{t-r}| + \delta_1 \sigma_{t-1} + \delta_2 \sigma_{t-2} + ... + \delta_q \sigma_{t-q}$$

The repeated back substitution of the lagged dependent variables allows to define QGARCH(r, q) process as a QARCH(∞), $\sigma_t = \alpha_0 + \sum_{h=1}^{\infty} \alpha_h |u_{t-h}|$. For instance the QGARCH(1,1) case becomes

$$\sigma_t = \frac{\gamma_0}{1 - \delta_1} + \gamma_1 (|u_{t-1}| + \delta_1 |u_{t-2}| + \delta_1^2 |u_{t-3}| + \delta_1^3 |u_{t-4}| ....)$$

where $\alpha_0 = \frac{\gamma_0}{1-\delta_1}$ and $\alpha_h = \gamma_1 \delta_1^{h-1}$. Following the two-step approach, the QARCH(∞) process can be estimated by

$$|\tilde{u}_t| = \alpha_0 + \sum_{h=1}^{\infty} \alpha_h |\tilde{u}_{t-h}|$$

6.7.1   Example of a conditional heteroskedastic model

Consider the three-month US inflation rate of Section 6.2.1 and the graph in Figure 6.13, which describes the behavior of the original series π_{3t} and of the residuals of the QAR(4) process as estimated at the median, in a sample of $n = 462$ observations. In this graph the series of residuals presents sudden jumps. Here the attempt is to model these jumps by means of a QARCH model. The estimated coefficients of the QAR(4) model are reported in Table 6.7. The residuals of this QAR(4) estimated model provide the first of the two steps needed to compute the QARCH(r) process. Once saved the residuals, their absolute value become the dependent variable of the auxiliary regression which provides the second step of the procedure. The explanatory variables are the lagged values of the dependent variable,

$$|\tilde{u}_t| = \gamma_0 + \gamma_1 |\tilde{u}_{t-1}| + \gamma_2 |\tilde{u}_{t-2}| + ... + \gamma_r |\tilde{u}_{t-r}|$$

Following an overfitting approach, a QARCH(3) model is estimated at the median, and this yields significant values only for the coefficients γ_0 and γ_1, as reported in the first two rows of Table 6.14. The conditional heteroskedasticity can thus be modeled by a QARCH(1), and these estimates are in the bottom two rows of Table 6.14.

Summarizing, the QAR(4) process for the three-month US inflation rate presents conditionally heteroskedastic errors of order $r = 1$ as estimated at the median. Figure 6.28 plots the original series and the residuals of the estimated QARCH(1) process, while Figure 6.29 compares QAR(4) and QARCH(1) residuals, where the QARCH(1) residuals are much smoother.

Table 6.14   Three-month US Treasury Bill rate, QARCH(3) and QARCH(1) estimates.

	γ_0	γ_1	γ_2	γ_3
QARCH(3)	0.385	0.225	0.041	0.078
se	(0.079)	(0.057)	(0.056)	(0.057)
QARCH(1)	0.498	0.221		
se	(0.055)	(0.049)		

Note: Standard errors in parenthesis

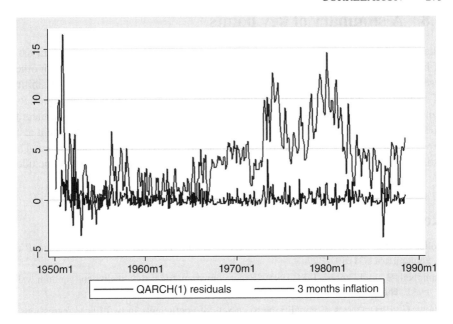

Figure 6.28   Three-month US inflation rate π_{3t} and, at the bottom, the residuals of the QARCH(1) process as estimated at the median.

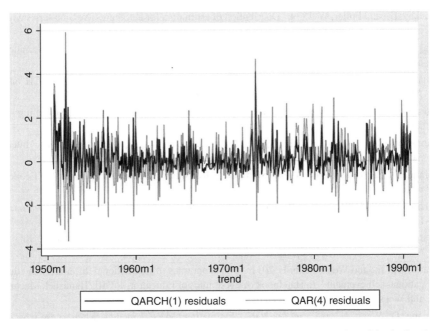

Figure 6.29   Comparison of QAR(4) and QARCH(1) estimated residuals in the three- month US inflation rate process π_{3t}. By modeling the error conditional heteroskedasticity, the sudden increases in the variability of the QAR(4) residuals are greatly reduced.

6.8   A summary of key points

The quantile regression estimates of an autoregressive process is discussed in case of both stationary and non-stationary series. A test for unit roots together with other tests closely related to the quantile regression model are considered. The possibility of a spurious regression is analyzed with simulated data, and the quantile regression cointegrated model is presented, together with the description of a test for cointegration. Next, a changing coefficient model is taken into account and two different tests to verify the presence of changing coefficients are discussed and implemented. The estimation of conditionally heteroskedastic errors in quantile regression concludes the chapter.

References

Bollerslev, T. 1986. "Generalized autoregressive conditional heteroskedasticity." Journal of Econometrics 31, 307–327.

Campbell, B., and Dufour, J. M. 1995. "Exact nonparametric orthogonality and random walk tests." The Review of Economics and Statistics 77, 1–16.

Chow, G. 1960. "Tests of equality between sets of coefficients in two linear regressions." Econometrica 28, 591–605.

de Jong, R., Amsler, C., and Schmidt, P. 2007. "A robust version of the KPSS test based on indicators." Journal of Econometrics 137, 311–33.

Dickey, D., and Fuller, W. 1979. "Distribution of estimators for autoregressive time series with a unit root." Journal of the American Statistical Association 74, 427–431.

Engle, R. 1982. "Autoregressive conditional heteroskedasticity with estimates of the variance of the United Kingdom inflation." Econometrica 50, 987–1007.

Francq, C., and Zakoian, J. 2004. "Maximum likelihood estimation of pure GARCH and ARMA-GARCH processes." Bernoulli 10, 605–637.

Furno, M. 2007. "Parameter instability and quantile regression." Statistical Modelling 7, 345–362.

Furno, M., and Vistocco, D. 2013. "Qu test for structural breaks in quantile regressions." International Journal of Statistics and Probability 2, 42–55.

Granger, C., and Newbold, P. 1974. "Spurious regressions in econometrics." Journal of Econometrics 2, 111–120.

Hansen, B. 1995. "Rethinking the univariate approach to unit root tests: how to use covariates to increase power." Econometric Theory 11, 1148–1171.

Hanushek, E. 2006. "School resources." In Handbook of Economics of Education vol. II (Hanushek and Welch eds.), North Holland.

Hanushek, E., and Woessmann, L. 2011. "The economics of international differences in educational achievement." In Handbook of Economics of Education vol. III (Hanushek Machin and Woessmann eds) North Holland.

Huang, D., Wang, H., and Yao, Q., 2008. "Estimating GARCH models: when to use what?" Econometrics Journal 11, 27–38.

Koenker, R., and Bassett, G. 1982. "Tests of linear hypotheses and l1-estimation." Econometrica 50, 1577–83.

Koenker, R., and Machado, J. 1999. "Goodness of fit and related inference for quantile regression." Journal of the American Statistical Association 94, 1296–1310.

Koenker, R., and Xiao, Z. 2004. "Unit root quantile autoregression inference." Journal of the American Statistical Association 99, 775–787.

Koenker, R., and Xiao, Z. 2006. "Quantile autoregression." Journal of the American Statistical Association 101, 980–990.

Koenker, R., and Zhao, Q. 1996. "Conditional quantile estimation and inference for Arch models." Econometric Theory 12, 793–813.

Kwiatkowski, D., Phillips, P., Schmidt, P., and Shin, Y. 1992. "Testing the null hypothesis of stationarity against the alternative of a unit root." Journal of Econometrics 54, 159–78.

Li, G., Li, Y., and Tsai, C. 2015. "Quantile correlations and quantile autoregressive modelling." Journal of the American Statistical Association, forthcoming.

Oka, T., and Qu, Z. 2011. "Estimating structural changes in regression quantiles." Journal of Econometrics 162, 248–267.

Qu, Z. 2008. "Testing for structural change in regression quantiles." Journal of Econometrics 146, 170–184.

Saikkonen, P. 1991. "Asymptotic efficient estimation of cointegrating regressions." Econometric Theory 7, 1–21.

So, B., and Shin, D. 2001. "An invariant sign test for random walk based on recursive median adjustment." Journal of Econometrics 102, 197–229.

Xiao, Z. 2009. "Quantile cointegrating regression." Journal of Econometrics 150, 248–260.

Xiao, Z., and Koenker, R. 2009. "Conditional quantile estimation for generalized autoregressive conditional heteroskedasticity models." Journal of the American Statistical Association 104, 1696–1712.

Appendix: Stata computer codes

A) QAR process:	
`gen ylag=y[_n-1]`	generate lagged value of dep. variable
`gen t=m(1950m2)+_n-1`	define time frequency(m=monthly)
`format t %tm`	with starting date (1950m2)=february 1950
`tsset t, monthly`	
`corrgram y`	compute correlations and partial correlations and their graphs
`ac y`	graph of correlations
`pac y`	graph of partial correlations
`gen ylag2=y[_n-2]`	generate y_{t-2}, lag 2
`gen ylag3=y[_n-3]`	generate y_{t-3}, lag 3
`gen ylag4=y[_n-4]`	generate y_{t-4}, lag 4
`qreg y ylag ylag2 ylag3`	compute QAR(3)

* QACF:

`qreg y ylag, q(.75)`	compute $.75^{th}$ quartile regression
`predict qres75, resid`	save residuals
`scalar teta=e(q)`	save quantile θ
`scalar tetasq=teta*teta`	compute θ^2
`scalar nobs=e(N)`	save sample size
`su qres75,d`	
`scalar meanres=r(mean)`	save mean of residuals
`scalar vares=r(Var)`	save variance of residuals

*compute denominator

`scalar deno=(teta-tetasq)*vares*nobs`	
`gen signres=teta if`	$\psi(\hat{e}_t)$
`    qres75>=0 &`	
`    qres!=.`	
`replace signres=teta-1 if`	
`    qres75 <=0 &`	
`    qres!=.`	
`gen qreslag=qres75[_n-1]`	compute lagged residuals
`gen num=signres*(qreslag-meanres)`	$\psi(\hat{e}_t)[\hat{e}_{t-1} - E(\hat{e}_t)]$
`gen qacf=sum(num)`	
`su qacf`	
`scalar qacfin=r(max)`	
`scalar qacfinal=qacfin/deno`	compute QACF at lag 1
`gen qreslag2=qres75[_n-2]`	compute lag two residuals
`gen num2=signres*(qreslag2-meanres)`	$\psi(\hat{e}_t)[\hat{e}_{t-2} - E(\hat{e}_t)]$
`gen qacf2=sum(num2)`	
`scalar qacfin2=qacf2 in 100`	
`scalar qacfinal2=qacfin2/deno`	
`scalar list qacfinal qacfinal2`	compute QACF at lag 2

* Augmented QAR processes

`gen dy=y-ylag1`	generate first difference $\Delta y_t = y_t - y_{t-1}$
`gen dy1=ylag1-ylag2`	generate first difference $\Delta y_{t-1} = y_{t-1} - y_{t-2}$
`gen dy2=ylag2-ylag3`	generate first difference $\Delta y_{t-2} = y_{t-2} - y_{t-3}$
`gen dy3=ylag3-ylag4`	generate first difference $\Delta y_{t-3} = y_{t-3} - y_{t-4}$
`corrgram dy`	
`ac dy`	
`pac dy`	
`qreg y ylag1`	unit root test
`qreg y ylag1 dy1 dy2 dy3,`	augmented Dickey Fuller test
`    q(.5)`	

B) Sign based tests

Campbell Dufour test

`gen ylag1=y[_n-1]`	lagged variable, lag one
`gen dy=y-ylag1`	first difference Δy_t
`gen rroo=dy*ylag1`	product $\Delta y_t y_{t-1}$
`gen dum=1 if rroo >=0`	excludes non-positive signs
`egen C=sum(dum)`	sum the # of positive signs
`list C in 1/1`	prints the result

Campbell Dufour test

`su xlag1,d`	
`scalar med=r(p50)`	median of y_{t-1}
`scalar varia=r(Var)`	variance of y_{t-1}
`scalar obs=r(N)`	# of observations
`gen difmed=y-med`	$y_t - median$
`gen dam=1 if difmed >=0`	+1 for $y_{t-1} - median > 0$
`replace dam=-1 difmed<=0`	-1 for $y_{t-1} - median < 0$
`replace dam=0 if difmed==0`	0 for $y_{t-1} - median = 0$
`gen sumsquares=(sum(dam))^2`	sum of squares
`scalar cott=1/(obs*obs*varia)`	$n^{-2}\sigma^{-2}$
`gen J=cott*sum(sumsquares)`	test function, the last value of the series
`list J in 460/460`	prints the result

So and Shin test	
`scalar nobs=obs+1`	stopping rule at n
`gen constant=1`	
`gen trend=sum(constant)`	generates a trend
`gen ics = 0 if trend==2`	initialize ics
`cap program drop doit`	iterations to compute
`program def doit`	the running median
`local i=3`	
`while 'i' < nobs {`	
`  replace ics=ylag1 if`	
`      trend=='i'`	
`  quietly su ics if`	
`      trend<='i',d`	
`  scalar rmed=r(p50)`	
`  replace constant=rmed if`	running median=med $_{t-1}$
`      trend=='i'`	
`  local i='i'+1`	
`}`	
`end`	
`qui doit`	end of iterations
`gen difmedbis=ylag1-constant`	y_{t-1} minus the running median
`gen product=dy*difmedbis`	
`gen dam=1 if`	$+1$ for $\Delta y_t(y_{t-1} - med_{t-1}) > 0$
`    product>=0 & difmed!=.`	
`replace dam=-1 if`	-1 for $\Delta y_t(y_{t-1} - med_{t-1}) < 0$
`    product<=0 & difmed!=.`	
`replace dam=0 if`	0 for $\Delta y_t(y_{t-1} - med_{t-1}) = 0$
`    product==0 & difmed!=.`	
`gen C=sum(dam)`	sum of signs
`gen CC=(C+obs)/2`	test function, the last value of the series
`list CC in 462/462`	prints the result

C) Cointegration

`qreg y x`	original model
`gen lagx=x[_n-1]`	x_{t-1}
`gen lagy=y[_n-1]`	y_{t-1}
`qreg y lagy`	unit root test on Y
`qreg x lagx`	unit root test on X
`gen dy=y-lagy`	Δy_t
`gen dx=x-lagx`	Δx_t
`gen lagdx=x[_n-1]-x[_n-2]`	Δx_{t-1}
`gen leadx=x[_n+1]-x`	Δx_{t+1}
`gen leadx2=x[_n+2]-x[_n+1]`	Δx_{t+2}
`qreg y x dx lagdx leadx leadx2`	cointegrating model
`predict res, resid`	store residuals
`gen reslag=res[_n-1]`	lagged residuals
`qreg res reslag`	to test residuals stationarity

D) Qu changing coefficient test

`qreg y x1 x2 x3 x4, q(.5) nolog`	compute median regression
`scalar obs=e(N)`	save sample size
`predict res, resid`	save residuals
`gen signres=-.5 if res<=0 & res!=.`	
`replace signres=.5 if` `    res>0 & res!=.`	sign of residuals
`gen const=1 if res!=.`	constant term
`gen trend=sum(cost) if res!=.`	partial sum of the constant
`gen lambda=trend/obs if res!=.`	compute lambda
`gen partialsum=sum(signres) if res!=.`	partial sum of sgn(residual)
`egen totalsum=sum(signres) if res!=.`	total sum of sgn(residual)
`matrix accum XX=x1 x2 x3 if res!=.`	compute matrix $X^T X$
`matrix XXinv=syminv(XX)`	compute $(X^T X)^{-1}$
`matrix A=cholesky(XXinv)`	Cholesky factorization
`gen a=el(A,4,4)`	selects the element in A for the intercept

*Qu of intercept

```
gen qo=(abs(partialsum-lambda*totalsum)*a)/.5 if res!=.
```

`drop a partialsum totalsum`	drops intermediate terms
`gen gradient=signres*x1 if res!=.`	gradient of x1
`gen partialsum=sum(gradient) if res!=.`	partial sum
`egen totalsum=sum(gradient) if res!=.`	total sum
`gen a=el(A,1,1)`	selects the element in A for x1

*Qu of x1

```
gen qo1=(abs(partialsum-lambda*totalsum)*a)/.5 if res!=.
```

`drop gradient partialsum totalsum a`	drops intermediate terms
`gen gradient=signres*x2 if res!=.`	gradient of x2
`gen partialsum=sum(gradient) if res!=.`	partial sum
`egen totalsum=sum(gradient) if res!=.`	total sum
`gen a=el(A,2,2)`	selects the element in A for x2

*Qu of x2

gen qo2=(abs(partialsum−lambda*totalsum)*a)/.5 if res!=.

`drop gradient partialsum totalsum a`	drops intermediate terms
`gen gradient=signres*x3 if res!=.`	gradient of x3
`gen partialsum=sum(gradient) if res!=.`	partial sum
`egen totalsum=sum(gradient) if res!=.`	total sum
`gen a=el(A,3,3)`	selects the element in A for x3

*Qu of x3

```
gen qo3=(abs(partialsum-lambda*totalsum)*a)/.5 if res!=.
```

`twoway (scatter qo trend)`
`    (scatter qo1 trend)`
`        ....`
`    (scatter qo3 trend)`
	plot Qu for each coefficient

`graph save filename`	save the graph
`su qo qo1 qo2 qo3`	find the max for each coefficient

E) robust CHOW test

`qreg y x1 x2 x3, q(.5) nolog`	constrained model
`scalar obs=e(N)`	# of observations
`scalar v=e(sum_adev)`	estimated objective function
`qreg y x1 x2 x3 in 1/50, q(.5)` `    nolog`	unconstrained model 1st subset
`scalar v1=e(sum_adev)`	estimated objective function
`qreg y x1 x2 x3 in 51/100,` `    q(.5) nolog`	unconstrained model 2nd subset
`scalar v2=e(sum_adev)`	estimated objective function
`scalar num=(v-v1-v2)/4`	numerator of the test
`scalar test=num*(obs-8)/(v1+v2)`	final value of the test
`scalar list test`	prints the results

F) QARCH(r) process

`gen ylag=y[_n-1]`	lag one of the dependent variable
`gen ylag2=ylag[_n-1]`	lag two of the dependent variable
`qreg y ylag ylag2, q(.5)`	estimate QAR(r) with $r = 2$
`predict res, resid`	save residuals
`gen resabs=abs(res)`	compute absolute value of residuals
`gen reslag=resabs[_n-1]`	lag one of absolute value of residuals
`gen reslag2=reslag[_n-1]`	lag two of absolute value of residuals
`gen reslag3=reslag2[_n-1]`	lag three of absolute value of residuals
`gen reslag4=reslag3[_n-1]`	lag four of absolute value of residuals
`qreg resabs reslag reslag2` `reslag3 reslag4, q(.5)`	estimate QARCH(r), $r = 4$

Index

Anscombe, F.
  data, 8, 9, 11, 18–20, 26, 30, 31, 34,
    35, 39, 52, 53, 61, 82, 83, 85,
    86, 88, 90, 91, 96
  model, 8, 19, 34, 52–54
AR (autoregressive), 233–239, 242,
    244, 252, 255, 269, 270
ARCH (autoregressive conditional
    heteroskedasticity model),
    269–271, 275
Artificial
  data set, 2, 22, 23, 26, 29, 72, 239
  objective function, 171, 173, 174,
    176, 195, 196, 199, 202
  problem, 170, 171, 176, 195, 196
  variable(s), 130, 131, 141, 143, 167,
    170–176, 189, 194–199, 202,
    205
Asymmetric Laplace distribution, 228,
    231
Asymptotic
  distribution, 81, 94, 100, 101, 243,
    260
  theory, 94, 97
Autocorrelation, 234, 237, 238, 240,
    241, 243
  partial, 237, 240
Autoregressive model(s), 233, 237, 239,
    275

Barrier method(s), 227
Basic
  solution(s), 141–148, 150–155,
    157–160, 162, 166, 168, 169,
    171, 172, 174, 176–178, 180,
    188, 190, 218
  variables, 142–144, 147–157, 160,
    162, 171, 172, 176, 180–182,
    184–186, 188, 189, 212
Basis, 148, 149, 151–168, 170, 172,
    174–178, 180–183, 185–189,
    195, 197–202, 206, 208–215,
    217, 218, 224, 225, 229
Bassett, G., xi, 82, 100, 101, 119, 265,
    274
Bayesian quantile regression, 228–231
Bootstrap, xvii, 89–94, 96–99, 101,
    108–110, 117–119, 121–123
Boscovich, R. J., xv, xvi, 192, 219, 228,
    229
Boundary, 133, 135, 145, 146, 227, 228

Canonical form, 141, 150–153, 155,
    157, 158, 160, 162–168,
    170–172, 174, 176, 177, 182,
    183, 194–196, 198, 199, 205,
    206
Cointegration, xviii, 233, 259, 261, 262,
    274, 279

Constraint(s), 128, 129, 131–134, 137–139, 141, 147, 150–153, 155–159, 163, 168–171, 173, 178, 189, 190, 193–196, 205, 211, 217, 219, 220, 227, 264, 265, 268

Contaminated distribution, 5–8, 17, 22–26, 39, 56, 64

Convergence, 39, 50, 51, 62, 95, 176, 177

Corner(s), 133–136, 138, 139, 141, 144–147, 150, 151, 153, 155, 157–159, 162, 164, 165–167, 170, 172–178, 180, 190, 199, 227

Correlation(s), xviii, 101, 119, 233–251, 254, 255, 260, 261, 265, 275
    partial, 236, 237, 240, 241, 243–246, 250, 251, 275

Counterfactual, 102, 108–111, 118, 119, 124

Cyclical(ity), 177, 178, 180, 181

Decomposition, 81, 101, 103–105, 107, 118, 119, 124, 217
    Chernozhukov et al., 92, 94, 95, 110–112, 118, 124
    Machado and Mata, 108, 110–113, 119, 124
    Oaxaca-Blinder, 104, 107, 119, 124
    QTE, 101, 107, 108, 110–112, 114, 117, 125

Degenerate, 146, 177, 178, 180, 218
    basis, 178
    corner(s), 176, 178, 180
    solution(s), 143, 146, 177, 180, 199

Diagnostic, 17–19, 22

Double median, 220, 224

Dual
    plane, 222, 223
    plot, xviii, 218, 220–227, 229

Dutch data, 14–17, 22, 45–48, 59, 64, 67, 68, 70–72, 91–94, 98, 99

Edge(s), 135, 136, 202, 220, 223–225, 227

Edgeworth, F. Y., xv, 119, 191, 192, 218, 220, 221, 228–230

Elemental set(s), xvi–xviii, 81–94, 119–121, 218, 221, 230

Epanechnikov kernel density, 7, 8, 23, 51, 251, 258

Expectile(s), xvi, xvii, 29–48, 59–65, 70, 73–75, 77, 228

Exterior-point, 227

Facet(s), 141, 223

Farebrother, 81, 118, 192, 218, 230

Feasible
    basic solution, 143, 145–147, 150, 151, 153–155, 162, 172, 176, 177, 180, 188, 190
    region, 128, 132–139, 141–147, 178, 190, 227
    set, 132, 133, 146, 171–174, 176, 195, 196, 227
    solution, 128, 132, 143, 144, 146, 156–158, 176, 199, 227

Fourier, xv, xvi

French data, 39, 41–45, 58, 64, 67, 70–72, 91–93, 97, 98

GARCH (generalized ARCH model), 270–273, 281

Gradient, 100, 234, 253, 254, 262, 263, 265, 266, 268, 280

Hampel, F., 17, 26, 51–55, 73, 76

Heteroskedasticity, 38, 119, 253, 271
    conditional heteroskedasticity, xviii, 269, 271–275

Huber, P., 18, 26, 49–65, 73, 76–78

Inequality(ies), xv, 94, 128–130, 132, 141, 155, 157, 167–170, 175–177, 193, 194, 217

Influence function, 1, 17, 19–21, 26, 49, 50, 59, 108

Interior-point, 227, 228, 231
Inverse Probability Weight (IPW),
112–114, 116, 124, 125

Koenker, R., xi, xiii, xv, xvi, xviii, 82,
100, 101, 119, 192, 201, 203,
217, 218, 220, 224–230, 234,
243, 248, 250–252, 260, 265,
270, 271, 274, 275

Laplace, xv, xvi, 192, 219, 220, 225,
228–231
Least absolute deviation(s), xv, 191,
192, 219, 229
Least squares, xv, 1, 29, 30, 38, 39,
49–51, 62, 72, 73, 75, 78, 84,
85, 119, 191
Legendre, xv, xvi
Level line(s), 132–137
Likelihood, 1, 5, 38, 39, 73, 113, 228,
270, 274
Linear equation(s), 129, 140, 151
Linear programming (LP), xv-xviii,
127–141, 144–147, 150–153,
155, 160, 167–169, 171, 173,
176–178, 180, 181, 183,
190–195, 204, 205, 210, 211,
217, 218, 227–231, 239
parametric, 227
Linear quantile mixed models, 228, 230
Loss function, xv, 217, 218, 228

Maximization problem, 128, 129, 137,
148
Median regression, xv, xvii, xviii, 1,
8–16, 18–26, 36, 38, 43, 45,
52–56, 58, 82–84, 86, 87,
89–94, 101, 121, 191–193,
201, 202, 204, 206, 209, 210,
217–225, 229, 235, 254, 256,
257, 259, 265, 268, 269, 279
M-estimator(s), xvi, xvii, 29, 49–62, 64,
72, 76, 78
Methode de situation, xv, 192, 219, 220,
225, 229

Minimax absolute deviation, 192
Minimization problem, 1, 128, 129,
156, 190, 191, 217, 223, 226
M-quantile(s), xvi, xvii, 29, 60–65,
67–73, 77, 78

Nonbasic variable(s), 142, 148, 149,
154–157, 160, 162, 171, 176,
181, 182, 184–186, 188, 189,
211, 212
Non-stationarity, xviii, 233, 242–244,
248–250, 259, 260, 274
Nonunique(ness), 1, 128, 141, 180,
202–204, 217, 221, 223

Objective function, xvii, 2, 29, 30, 32,
49, 60, 72, 100, 111, 128, 129,
131–144, 147–149, 151, 153,
157–168, 170–177, 188–190,
193, 195–204, 206–209, 211,
214, 216, 220, 223, 224, 227,
234, 264, 265, 268, 270, 281
Operation research, 192
Optimal
quantization, 228, 229
solution, 128, 131–141, 146–149,
153, 154, 159, 170–172,
175–178, 188, 190, 202, 209,
210, 213, 219, 222–226
value, 128, 136, 162, 176, 195, 224
Ordinary least squares, 1–5, 8, 10–25,
30, 34, 37–39, 45, 47, 49–50,
52–57, 59, 61, 82–94,
104–107, 234, 243, 248,
251–253, 255–257, 259
Outlier(s), xvii, 1–27, 29, 34–42, 44, 45,
47–54, 56–60, 62–65, 68, 69,
72, 76, 84, 85, 87, 89, 119

Pivot element, 151, 161–167, 173–175,
183, 189, 197, 200, 207–209
Pivoting, 151–154, 157, 158, 161–167,
172, 173, 181, 183, 185–187,
189, 196–198, 200, 207–209,
213

Plural median, xv, 192
Point/line duality, 191
Potential
  outcome, 113, 116
  output, 113–116, 125
Primal
  plane, 222
  plot, 221–223
Propensity score(s), 108, 112–116, 119,
    125

QAR (quantile autoregressive), 234,
    235, 237–240, 242–244,
    246–248, 253, 260, 270, 272,
    273, 275, 281
QARCH (quantile regression ARCH),
    270–273, 281
Quantile(s)
  conditional, 32, 108, 118, 227–229,
    275
  crossing, 36, 38, 47, 48, 73, 228
  extremal, 92, 94, 95, 122
  plot, 39–43, 45–47, 64–67, 74, 76,
    121
  regression process, xviii, 218, 226,
    227
  treatment effect, xvii, 101, 107, 118,
    124
  unconditional, 108
Quartile(s), 31, 35, 36, 38, 84, 91, 234,
    242, 244, 251, 252, 276

Reduced cost(s), 148, 149, 154,
    156–162, 175, 181, 182,
    184–189, 200, 206, 208, 209,
    211–214, 217
Resampling, xvii, 81, 91, 94–96, 109,
    119
Residual(s), 17–19, 21–25, 27, 32, 39,
    49, 56, 60, 62, 75, 77–79, 193,
    194, 201–205, 209, 217–221,
    223, 224, 228, 229, 237–240,
    242, 246, 247, 256, 257,
    259–262, 269, 270, 272, 273,
    276, 279, 281

Robustness, xvi, xvii, 1, 4, 14, 17, 21,
    26, 29, 38, 45, 49, 59, 61, 62,
    72, 84, 85, 92, 253
Robust regression, xvii, 2, 55, 57, 60,
    62, 77–79

Sample influence function (SIF), 20–22,
    25, 26, 59
SHARE, 12, 103, 114
Shift, 262, 264, 265
Simplex, xi, xv, 82, 176, 180, 181, 199,
    202, 204, 206, 214, 220,
    227–229
  algorithm, xviii, 130, 134, 141, 151,
    153, 154, 159, 162, 167–170,
    176, 179–181, 190, 191,
    195–197, 202, 205, 206, 210,
    217, 224, 225, 229
  method, xi, xv, xvii, xviii, 141, 146,
    176, 190, 192, 204, 211, 228
  revised, 181–183, 188–190
  rule, 214
  standard, xviii, 181, 189, 217
  tableau, 211, 218
  transformation, 211, 212, 215, 216
Small-area estimation, 29, 70, 228
Spurious regression, xviii, 233, 254,
    256, 259–261, 274
Standard form, 129–131, 141, 143,
    146–148, 150, 151, 155, 160,
    163, 167, 170, 190
Stationarity, 233, 242, 243, 249, 254,
    257, 262, 275, 276
Stigler, S. M., 192, 219, 230
Subsampling, xvii, 81, 92, 95, 118

Tableau, 159–168, 170, 171, 174, 175,
    177–183, 189, 193, 196–209,
    211–216, 218, 224, 225, 228
Test(s), 19, 92, 94, 147–148, 156–157,
    159, 185, 188
  for ARCH, 270–271
  for changing coefficients, 262–269
  for cointegration, 261–262
  for non-stationarity, 249–254, 260

Transformation, 129, 130, 151, 155,
      170, 171, 192, 211–213, 215,
      216, 253
Treatment effect, 81, 101, 102, 107–109,
      112, 113, 116–118, 124, 125
Tukey, J., 19, 26, 51–59, 61–65, 73, 76,
      78
Two-phase, 170–176, 189, 195–203,
      225

Unit root, 248–262, 269, 274, 275, 277,
      279
Unrestricted in sign, 130, 194, 205, 228

Variable(s)
   artificial, 142–144, 147–157, 167,
      171, 172, 176, 180–182,
      184–186, 188, 189, 212
   decisional, 128–135, 138, 139, 141,
      144, 145, 159, 168, 190, 192,
      194, 201, 228
   dependent, 2–5, 9–14, 17–24, 29, 30,
      32, 34, 35, 39, 40, 44, 45, 49,
      52, 53, 58, 64, 89, 95, 102,
      104, 105, 108, 109, 249, 254,
      257, 263, 265, 267–272, 281

   entering, 148–150, 154–156, 158,
      160–162, 164–168, 172,
      174–176, 181–187, 189,
      197–200, 202, 206–209, 212,
      213, 215, 216
   explicative, 192, 194, 217
   leaving, 148, 150, 154–156, 158,
      161–168, 172, 174–178, 181,
      184–187, 189, 197–200, 202,
      206–209, 211–216
   response, 127, 192, 195
   slack, 130, 141, 143, 155, 160, 163,
      168, 170–172, 176, 177, 183,
      189, 194–196, 201, 205
   surplus, 130

Xiao, Z., 234, 243, 248, 250–252, 259,
      260, 262, 271, 275

Zhao, Q., 270, 271, 275